Geophysical Monograph Series

Including

IUGG Volumes
Maurice Ewing Volumes
Mineral Physics Volumes

GEOPHYSICAL MONOGRAPH SERIES

Geophysical Monograph Volumes

1. Antarctica in the International Geophysical Year   *A. P. Crary, L. M. Gould, E. O. Hulburt, Hugh Odishaw, and Waldo E. Smith (Eds.)*
2. Geophysics and the IGY   *Hugh Odishaw and Stanley Ruttenberg (Eds.)*
3. Atmospheric Chemistry of Chlorine and Sulfur Compounds   *James P. Lodge, Jr. (Ed.)*
4. Contemporary Geodesy   *Charles A. Whitten and Kenneth H. Drummond (Eds.)*
5. Physics of Precipitation   *Helmut Weickmann (Ed.)*
6. The Crust of the Pacific Basin   *Gordon A. Macdonald and Hisahi Kuno (Eds.)*
7. Antarctica Research: The Matthew Fontaine Maury Memorial Symposium   *H. Wexler, M. J. Rubin, and J. E. Caskey, Jr. (Eds.)*
8. Terrestrial Heat Flow   *William H. K. Lee (Ed.)*
9. Gravity Anomalies: Unsurveyed Areas   *Hyman Orlin (Ed.)*
10. The Earth Beneath the Continents: A Volume of Geophysical Studies in Honor of Merle A. Tuve   *John S. Steinhart and T. Jefferson Smith (Eds.)*
11. Isotope Techniques in the Hydrologic Cycle   *Glenn E. Stout (Ed.)*
12. The Crust and Upper Mantle of the Pacific Area   *Leon Knopoff, Charles L. Drake, and Pembroke J. Hart (Eds.)*
13. The Earth's Crust and Upper Mantle   *Pembroke J. Hart (Ed.)*
14. The Structure and Physical Properties of the Earth's Crust   *John G. Heacock (Ed.)*
15. The Use of Artificial Satellites for Geodesy   *Soren W. Henricksen, Armando Mancini, and Bernard H. Chovitz (Eds.)*
16. Flow and Fracture of Rocks   *H. C. Heard, I. Y. Borg, N. L. Carter, and C. B. Raleigh (Eds.)*
17. Man-Made Lakes: Their Problems and Environmental Effects   *William C. Ackermann, Gilbert F. White, and E. B. Worthington (Eds.)*
18. The Upper Atmosphere in Motion: A Selection of Papers With Annotation   *C. O. Hines and Colleagues*
19. The Geophysics of the Pacific Ocean Basin and Its Margin: A Volume in Honor of George P. Woollard   *George H. Sutton, Murli H. Manghnani, and Ralph Moberly (Eds.)*
20. The Earth's Crust: Its Nature and Physical Properties   *John G. Heacock (Ed.)*
21. Quantitative Modeling of Magnetospheric Processes   *W. P. Olson (Ed.)*
22. Derivation, Meaning, and Use of Geomagnetic Indices   *P. N. Mayaud*
23. The Tectonic and Geologic Evolution of Southeast Asian Seas and Islands   *Dennis E. Hayes (Ed.)*
24. Mechanical Behavior of Crustal Rocks: The Handin Volume   *N. L. Carter, M. Friedman, J. M. Logan, and D. W. Stearns (Eds.)*
25. Physics of Auroral Arc Formation   *S.-I. Akasofu and J. R. Kan (Eds.)*
26. Heterogeneous Atmospheric Chemistry   *David R. Schryer (Ed.)*
27. The Tectonic and Geologic Evolution of Southeast Asian Seas and Islands: Part 2   *Dennis E. Hayes (Ed.)*
28. Magnetospheric Currents   *Thomas A. Potemra (Ed.)*
29. Climate Processes and Climate Sensitivity (Maurice Ewing Volume 5)   *James E. Hansen and Taro Takahashi (Eds.)*
30. Magnetic Reconnection in Space and Laboratory Plasmas   *Edward W. Hones, Jr. (Ed.)*
31. Point Defects in Minerals (Mineral Physics Volume 1)   *Robert N. Schock (Ed.)*
32. The Carbon Cycle and Atmospheric CO_2: Natural Variations Archean to Present   *E. T. Sundquist and W. S. Broecker (Eds.)*
33. Greenland Ice Core: Geophysics, Geochemistry, and the Environment   *C. C. Langway, Jr., H. Oeschger, and W. Dansgaard (Eds.).*
34. Collisionless Shocks in the Heliosphere: A Tutorial Review   *Robert G. Stone and Bruce T. Tsurutani (Eds.)*
35. Collisionless Shocks in the Heliosphere: Reviews of Current Research   *Bruce T. Tsurutani and Robert G. Stone (Eds.)*
36. Mineral and Rock Deformation: Laboratory Studies   The Paterson Volume   *B. E. Hobbs and H. C. Heard (Eds.)*
37. Earthquake Source Mechanics (Maurice Ewing Volume 6)   *Shamita Das, John Boatwright, and Christopher H. Scholz (Eds.)*

38 Ion Acceleration in the Magnetosphere and Ionosphere *Tom Chang (Ed.)*

39 High Pressure Research in Mineral Physics (Mineral Physics Volume 2) *Murli H. Manghnani and Yasuhiko Syono (Eds.)*

40 Gondwana Six: Structure, Tectonics, and Geophysics *Garry D. McKenzie (Ed.)*

41 Gondwana Six: Stratigraphy, Sedimentology, and Paleontology *Garry D. McKenzie (Ed.)*

42 Flow and Transport Through Unsaturated Fractured Rock *Daniel D. Evans and Thomas J. Nicholson (Eds.)*

43 Seamounts, Islands, and Atolls *Barbara H. Keating, Patricia Fryer, Rodey Batiza, and George W. Boehlert (Eds.)*

44 Modeling Magnetospheric Plasma *T. E. Moore and J. H. Waite, Jr. (Eds.)*

45 Perovskite: A Structure of Great Interest to Geophysics and Materials Science *Alexandra Navrotsky and Donald J. Weidner (Eds.)*

46 Structure and Dynamics of Earth's Deep Interior (IUGG Volume 1) *D. E. Smylie and Raymond Hide (Eds.)*

47 Hydrological Regimes and Their Subsurface Thermal Effects (IUGG Volume 2) *Alan E. Beck, Grant Garven, and Lajos Stegena (Eds.)*

48 Origin and Evolution of Sedimentary Basins and Their Energy and Mineral Resources (IUGG Volume 3) *Raymond A. Price (Ed.)*

49 Slow Deformation and Transmission of Stress in the Earth (IUGG Volume 4) *Steven C. Cohen and Petr Vaníček (Eds.)*

50 Deep Structure and Past Kinematics of Accreted Terranes (IUGG Volume 5) *John W. Hillhouse (Ed.)*

51 Properties and Processes of Earth's Lower Crust (IUGG Volume 6) *Robert F. Mereu, Stephan Mueller, and David M. Fountain (Eds.)*

52 Understanding Climate Change (IUGG Volume 7) *Andre L. Berger, Robert E. Dickinson, and J. Kidson (Eds.)*

53 Plasma Waves and Istabilities at Comets and in Magnetospheres *Bruce T. Tsurutani and Hiroshi Oya (Eds.)*

54 Solar System Plasma Physics *J. H. Waite, Jr., J. L. Burch, and R. L. Moore (Eds.)*

55 Aspects of Climate Variability in the Pacific and Western Americas *David H. Peterson (Ed.)*

Maurice Ewing Volumes

1 Island Arcs, Deep Sea Trenches, and Back-Arc Basins *Manik Talwani and Walter C. Pitman III (Eds.)*

2 Deep Drilling Results in the Atlantic Ocean: Ocean Crust *Manik Talwani, Christopher G. Harrison, and Dennis E. Hayes (Eds.)*

3 Deep Drilling Results in the Atlantic Ocean: Continental Margins and Paleoenvironment *Manik Talwani, William Hay, and William B. F. Ryan (Eds.)*

4 Earthquake Prediction—An International Review *David W. Simpson and Paul G. Richards (Eds.)*

5 Climate Processes and Climate Sensitivity *James E. Hansen and Taro Takahashi (Eds.)*

6 Earthquake Source Mechanics *Shamita Das, John Boatwright, and Christopher H. Scholz (Eds.)*

IUGG Volumes

1 Structure and Dynamics of Earth's Deep Interior *D. E. Smylie and Raymond Hide (Eds.)*

2 Hydrological Regimes and Their Subsurface Thermal Effects *Alan E. Beck, Grant Garven, and Lajos Stegena (Eds.)*

3 Origin and Evolution of Sedimentary Basins and Their Energy and Mineral Resources *Raymond A. Price (Ed.)*

4 Slow Deformation and Transmission of Stress in the Earth *Steven C. Cohen and Petr Vaníček (Eds.)*

5 Deep Structure and Past Kinematics of Accreted Terranes *John W. Hillhouse (Ed.)*

6 Properties and Processes of Earth's Lower Crust *Robert F. Mereu, Stephan Mueller, and David M. Fountain (Eds.)*

7 Understanding Climate Change *Andre L. Berger, Robert E. Dickinson, and J. Kidson (Eds.)*

8 Evolution of Mid Ocean Ridges *John M. Sinton (Ed.)*

Mineral Physics Volumes

1 Point Defects in Minerals *Robert N. Schock (Ed.)*

2 High Pressure Research in Mineral Physics *Murli H. Manghnani and Yasuhiko Syono (Eds.)*

Geophysical Monograph 59
IUGG Volume 9

Variations in Earth Rotation

Dennis D. McCarthy
William E. Carter
Editors

American Geophysical Union
International Union of Geodesy and Geophysics

Geophysical Monograph/IUGG Series

Library of Congress Cataloging-in-Publication Data

Variations in Earth rotation / Dennis D. McCarthy, William E. Carter, editors.
    p. cm. — (Geophysical monographs ; 59)   (IUGG ; v. 9)
  ISBN 0-87590-459-9
    1. Earth—Rotation.   I. McCarthy, Dennis D.   II. Carter, William E. (William Eugene), 1939–   . III. Series.   IV. Series: IUGG (Series) ; v. 9.
  QB633.V37   1990
  525'.35—dc20   90-1027
                                                                                  CIP

Copyright 1990 by the American Geophysical Union, 2000 Florida Avenue, NW, Washington, DC 20009

Figures, tables, and short excerpts may be reprinted in scientific books and journals if the source is properly cited.

    Authorization to photocopy items for internal or personal use, or the internal or personal use of specific clients, is granted by the American Geophysical Union for libraries and other users registered with the Copyright Clearance Center (CCC) Transactional Reporting Service, provided that the base fee of $1.00 per copy plus $0.10 per page is paid directly to CCC, 21 Congress Street, Salem, MA 10970. 0065-8448/89/S01. + .10.
    This consent does not extend to other kinds of copying, such as copying for creating new collective works or for resale. The reproduction of multiple copies and the use of full articles or the use of extracts, including figures and tables, for commercial purposes requires permission from AGU.

Printed in the United States of America.

CONTENTS

Preface

International Cooperation in the Study of the Rotation of the Earth
  G. A. Wilkins   1

Variational Calculation of Wobble Modes of the Earth
  D. E. Smylie   5

On the Complex Eigenfrequency of the "Nearly Diurnal Free Wobble" and its Geophysical Interpretation
  Jürgen Neuberg, Nacques Hinderer, and Walter Zürn   11

Numerical Solution for the Rotation of a Rigid Model Earth
  Joachim Schastok, Michael Soffel, and Hanns Ruder   17

The Long Period Elastic Behavior of the Earth
  Bernd Richter   21

The Earth's Differential Rotation: Hydrospheric Changes
  Nils-Axel Mörner   27

The Influence of Ocean and Solid Earth Parameters on Oceanic Eigenoscillations, Tides and Tidal Dissipation
  Wilfried Zahel   33

Secular Tidal and Nontidal Variations in the Earth's Rotation
  Milan Burša   43

Tidal Deceleration of the Earth
  Peter Brosche   47

Effects of the Tidal Dissipation on the Moon's Orbit and the Earth's Rotation
  M. Ooe, H. Sasaki, and H. Kinoshita   51

The Pole Tide in Deep Oceans
  S. R. Dickman   59

Tidal Parameters and Nutation: Influence From the Earth Interior
  Veronique Dehant   69

The Earth's Forced Nutations: Geophysical Implications
  J. M. Wahr and D. de Vries   79

Study of Fluid-Solid Earth Coupling Process Using Satellite Altimeter Data
  Wooil M. Moon, Roger Tang, and H. H. Choi   85

Orthogonal Stack of Global Tide Gauge Sea Level Data
  A. Trupin and J. Wahr   111

Atmospheric Excitation of the Earth's Rotation Rate
  J. B. Merriam   119

Earth Rotation and Climatic Periodicities
  J. P. Rozelot and D. Spaute   127

Enso-Related Signals in Earth Rotation, 1962–87
  Martine Feissel and Jean Gavoret   133

Forecasting Atmospheric Angular Momentum and Length-of-Data Using Operational Meteorological Models
  R. D. Rosen, D. A. Salstein, T. Nehrkorn, J. O. Dickey, T. M. Eubanks, J. A. Steppe, M. R. P. McCalla and A. J. Miller   139

Forecasting Short-Term Changes in the Earth's Rotation Using Global Numerical Weather Prediction Models
  Raymond Hide   145

Global Water Storage and Polar Motion
  John W. Kuehne and Clark R. Wilson   147

Maximum Likelihood Estimates of Polar Motion Parameters
  Clark R. Wilson and R. O. Vicente   151

Interannual and Decade Fluctuations in the Earth's Rotation
  Jean O. Dickey, T. Marshall Eubanks, and Raymond Hide   157

Short Period UT1 Variations From Iris Daily VLBI Observations
  D. S. Robertson, W. E. Carter, and F. W. Fallon   163

Daily Pole Positions Monitored by Very Long Baseline Interferometry
  A. Nothnagel, G. D. Nicolson, H. Schuh, J. Campbell, and R. Kilger   171

Error Analysis for Earth Orientation Recovery From GPS Data
  N. Zelensky, J. Ray, and P. Liebrecht   177

Simulations to Recover Earth Rotation Parameters With GPS System
  P. Paquet and L. Louis   185

Station Coordinates and Earth Rotation Parameters 1986
  H. Hauck   189

Reference Frame of LLR
  Jin Wen-Jing and Wang Qiang-guo   193

Definition and Realization of Terrestrial Reference Systems for Monitoring Earth Rotation
  Claude Boucher   197

A Correlation Study of the Earth's Rotation with El Nino/Southern Oscillation
  B. Fong Chao   203

Statistical Investigations on Atmospheric Angular Momentum Functions and on Their Effects Polar Motions
  Aleksander Brzezinski   205

PREFACE

As part of the Nineteenth General Assembly of The International Union of Geodesy and Geophysics Symposium (IUGG) in Vancouver, Canada, Union Symposium U4, "Variations in Earth Rotation" was held August 18–19 1987. The Convenor was Dennis D. McCarthy, U.S. Naval Observatory with P. Paquet, Observatoire Royal de Belgique and M. G. Rochester, St. Johns University serving as co-convernors. In a session on internal structure of the Earth papers dealt with the geophysical effects on Earth rotation parameters. Mantle anelasticity increases the free core nutation (FCN) period by a few days. The period of the FCN and the amplitudes of the main nutation components are sensitive to the ellipticity of the core-mantle boundary (CMB), and a non-hydrostatic increase of 400 m in the flattening of the CMB is a possible explanation of the discrepancies from theory. An alternative suggestion rests on the subseismic description of the nutation spectrum of the stratified liquid core. Evidently new models will have to take into account contributions from the oceans, mantle anelasticity, non-hydrostatic pre-stress, CMB topography and internal core structure.

The relationship of Earth rotation parameters to tides and oceans was discussed in a session. Papers showed that changes in continentality and bathymetry over a few tens of millions of years alter the eigenperiod spectrum of the world ocean enough to change the tidal torque by a factor of two. Tide gauge data from southeast Asian waters were used to constrain numerical models of the angular momentum and energy balances for this topographically complex region. Tides were computed in a hemispherical ocean bounded by meridians, to study the effects of self-gravitation, bottom friction and loading. SEASAT altimetry data gave determinations of the bottom friction coefficient. Stacking of ocean tide data against appropriate low degree spherical harmonics can be used to reduce noise and test the equilibrium response of the tides at Chandler and 18.5 year periods.

A session on terrestrial and celestial reference systems reviewed several approaches based on classical optical astrometry (FK4, FK5) and very long baseline interferometry (VLBI), whose precisions are on the order of ±0.01 and ±0.001 respectively. The need to connect optical catalogs with radio catalogs was emphasized. New space programs for star positioning (HIPPARCOS) are of high interest, but to fully benefit from the high quality of proper motion measurements 'follow-on' programs will be needed.

Refinements in the terrestrial reference frames result from the requirements of the International Earth Rotation Service (IERS). Since 1986, station coordinates have been given together with their motion as deduced from the Minster-Jordan model of plate movements. The motions deduced from laser tracking of LAGEOS are consistent with the Minster-Jordan model at the level of a few cm/year, while the different terrestrial systems agree to within ±10 cm.

Studies of the now well-known relation between the length of day (LOD) and atmospheric angular momentum (AAM) were discussed in a session which identified several directions for further work. A better understanding of fluctuations in the range 30–60 days and at interannual periods is needed, the computation of AAM needs to be extended to higher altitudes, and the details of the atmosphere-mantle coupling mechanisms (of which mountain pressure torque is one) require study. Oceanic angular momentum must be taken into account. Several papers were dedicated to the correlation between abrupt changes of Earth rotation parameters and El Niño phenomena. The strong LOD signal tied to the 1982–83 El Niño is not evident in other El Niño years since 1962.

In a session on polar motion daily determinations of atmospheric wind and pressure were compared with (VLBI) and satellite laser ranging (SLR) observations of polar motion, but coherence studies still leave atmospheric excitation of the Chandler wobble as "not proven", though verifying the semi-annual wobble and higher-frequency irregular fluctuations. The latter are not well accounted for by the inverted barometer model of equilibrium ocean response. Effects of the ocean pole tide on wobble can be modelled by applying the Laplace tidal equations with dissipation, continentality and bathymetry. Future studies of polar motion excitation will need better hydrological data.

In a session devoted to decade and long-term fluctuations of Earth rotation parameters it was shown that coupling torques due to seismically-inferred core-mantle boundary (CMB) topography can be estimated by estimating the toroidal part of the core velocity field near the CMB from the geostrophic approximation and the frozen flux theorem, and have the right sign and magnitude for the current decade change in the LOD. Several papers discussed long-term tidal and non-tidal changes of the spin rate. Modelling tidal friction in the distant past, including re-arrangements of continentality by plate motions, suggests the Gerstenkorn event could be displaced to a sufficiently early time in Earth-Moon history. The geographical distribution of ice sheets during the last glacial maximum is constrained by requiring current polar wander, sea level change and regional effects of postglacial isostatic adjustment to fit the same radial profile of viscoelasticity in the Earth.

In the final session on new methods of measurement and prediction of Earth rotation parameters it was reported that the new IERS, based on international networks using VLBI, SLR and lunar laser ranging, supplemented by regular frequent determinations of AAM, will also maintain the necessary terrestrial and celestial reference systems. Several papers reviewed the precision of current determinations and short-range predictions.

This volume includes many of the papers dealing with these topics. It does not include all as some have been published elsewhere. They were edited for compliance with style standards but the responsibility for scientific content and quality rest with the authors. Their cooperation and patience is gratefully acknowledged.

D. McCarthy
P. Paquet
M. Rochester
Co-Convenors

Geophysical Monograph 59

INTERNATIONAL COOPERATION IN THE STUDY OF THE ROTATION OF THE EARTH

G.A. Wilkins

Royal Greenwich Observatory, Herstmonceux Castle,
Hailsham, East Sussex, BN27 1RP, U.K.

Abstract. During the nine years since the formation of the IAU/IUGG Working Group on the Rotation of the Earth there has been a dramatic increase in the extent of the international cooperation in the study of the rotation of the Earth. The operational arrangements that were brought into use for the MERIT Main Campaign in 1983/4 have been continued and will form the model for the new international Earth rotation service. This will be based on the use of international networks for the new techniques of very-long-baseline radio interferometry (VLBI) and of satellite and lunar laser ranging, supplemented by the regular determination of the angular momentum of the atmosphere. The service will also be responsible for the establishment and maintenance of conventional terrestrial and celestial reference systems, and for providing information about their relationships with the implicit systems of the separate techniques used in the determination of Earth rotation parameters.

The observational data are made available for analysis to research groups in many countries, and the exchange of other information and software is well developed. This has already led to significant increases in our knowledge of the characteristics of the variations in the rate of rotation and in the motions of the axis of rotation within the Earth and in space. In turn, this has given new data on the properties of the interior of the Earth and about the interactions between the crust, oceans and atmosphere of the Earth. The observational data are also being used to improve the terrestrial and celestial reference systems and to determine the current motions of tectonic plates.

Introduction

International cooperation in the monitoring of the rotation of the Earth was the subject of a review paper [Wilkins, 1985] presented at the Longitude Zero Symposium held at Greenwich in 1984

Copyright 1990 by
International Union of Geodesy and Geophysics
and American Geophysical Union.

to mark the centenary of the adoption of the recommendations that the prime meridian passes through Greenwich, and that the universal day begins for all the world at the moment of mean midnight at Greenwich. That paper described the development of the international services for polar motion and time up to 1984 and drew attention to other related activities. The aims of this paper are to review briefly the subsequent activities and to describe the principal scientific results that have been obtained in the past few years. Further details of the organizational aspects of international cooperation in this field have been given in my separate report to IAG Commission VIII on International Coordination of Space Techniques for Geodesy and Geodynamics (CSTG) at this General Assembly. Further and more up-to-date details of the scientific research are given in other papers in this volume.

Observational and Data Processing Activities

The current arrangements for international cooperation in the monitoring of the rotation of the Earth have developed from the MERIT program of international collaboration to Monitor Earth Rotation and Intercompare the Techniques of observation and analysis; this program was proposed and organized by the IAU/IUGG Working Group on the Rotation of the Earth, which was set up in 1978 [Wilkins, 1980]. The MERIT Main Campaign took place during the period 1 September 1983 to 31 October 1984, and the arrangements then introduced have continued with only minor changes to the present time. The improvements in recent years in accuracy of the measurements are so great that the terrestrial and celestial reference frames used to define the rotation must be specified much more precisely than the currently adopted definitions allow. The MERIT activities from 1982 onwards were organized jointly with the IAG/IAU Working Group on the establishment and maintenance of a new Conventional Terrestrial System (COTES), and additional observations were made to determine as accurately as possible the relationships between the reference frames of different techniques of observation.

The activities and initial scientific results of the MERIT Main Campaign were discussed at an international conference held at Columbus, Ohio in 1985 [Mueller, 1985] and further results have been presented at subsequent national and international conferences. A catalog of the observational results on Earth rotation and reference systems has been published [Feissel, 1986] and a general description of the MERIT-COTES data base has been given by Boucher et al [1987].

The discussions at the MERIT-COTES workshop and committee meeting that were held in conjunction with the Columbus conference were the basis of the Joint Summary Report [Wilkins & Mueller, 1986], which was prepared for consideration at the IAU General Assembly at New Delhi in December 1985. The recommendations were adopted by both IAU and IAG and a Provisional Directing Board was set up to prepare detailed proposals for a new International Earth Rotation Service (IERS), which will come into operation on 1988 January 1. The recommendations of the Board have been endorsed by IAG and IUGG at this General Assembly.

Scientific Results

Since 1979 the MERIT-COTES program stimulated improvement of the techniques for monitoring the rotation of the Earth and for determining the positions of points on the Earth's surface. The slowly varying Earth rotation parameters are presented as the difference between universal time (UT) and atomic time (UTC or TAI), and the coordinates of the pole of rotation (celestial ephemeris pole, CEP) with respect to a conventional origin. (The variations in UT are often represented by the variations of the length of day (LOD), which are inversely proportional to the variations in the rate of rotation of the Earth.) The accuracy of measurement has improved by an order of magnitude, so that the length of day is now known to better than 0.1 ms and the direction of the axis better than 1 mas. Correspondingly, there is also evidence that detectable changes in the rate of rotation take place over intervals of a few days, and so the tabulation of the parameters at intervals of 5 days, as has been customary, is no longer adequate for some purposes.

These improvements have come largely from the development of a worldwide network of satellite laser-ranging (SLR) stations and of intercontinental very-long-baseline radio-interferometric (VLBI) networks. The contributions of optical astrometry, Doppler tracking of satellites and connected-element radio interferometers were important during the early phases of the MERIT campaign, but are now much less significant, and these techniques will not be used in IERS. Lunar laser ranging (LLR) has provided valuable data on UT since 1972, but a fully operational worldwide network has not yet been developed; it is hoped, however, that LLR will make a significant contribution to IERS. The improved data have led to important advances in our knowledge about the Earth.

The short-period variations in the length of day have been shown to be very strongly correlated with the variations in the angular momentum of the atmosphere (AAM); the latter quantity is now evaluated from observed data on a routine basis by four meteorological centers and from forecast data by three centers. It may be possible to use the meteorological forecasts to improve the prediction of the variations of UT, for which accurate values are required for the navigation of spacecraft and other purposes. It was noticed that the El-Nino phenomenon of the equatorial Pacific was associated with a significant change in LOD in early 1983 and it is now realized that the effect can also be traced in early LOD records. By removing the AAM contribution from the variations in LOD it will be possible to see if there are any other short-period contributions to LOD due to oceanic effects or other causes. It will also be possible to determine more precisely the characteristics of the decade fluctuations in LOD. These are believed to be due largely to interactions between the mantle and the core; the presence or absence of correlations with the changes in the secular variation of the geomagnetic field would be useful in establishing the nature of the interactions at the core-mantle boundary. The characteristics of the motion of the pole are now much clearer; the motion is comparatively smooth and largely free from sudden changes. It is not clear, however, whether the interaction between the crust and atmosphere is the major cause of excitation of the Chandler wobble. The improved data have also made it worthwhile to look for small departures from the adopted theory of the nutation of the axis in space, and it is claimed that results indicate that the current model of the core must be changed.

The coordinates of the stations used in each technique have been derived and intercompared in order to determine the systematic differences in orientations of the corresponding reference frames as a first step to the establishment of a new conventional terrestrial reference system. In addition, the coordinates of all stations were determined by the Doppler technique during the MERIT Main Campaign, and special efforts were made to collocate instruments of different techniques at selected stations, so that the differences in the coordinates for those stations may be compared with the values obtained by local surveys. In some cases, the collocation is permanent, but in others mobile systems have been used. Sets of station coordinates, and baseline lengths between pairs of stations, have also been determined at different times in order to determine the relative motions, which are due to plate motions and local deformation. In general, good agreement with the Minster-Jordan models, which are based on studies of the geological record, has been found. This work represents a major step towards the establishment of a new conventional terrestrial reference system based on the adopted coordinates and motions of a worldwide network of reference stations.

The MERIT-COTES activities have been of direct benefit to other studies of the Earth and to astronomy. For example, the worldwide regular observations of LAGEOS have been vital to the success of regional projects for the determination of crustal motions using mobile SLR systems. The analysis of the LAGEOS data has also revealed a time variation of the J_2-coefficient in the gravitational field of the Earth; this has been ascribed to the isostatic uplift of the North American shield and gives rise to a non-tidal component in the secular change in LOD. The comparison of the results from different techniques has also been used to look for differences between their celestial reference systems. It is recognized that the new Service should refer the earth-rotation parameters to a celestial reference system defined by a catalogue of VLBI (extragalactic) radio sources, and special efforts will be required to link this system as precisely as possible to the stellar systems of FK5 and HIPPARCOS and to the dynamical systems of the satellites and of the planets.

Conclusion

The MERIT-COTES program has stimulated technical developments and increased international cooperation in studies of the rotation of the Earth and of terrestrial and celestial reference systems. High-precision data are now obtained regularly by new techniques and much valuable information about the Earth and its environment has been derived. Further scientific results and practical benefits may be expected from the analysis and use of the data on Earth rotation and reference systems that will be obtained by the new International Earth Rotation Service.

References

C. D. Boucher, M. Feissel and G. A. Wilkins, The MERIT/COTES database on the rotation of the Earth and terrestrial reference systems, in P. S. Glaeser (ed.), Computer Handling and Dissemination of Data, Elsevier Science Publishers (North-Holland), 1987.

M. Feissel (ed.), Reports on the MERIT-COTES Campaign on Earth Rotation and Reference Systems, Part III, Observational Results, Bureau International de l'Heure, Paris, 1986.

I. I. Mueller (ed.), Reports on the MERIT-COTES Campaign on Earth Rotation and Reference Systems, Part II, Proceedings of the International Conference on Earth Rotation and the Terrestrial Reference Frame, Columbus, Ohio, July 31-August 2, 1985. Department of Geodetic Science, Ohio State University, Columbus, Ohio, USA, 1985.

G. A. Wilkins (ed.), A review of the techniques to be used during Project MERIT to monitor the rotation of the Earth. Published jointly by Royal Greenwich Observatory, Herstmonceux, UK, and Institut für Angewandte Geodäsie, Frankfurt, GFR, 1980.

G. A. Wilkins, International cooperation in the monitoring of the rotation of the Earth. Vistas in Astronomy, 28, 329-335, 1985.

G. A. Wilkins and I. I. Mueller, Joint Summary Report of the IAU/IUGG Working Groups on the Rotation of the Earth and the Terrestrial Reference System, in J.P. Swings (ed.), Highlights of Astronomy, 7, 771-788, Reidel, Dordrecht, Holland, 1986. See also Bull. Geod., 60, 85-100, 1986.

VARIATIONAL CALCULATION OF WOBBLE MODES OF THE EARTH

D. E. Smylie

Department of Earth and Atmospheric Science
York University North York, Ontario M3J 1P3 Canada

Abstract. The theory allowing variational calculation of the wobble modes of the Earth arising from the presence of the fluid core is developed. Included are the effects of shell elasticity, core stratification and compressibility, and gravitational coupling. Numerical implementation of this theory should allow computation of the full suite of Earth's wobble modes.

Introduction

Fundamental to an understanding of the effect of the fluid outer core on Earth's rotational dynamics are those normal modes of the core which can exchange equatorial angular momentum with the shell. They therefore constitute a suite of wobble modes of the whole Earth and the Earth's rotational response to tidal and other torques can be described in terms of their superposition.

In the classical description of the modes of a rotating fluid (Poincaré, 1885; Greenspan, 1969), the pressure field at the surface is a single spherical harmonic for each individual mode. Applied to an Earth model with a rigid, ellipsoidal outer core boundary, a uniform, incompressible fluid core with no inner solid body, this theory (Kudlick, 1966) yields only one wobble mode which is capable of exchanging equatorial angular momentum with the shell. The reason that there is only one such wobble mode in this case is that the exchange of equatorial angular momentum is entirely through that part of the pressure field at the boundary which is a spherical harmonic of azimuthal number unity, zonal number two, and in turn, in the classical description, only one mode exists with the required surface pressure field (Smith, 1977).

The real Earth, however, differs markedly from the classical description in that the shell has finite elasticity, the fluid outer core is stratified and compressible and there is a solid inner core. In addition, the forces of self gravitation are non-negligible and the gravitational coupling of the shell and core must be included along with the pressure coupling. Each of these departures of the real Earth from the classical description produces a violation of the simple theory which predicts only one wobble mode.

In this paper, we formulate the theory required to calculate the wobble modes of realistic Earth models and implement it through a variational principle for the subseismic wave equation governing long period core dynamics (Smylie, 1988).

Application of the Variational Method to Wobble Mode Calculation

At frequencies below seismic frequencies ($< 300\mu$Hz), the appropriate governing equation for oscillations of the fluid core is the subseismic equation,

$$\sigma^2 \nabla^2 \chi - \frac{\partial^2 \chi}{\partial z^2} - \frac{A}{B}\mathbf{C} \cdot \nabla \chi - \mathbf{C}^* \cdot \nabla \left(\frac{\mathbf{C} \cdot \nabla \chi}{B}\right) = 0, \quad (1)$$

where

$$A = \frac{\omega^2}{\beta}\left(\sigma^2 - 1\right) + \sigma^2\left(4\pi G \rho_0 - 2\Omega^2\right) + \left(\hat{\mathbf{k}} \cdot \nabla\right)\hat{\mathbf{k}} \cdot \mathbf{g}_0, \quad (2)$$

$$B = \frac{\alpha^2 \omega^2}{\beta}\left(\sigma^2 - 1\right) + \sigma^2 g_0^2 - \left(\hat{\mathbf{k}} \cdot \mathbf{g}_0\right)^2, \quad (3)$$

and

$$\mathbf{C} = \left(\hat{\mathbf{k}} \cdot \mathbf{g}_0\right)\hat{\mathbf{k}} + i\sigma\hat{\mathbf{k}} \times \mathbf{g}_0 - \sigma^2 \mathbf{g}_0. \quad (4)$$

The scalar variable χ which this equation governs is defined by

$$\chi = \frac{p_1}{\rho_0} - V_1 \quad (5)$$

with p_1 the flow pressure perturbation, ρ_0 the equilibrium density and V_1 the negative change in gravitational potential. The reference frame is taken to be rotating uniformly with the mean angular speed Ω of Earth's rotation about a fixed spatial direction aligned with the z-axis. ω is the angular frequency of oscillation and σ is its dimensionless measure against twice the rotation speed; that is

$$\sigma = \frac{\omega}{2\Omega}. \qquad (6)$$

$\hat{\mathbf{k}}$ is the unit vector in the direction of the mean rotation axis and $\mathbf{g}_0$ is the equilibrium gravity vector, g_0 its scalar magnitude. G is the universal constant of gravitation, α^2 the square of the P-wave velocity and β is a dimensionless stability factor introduced by Pekeris and Accad (1972) and related to the seismic stratification parameter η (the ratio of the actual density lapse rate to the adiabatic lapse rate) by

$$\beta = 1 - \eta. \qquad (7)$$

$\mathbf{C}^*$ is the complex conjugate of the vector $\mathbf{C}$.

The variational solution of the subseismic equation is the subject of another paper in this volume (Smylie, 1988) and we omit details here. For present purposes we require only the surface integral contribution to the functional,

$$\int (\chi^* \mathbf{u} + \chi \mathbf{u}^*) \cdot \hat{\mathbf{n}} dS. \qquad (8)$$

Here $\mathbf{u}$ is the vector displacement field on the bounding surfaces of the fluid core and $\hat{\mathbf{n}}$ is the unit outward normal vector. Because the reference frame we have chosen points in a fixed spatial direction, the normal displacement to be used in (8) in a wobble mode calculation must include contributions from both the wobble motion and the deformation of the shell.

For modes of angular frequency ω, the Liouville equations (Munk and MacDonald, 1960), written in the uniformly rotating space-fixed frame presently employed, provide the equation of motion

$$i\left[\omega - \omega_0 + (1 + e_0)\Omega\right]\tilde{m} = \frac{\tilde{\Gamma}}{A_0\Omega} - i\frac{(\omega + \Omega)}{A_0}\left(\frac{\tilde{l}}{\Omega} + \tilde{c}_0\right), \qquad (9)$$

where $\tilde{m}$, $\tilde{\Gamma}$, $\tilde{l}$ and $\tilde{c}_0 = c_{0_{13}} + ic_{0_{23}}$ are the familiar complex phasor representations of dimensionless wobble angular velocity, equatorial torque, relative angular momentum and off-diagonal inertia tensor components. ω_0 is the resonant Chandler angular frequency of the shell alone, A_0, C_0 are the equatorial and axial moments of inertia of the shell, and $e_0 = (C_0 - A_0)/A_0$ is its dynamical ellipticity. Since the shell reference frame is inclined by a small angle $\delta\Theta_d$ (see equation (15)) to the frame in uniform rotation about a fixed spatial direction used here, there is a relative angular momentum given by

$$\tilde{l} = -\tilde{m}\frac{\Omega}{\Omega + \omega}A_0(1 + e_0)\Omega. \qquad (10)$$

For the single complex harmonic motions contemplated here, (9) may be solved for the wobble angular velocity to yield

$$\tilde{m} = \frac{1}{A_0\Omega}\frac{1}{(\omega_0 - \omega)}\left[(\omega + \Omega)\Omega\tilde{c}_0 + i\tilde{\Gamma}\right]. \qquad (11)$$

Thus, in order to specify the wobble of the shell we must be able to specify both $\tilde{c}_0$ and $\tilde{\Gamma}$.

As well as the direct contribution shell wobble makes to the normal displacement field to be used in evaluating the surface integral (8), there is an indirect contribution through the rotational deformation of the shell which wobble induces.

Due to rotational deformation, the figure axis of the shell shifts with respect to the shell's material elements and thus the symmetry axis of the ellipsoidal core-mantle boundary is similarly rotated which, because of the slight ellipticity of the boundary, produces a contribution to the normal displacement field there. Changes in the off-diagonal components of the shell inertia tensor from non-rotational sources also contribute and the net angular rotation $\delta\Theta$ can be found by inverse similarity transformation to be determined by

$$\delta\Theta \times \hat{\mathbf{k}} = \frac{k_2}{k_{2_S}}\mathbf{m} + \frac{\mathbf{c}_0}{A_0 e_0}, \qquad (12)$$

with

$$\mathbf{m} = m_1\hat{\mathbf{i}} + m_2\hat{\mathbf{j}},$$
$$\mathbf{c}_0 = c_{0_{13}}\hat{\mathbf{i}} + c_{0_{23}}\hat{\mathbf{j}},$$

and where k_2 is a 'tidal effective Love number', k_{2_S} it's secular or long term value (Munk and MacDonald, 1960). The ratio k_2/k_{2_S} is given by the defect from unity of the ratio of the Chandler angular frequency of an isolated elastic shell to its rigid-body angular frequency or

$$\frac{k_2}{k_{2_S}} = 1 - \frac{\omega_0}{e_0\Omega}. \qquad (13)$$

The contribution of the rotation (12) to the normal displacement of a point on the core-mantle boundary with co-ordinates (x, y, z) is then

$$-2f_1\frac{z}{b}\mathbf{R}\cdot\left(\frac{k_2}{k_{2_S}}\mathbf{m} + \frac{\mathbf{c}_0}{A_0 e_0}\right) \qquad (14)$$

with $\mathbf{R} = \hat{\mathbf{i}}x + \hat{\mathbf{j}}y$, f_1 the flattening of the core-mantle boundary and b its mean radius.

The direct contribution of shell wobble to the normal displacement field to be used in evaluating the surface integral (8) can also be derived from a net angular rotation, $\delta\Theta_d$, obeying

$$\delta\Theta_d \times \hat{\mathbf{k}} = -\mathbf{m}\frac{\Omega}{\Omega + \omega}. \qquad (15)$$

The total contribution of bodily motion of the core-mantle boundary to the normal displacement field there is given by

$$-2f_1 \frac{z}{b} \mathbf{R} \cdot \left[\left(\frac{k_2}{k_{2s}} - \frac{\Omega}{\Omega + \omega} \right) \mathbf{m} + \frac{\mathbf{c}_0}{A_0 e_0} \right]. \quad (16)$$

Substituting this expression into the surface integral (8), we find that the wobbling elastic shell selects only the azimuthal number unity, zonal number two parts of the χ field in (8) to give

$$\frac{8}{5} \pi f_1 b^3 \left[\tilde{\chi}_2^{1*} W(\omega) + \tilde{\chi}_2^1 W^*(\omega) \right] \quad (17)$$

where

$$W(\omega) = \left(\frac{k_2}{k_{2s}} - \frac{\Omega}{\Omega + \omega} \right) \tilde{m} + \frac{\tilde{c}_0}{A_0 e_0} \quad (18)$$

and

$$\tilde{\chi}_2^1 = \chi_{2_C}^1 + i \chi_{2_S}^1.$$

Note that only the parts

$$\chi_{2_C}^1 P_2^1(\cos\theta)\cos\phi,$$
$$\chi_{2_S}^1 P_2^1(\cos\theta)\sin\phi$$

of χ on the core-mantle boundary are involved.

Before we can complete the specification of the term (17) in the functional, we must be able to calculate $\tilde{c}_0$ and $\tilde{\Gamma}$ and, by the use of (11), $\tilde{m}$ in terms of the χ field.

Pressure and Gravitational Couples

The slight ellipticity of the core-mantle boundary produces a pressure couple on the shell which is a result solely of the azimuthal number unity, zonal number two parts of the pressure field at the boundary. In the y-notation of free oscillations theory, these are the negatives of the corresponding parts of the normal stress field,

$$-y_{2_C}^1(b) P_2^1(\cos\theta)\cos\phi,$$
$$-y_{2_S}^1(b) P_2^1(\cos\theta)\sin\phi.$$

Correct to first order in the flattening, the pressure couple is

$$\tilde{\Gamma}_P = i \frac{2}{3} f_1 b^3 \int_0^{2\pi} \int_0^{\pi} p P_2^1(\cos\theta) e^{i\phi} \sin\theta \, d\theta \, d\phi, \quad (19)$$

where p is the pressure perturbation at the deformed boundary. The orthogonality relations among spherical harmonics then yield the final expression for the pressure couple,

$$\tilde{\Gamma}_P = -i \frac{8}{5} \pi f_1 b^3 \tilde{y}_2^1(b), \quad (20)$$

with $\tilde{y}_2^1(b) = y_{2_C}^1(b) + i y_{2_S}^1(b)$.

In addition, the small misalignment of the symmetry axis of the core-mantle boundary and the figure axis of the core produces a couple due to centrifugal forces. This couple is most conveniently treated in conjunction with the gravitational couple arising from the misalignment of the figure axes of the shell and core. Indeed, such a combined treatment has already been given (Szeto and Smylie, 1984) in connection with the gravity (centrifugal plus gravitational) restoring torque acting on the inner core. The analysis, based on the 'uniform method of Wavre' (Wavre, 1932; Jardetzkey, 1958), can be carried over directly to the present problem.

By extension of the arguments leading to expressions (12) and (15), the angular displacement of the figure axis of the shell over that of the core is represented by the complex phasor

$$\delta \tilde{\theta} = i \left[W(\omega) - \frac{\tilde{c}_1}{A_1 e_1} \right], \quad (21)$$

where $\tilde{c}_1 = c_{1_{13}} + i c_{1_{23}}$ represents the off-diagonal components of the inertia tensor of the core and A_1, e_1 are its equatorial moment of inertia and dynamical ellipticity, respectively. The total gravity couple which the core exerts on the mantle is then given by

$$\tilde{\Gamma}_G = -\gamma' \delta \tilde{\theta}, \quad (22)$$

with

$$\gamma' = \frac{2}{5} A_1 e_1 \left[\frac{GM_1}{b^3}(2f_1 + bf_1') + \frac{GM_1'}{b^3}(3f_1) \right]. \quad (23)$$

M_1 is the mass of the core and M_1' is the mass it would have if its density throughout was the same as that at its surface. f_1' is the radial derivative of flattening at the core surface. The quantities required for the evaluation of this expression are given by the integration of the Clairaut equation as illustrated in Table 1 for earth model 1066A of Gilbert and Dziewonski (1975).

Shell Deformation and Extension of MacCullagh's Formula

MacCullagh's formula relates the coefficients of the zonal number two spherical harmonics in the expansion of the gravitational potential external to an arbitrary body to the components of the inertia tensor. Applied to the off-diagonal components appearing here, it yields

$$\tilde{y}_{5_2}^1(d) = -\frac{G}{d^3}(\tilde{c}_0 + \tilde{c}_1), \quad (24)$$

where $\tilde{y}_{5_2}^1(d) = y_{5_{2_C}}^1(d) + i y_{5_{2_S}}^1(d)$ and the changes in gravitational potential at the surface (mean radius d) involved are

$$-y_{5_{2_C}}^1(d) P_2^1(\cos\theta) \cos\phi$$

and

$$-y_{5_{2_S}}^1(d) P_2^1(\cos\theta) \sin\phi.$$

TABLE 1. Integration of the Clairaut Equation for Earth Model
1066A of Gilbert and Dziewonski (1975).

r_0 $10^3 m$	ρ_0 $10^3 kg \cdot m^{-3}$	$1/f$	f' $10^{-11} m^{-1}$	$C_1 - A_1$ $10^{32} kg \cdot m^2$	M_1 $10^{22} kg$	I_1 $10^{34} kg \cdot m^2$
0	13.421	418.67	0.473	0	0	0
204.9	13.390	418.46	0.742	–	–	–
409.8	13.336	418.13	1.108	0.01	0.4	0.03
614.8	13.251	417.64	1.598	0.05	1.3	0.20
819.7	13.170	417.03	1.754	0.20	3.1	0.82
1024.6	13.093	416.41	1.763	0.60	5.9	2.49
1229.5	13.021	415.78	1.743	1.48	10.2	6.16
1229.5	12.153	415.78	3.115	1.48	10.2	6.16
1605.3	11.931	411.48	8.326	5.34	21.7	21.98
1981.1	11.680	406.79	6.641	15.05	39.7	61.22
2356.9	11.337	403.01	5.774	35.35	65.3	142.52
2732.7	10.916	399.57	5.690	72.77	99.4	290.87
3108.5	10.456	396.10	6.044	135.56	142.5	537.23
3484.3	9.914	392.41	6.622	233.75	194.8	917.95

Before MacCullagh's formula can be applied to the core, the contribution of the shell to the potential at the core-mantle boundary must be subtracted off. Seen from the interior, it can be shown (Szeto and Smylie, 1984) that this contribution has the second degree term

$$\frac{1}{3}\frac{\gamma b^2}{A_1 e_1} P_2(\cos\theta'), \quad (25)$$

where θ' is the colatitude measured with respect to the shell figure axis and

$$\gamma = \gamma' - A_1 e_1 \Omega^2. \quad (26)$$

Misalignment of the shell figure axis with respect to the reference axis results, in turn, in a contribution to the azimuthal number unity, zonal number two potential which can be calculated from the addition theorem for Legendre functions. The appropriate extension of MacCullagh's formula to the core then becomes

$$\tilde{c}_1 = -\frac{b^3}{G}\left[\tilde{y}_{5_2}^1(b) - \frac{1}{3}\frac{\gamma b^2}{A_1 e_1} W(\omega)\right]. \quad (27)$$

Substitution of the couples (20) and (22) for the torque in equation (11) for shell wobble, and the use of (13), provides

$$W(\omega) = \frac{1}{D(\omega)}\{N_1(\omega)\tilde{c}_0 \\
- N_2(\omega)\left[\frac{\gamma'}{A_1 e_1 \Omega^2}\tilde{c}_1 - \frac{8}{5}\pi f_1 \frac{b^3}{\Omega^2}\tilde{y}_{2_2}^1(b)\right]\}, \quad (28)$$

where

$$N_1(\omega) = \Omega(\Omega+\omega)\{e_0\Omega\omega - \omega(\Omega+\omega_0)\}, \quad (29)$$

$$N_2(\omega) = \Omega^2\{e_0\Omega\omega - \omega_0(\Omega+\omega)\}, \quad (30)$$

and

$$D(\omega) = A_0 e_0 \Omega^2 (\omega_0 - \omega)(\Omega+\omega) \\
- \gamma'\{e_0\Omega\omega - \omega_0(\Omega+\omega)\}. \quad (31)$$

Using (24) and (27) to replace $\tilde{c}_0$ and $\tilde{c}_1$, we find that $W(\omega)$ can be expressed entirely in terms of quantities describing the deformation of the shell. We have

$$W(\omega) = \frac{1}{D'(\omega)}\left\{N_1(\omega)\frac{(b^3\tilde{y}_{5_2}^1(b) - d^3\tilde{y}_{5_2}^1(d))}{G} \\
+ N_2(\omega)\left[\frac{\gamma'}{A_1 e_1 \Omega^2}\frac{b^3}{G}\tilde{y}_{5_2}^1(b) + \frac{8}{5}\pi f_1 \frac{b^3}{\Omega^2}\tilde{y}_{2_2}^1(b)\right]\right\}, \quad (32)$$

where

$$D'(\omega) = D(\omega) + \frac{\gamma b^5}{3GA_1 e_1}\left(N_1(\omega) + \frac{\gamma'}{A_1 e_1 \Omega^2}N_2(\omega)\right). \quad (33)$$

In the accompanying paper in this volume (Smylie, 1988), it is shown how non-rotational shell deformation arising from pressure and gravitational interaction with the core can be described in terms of Love-like coefficients. For a particular degree n we may write

$$\tilde{y}_1(b) = \frac{1}{g_0(b)} \left[h_n^1 \tilde{\chi}_n(b) + b h_n^2 \frac{d\tilde{\chi}_n}{dr}(b) \right], \quad (34)$$

$$\tilde{y}_2(b) = \rho_0(b^-) \left[j_n^1 \tilde{\chi}_n(b) + b j_n^2 \frac{d\tilde{\chi}_n}{dr}(b) \right], \quad (35)$$

$$\tilde{y}_5(b) = -\tilde{\chi}_n(b), \quad (36)$$

$$\tilde{y}_5(d) = k_n^1 \tilde{\chi}_n(b) + b k_n^2 \frac{d\tilde{\chi}_n}{dr}(b), \quad (37)$$

where h_n^1, h_n^2; j_n^1, j_n^2 and k_n^1, k_n^2 are coefficients which are easily found by integration of the the sixth order spheroidal system of differential equations describing shell deformation, and where $\rho_0(b^-)$ is the equilibrium density at the top of the core. As before, the complex phasor

$$\tilde{\chi}_n(r) = \chi_{n_C}(r) + i \chi_{n_S}(r)$$

has been formed from the coefficients of the parts

$$\chi_{n_C}(r) P_n^m(\cos\theta) \cos m\phi,$$

$$\chi_{n_S}(r) P_n^m(\cos\theta) \sin m\phi$$

of the scalar χ field. Only its value and that of its radial derivative at the core-mantle boundary are required.

Shell deformation contributes directly to the integral (8) through the radial displacement described by (34), while the contributions given by (17) are determined by (32) through equations (35), (36) and (37). These may all be combined in the single expression

$$\frac{4\pi b^2}{g_0(b)} \sum_{n=1} \frac{1}{2n+1} \frac{(n+1)!}{(n-1)!} \left\{ h_n'^1 \tilde{\chi}_n \tilde{\chi}_n^* + b h_n'^2 \frac{d}{dr}\left(\frac{\tilde{\chi}_n \tilde{\chi}_n^n}{2}\right) \right\}, \quad (38)$$

with $h_n'^1 = h_n^1 + \delta_n^2 H_1(\sigma)$, $h_n'^2 = h_n^2 + \delta_n^2 H_2(\sigma)$. The functions $H_1(\sigma)$, $H_2(\sigma)$ may be regarded as wobble admittances defined by

$$W = \frac{3}{2} \frac{1}{g_0(b) b f_1} \left[H_1 \tilde{\chi}_2^1 + b H_2 \frac{d\tilde{\chi}_2^1}{dr} \right]. \quad (39)$$

Comparison with the result of substituting (35), (36) and (37) into (32) yields the expressions

$$H_1 = -\frac{2}{3} \frac{g_0(b) b^4 f_1}{D'G} \left\{ \left[1 + \left(\frac{d}{b}\right)^3 k_2^1 \right] N_1 + \frac{1}{\Omega^2} \left[\frac{\gamma'}{A_1 e_1} - \frac{8}{5} \pi f_1 G \rho_0(b^-) j_2^2 \right] N_2 \right\}, \quad (40)$$

$$H_2 = -\frac{2}{3} \frac{g_0(b) b^4 f_1}{D'G} \left\{ \left(\frac{d}{b}\right)^3 k_2^2 N_1 - \frac{8}{5} \pi f_1 \frac{G \rho_0(b^-)}{\Omega^2} j_2^2 N_2 \right\}. \quad (41)$$

The wobble admittances are shown plotted against dimensionless angular frequency in Figure 1.

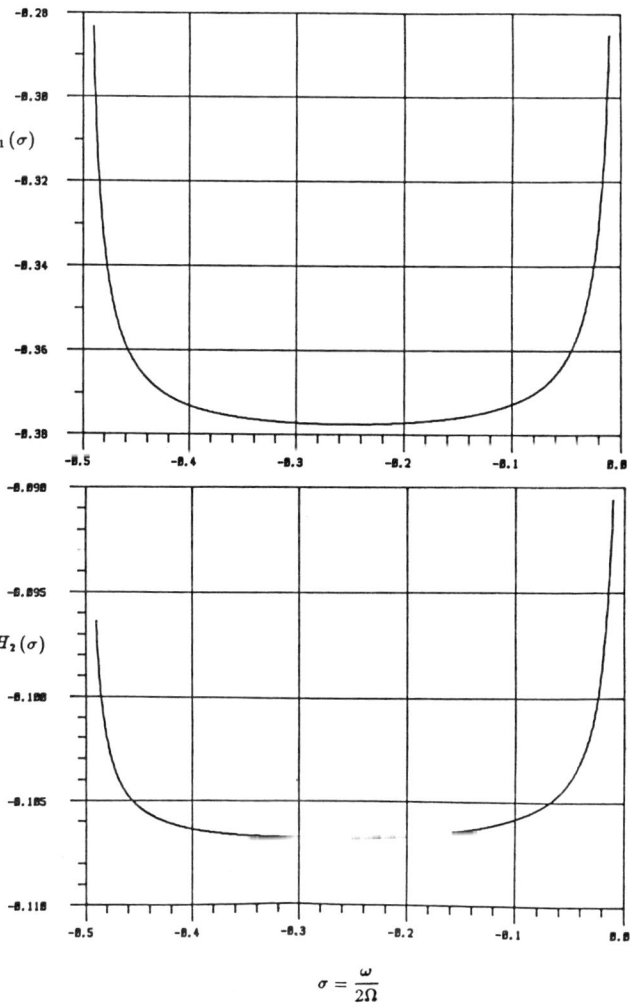

Fig. 1. Wobble admittances as functions of dimensionless angular frequency.

Apart from the requirement of continuity in the radial displacement at the core-mantle boundary, as described by (34), the trial functions are unspecified. They are then converted to spherical harmonics at the boundary to apply continuity in the radial displacement and to complete the specification of the surface integral in the functional through expression (38). Because there is no coupling across azimuthal number, all the wobble modes have the same azimuthal number unity dependence on longitude and the problem becomes two-dimensional. There is also a separation of modes even in the equatorial plane and modes odd in the equatorial plane, allowing solutions to be made in a quarter circle domain. Numerical implementation of the foregoing theory is under way using piecewise bicubic splines as trial functions.

Discussion

The present theory shows that realistic Earth properties produce significant divergences from the predictions of classical theory.

Perhaps the most significant of these is that there appears to be a whole suite of wobble modes of the Earth, in contrast to the single mode which classical theory predicts. Thus, for the full interpretation of the rotational response of the Earth to tidal and other forces one must sum over the full suite of wobble modes. Before this can be done, of course, reliable computational techniques must be developed to find the eigenfunctions and eigenvalues of these modes. This work is now in progress.

It is also important to note that the figure-figure gravitational interaction of the shell and core is significant. This interaction would be missed completely in a theory which is entirely first order in the flattening or which takes the internal level surfaces to all have the same flattening.

Acknowledgments. The author is grateful to Dr. A.M.K. Szeto and to Dr. Kachishige Sato for discussions and for pointing out errors in earlier versions of the manuscript.

References

Gilbert, F. and Dziewonski, A.M., An application of normal mode theory to the retrieval of structural parameters and source mechanisms from seismic spectra, *Phil. Trans. R. Soc. Lond.* A, **278**, 187-269, 1975.

Greenspan, H.P., *The Theory of Rotating Fluids*, Cambridge University Press, Cambridge, 1969.

Jardetzkey, W.S., *Theories of Figures of Celestial Bodies*, Interscience, N.Y., 1958.

Kudlick, M.D., On transient motions in a contained, rotating fluid, *Ph.D. thesis*, M.I.T., 1966.

Munk, W.H. and MacDonald, G.J.F., *The Rotation of the Earth*, Cambridge University Press, Cambridge, 1960.

Pekeris, C.L. and Accad, Y., Dynamics of the liquid core of the Earth, *Phil. Trans. R. Soc. Lond.* A, **273**, 237-260, 1972.

Poincaré, H., Sur l'équilibre d'une masse fluide animée d'un mouvement de rotation, *Acta Math.*, **7**, 259-380, 1885.

Smith, M.L., Wobble and nutation of the Earth, *Geophys. J. R. astron. Soc.*, **50**, 103-140, 1977.

Smylie, D.E., Variational calculation of core modes in realistic Earth models, (this issue) 1988.

Szeto, A.M.K. and Smylie, D.E., Coupled motions of the inner core and possible geomagnetic implications, *Phys. Earth Planet. Int.*, **36**, 27-42, 1984.

Wavre, R., *Figures Planétaires et Géodésie*, Gauthier-Villars, Paris, 1932.

ON THE COMPLEX EIGENFREQUENCY OF THE "NEARLY DIURNAL FREE WOBBLE"
AND ITS GEOPHYSICAL INTERPRETATION

Jürgen Neuberg[1], Jacques Hinderer[2], and Walter Zürn[3]

Abstract. Different tidal measurements from six tidal stations in Central Europe have been used to investigate the resonace effect in the diurnal tidal band. This rotational eigenmode of the Earth, commonly called the "nearly diurnal free wobble" is caused by different coupling mechanisms between the mantle and the outer core of the Earth. By the use of a stacking method it was possible to determine the quality factor of this eigenmode and its eigenfrequency, which is significantly higher than predicted by theories. This frequency shift - in good agreement with VLBI observations - allows one to draw conclusions for the coupling mechanisms, and thus provides a new constraint for Earth models in the vicinity of the core-mantle boundary.

Introduction

The so-called "nearly diurnal free wobble" (NDFW) can be described as a retrograde motion of the instantaneous rotation axis of the Earth relative to the figure axis with an eigenperiod of nearly 1 sidereal day. The associated motion, as seen from the inertial space, consists of a relative rotation of the instantaneous rotation axis in respect to the direction of angular momentum: the "free core nutation" (FCN). Both motions are strongly related by conservation of angular momentum and form together an eigenmode of the rotating Earth.

While the FCN can be investigated by VLBI observations, the NDFW can be studied by using Earth tides in the diurnal frequency band as an excitation mechanism. Tidal waves in the diurnal band - in the spectral vicinity of that eigenfrequency - cause a resonant response of the Earth observed in the corresponding deformation field. Thus, the tidal forcing near the FCN and NDFW eigenfrequency leads to a resonant behavior of the forced nutations and Earth tides, respectively. The investigations of the two aspects of the same eigenmode (VLBI and Earth tide measurements) leads to results, which can be compared directly.

Data, Model, and Results

Tidal gravity measurements from six stations in Central Europe were used to determine resonance parameters of the NDFW: the quality factor and the eigenfrequency. In the following these quantities are expressed as a complex eigenfrequency

$$\tilde{\sigma}_{ND} = \sigma_{ND}\left(1 + \frac{i}{2Q}\right). \qquad (1)$$

To obtain these parameters the tidal admittances of the resonant constituents P1, K1, Ψ1, and Φ1 are compared with a model of a driven damped harmonic oscillator. Searching for global parameters, whose values should be the same for any tidal station, we used a stacking procedure in a linearized least squares fit. The data reduction, the stacking method, and the model function are described in more detail in Neuberg, Hinderer, and Zürn [1987]. Figure 1 shows the distribution of used tidal stations.

The formal errors of the resonance parameters obtained by the least squares estimation are superimposed by systematic errors due to the uncertainty in the ocean load corrections. These errors enter the analysis via data reduction (ocean load correction). To assess their possible influence on the resulting resonance parameters we chose a kind of Monte-Carlo method and varied the given correction values randomly within a range of ± 40 percent (see inset Figure 2). The variation of correction values causes a corresponding spreading of resulting resonance parameters, as plotted in the Q/σ_{ND} - plane (Figure 2). In spite

[1]Geophysikalisches Institut, Universität Karlsruhe, Hertzstraße 16, 7500 Karlsruhe 21, West Germany
[2]Institut de Physique du Globe, Rue Rene Descartes, 67084 Stasbourg, France
[3]Observatorium Schiltach, Heubach 206, 7620 Wolfach, West Germany

Copyright 1990 by
International Union of Geodesy and Geophysics
and American Geophysical Union.

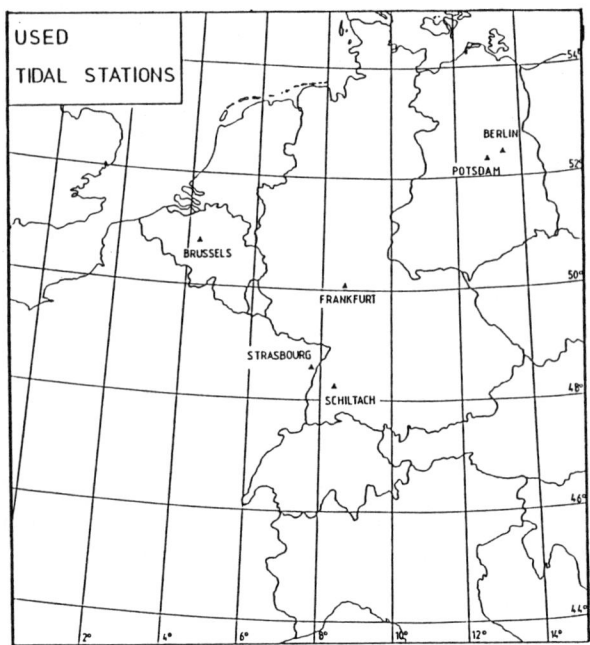

Fig. 1. Used tidal station in Central Europe

of the FCN we get 434 ± 7 days, which is in good agreement with VLBI results [Herring et al. 1986; Eubanks et al. 1986]. Our Q-estimate of about 2800 ± 500 (corresponding to a decay time of $\tau = 2.4$ years) differs from the VLBI determinations ($\tau = 19$ years ± 50 percent)

Interpretation

We base our interpretation on the theory of Sasao et al.[1980] and Hinderer et al.[1982], which provides analytical expressions for the NDFW eigenfrequency, where physical properties are explicitly involved. In turn, that formulation allows one to estimate the influence of the individual properties of the Earth. The complex eigenfrequency of the NDFW is given by, e.g., Neuberg et al.[1987],

$$\tilde{\sigma}_{ND} = -\Omega(1+A/A^m)(\alpha^c - q_o h^c/2 + K' + iK) \qquad (2)$$

where Ω is the Earth's rotation rate, A, A^m are the equatorial moments of inertia of the whole Earth and mantle, respectively; α^c is the dynamical core ellipticity, h^c describes the elastic behavior of the core-mantle boundary (CMB) in response to the pressure distribution caused by the wobble, and q_o is the ratio of gravitational force to centrifugal force at the Earth's surface; K and K' are dimensionless visco-magnetic coupling constants.

In the following we take the observed frequency shift $\Delta\sigma_{ND}$ and the damping $\sigma_{ND}/2Q$ to be 100 percent

of the spreading the obtained eigenfrequency differs significantly from the theoretical values by Wahr [1981] and Sasao et al.[1980]. Instead of 466 sidereal days for the associated eigenperiod

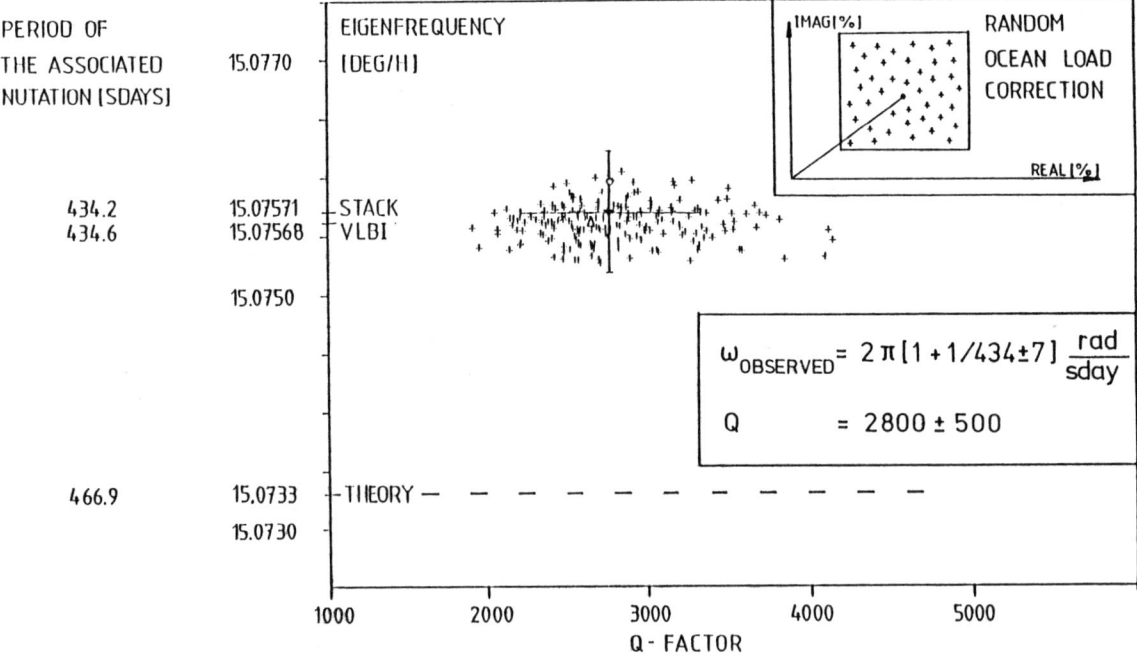

Fig. 2. Resulting resonance parameters in Q/σ_{ND}- parameter plane; eigenfrequency from theory [Sasao et al., 1980] and VLBI observations [Herring et al.,1986] are indicated. The crosses are results from the Monte-Carlo method (inset).

and normalize the different contributions to frequency shift and damping to these quantities. In order to get upper bounds for each influence we try to explain the whole amount of frequency shift/damping by the influence of one parameter alone. In a second approach we adopt values from literature for involved parameters to estimate more realistic contributions of their influence.

Inner Core

The change in the eigenfrequency resulting from a variation ΔA_c in the core moment of inertia ($A^C = A - A^m$) is given by

$$\Delta \tilde{\sigma}_{ND} = \frac{-A\Omega}{(A-A^C)^2}(\alpha^C - q_o h^C/2 + K' + iK)\Delta A^C. \quad (3)$$

If the maximal change in the core inertia comes from the lack of participation of the solid inner core in the nearly diurnal wobbling motion, then $\Delta A^C = -A^{iC}$ (moment of inertia of the inner core). Setting $A^{iC}/A = 8 \; 10^{-4}$ [Smith and Dahlen, 1981], the relative frequency shift is then -1.1 percent; thus, the amplitude of the shift is not only too small, but leads into the wrong direction.

Visco-Magnetic Coupling

This kind of coupling between the core and the mantle results from the tangential stresses of viscous and electromagnetic origin generated at the CMB (or rather within a thin boundary layer). The complex coupling constants depend essentially on the value of the kinematic viscosity ν of the outer fluid core and on the value of the electrical conductivity of the lower mantle β^m in the following dimensionless form (for more details see Loper [1975], Rochester [1976]):

$$K_V = 2.6 \; E^{1/2} \quad (4)$$

$$K_V' \cong K_V/10$$

$$K_m = 2.5 \; B^2/F^{1/2} \quad (5)$$

$$K_m' \cong K_m$$

where $E = \nu/\Omega \; b^2$ is the core Ekman number, b the core radius and $F = (\mu \beta^m \Omega b^2)^{-1}$ the lower mantle magnetic Ekman number (μ magnetic permeability); $B = B_o(\mu \rho^C)^{-1/2}(\Omega b)^{-1}$, where B_o is the mean radial part of the magnetic field (in the axial dipolar approximation) at the CMB, ρ^C is the outer core density.

Setting $K = K_V + K_m$ and $K' = K_V' + K_m'$ the resulting change in the complex eigenfrequency $\Delta \tilde{\sigma}_{ND}$ due to visco-magnetic coupling is:

$$\Delta \tilde{\sigma}_{ND} = -\frac{A\Omega}{A^m}(\Delta K' + i\Delta K). \quad (6)$$

To explain the observed damping by viscous interaction only (in the laminar approximation) implies an Ekman number E of order 10^{-9} corresponding to an upper bound for the viscosity $\nu = 3.3 \; m^2/s$. The induced contribution to the frequency shift is then only 11 percent. Notice that the required value for the viscosity is about 10^6 larger than the theoretical estimate by Gans [1972]. It is, however, quite close to the upper bound proposed by Toomre [1974] inferred from the possible phase change in the 18.6 year principle nutation. It seems that viscous friction probably plays only a negligible role in the coupling process and is unable to explain the observational results. Nevertheless, this conclusion possibly does not hold if the boundary layer is hydrodynamically unstable leading to turbulent coupling as suggested by Toomre [1966].

Similarly, to explain the damping in terms of magnetic coupling alone, requires a magnetic Ekman number of order 10^{-12} implying a very high value of the lower mantle conductivity $\beta^m \cong 4 \; 10^8 (\Omega m)^{-1}$. Using a reasonable value of β^m between 10^2 and 10^3 we obtain relative contributions to the frequency shift and damping between 0.018 percent and 0.18 percent, respectively. Visco-magnetic damping seems, therefore, completely inappropriate to explain the discrepancy between theory and observations (for similar conclusions see, e.g., Gwinn et al. [1986]).

Elasticity and Anelasticity

The perturbation in the eigenfrequency caused by changes in the elastic parameters is:

$$\Delta \tilde{\sigma}_{ND} = \Omega \; \frac{A}{A^m} \; \frac{q_o}{2} \; \Delta h^C. \quad (7)$$

In the simplified case of an incompressible two layer Earth model (homogeneous core and mantle with fixed density contrast) studied elsewhere [Hinderer et al., 1987], it can be shown that h^C is a simple function of the mantle mean shear modulus μ:

$$\frac{\Delta h^C}{h^C} \cong -0.57 \; \frac{\Delta \mu}{\mu} \quad (8)$$

To explain the frequency shift by a change in the mantle elastic rigidity would require a very large increase in mantle stiffness $\Delta \mu/\mu \cong 26$ percent; this can be excluded with confidence according to a small variation in the seismic velocities (and hence in Lame parameters) between different Earth models, and, in general, $\Delta h^C/h^C$ is less than 2 percent [Gwinn et al., 1986]. With this last value the relative frequency shift is then about 9 percent.

In the case of an anelastic medium the rheological equation between stress and deformation is more complicated than in the previous purely elastic case (see, e.g., Wahr and Bergen, [1986]) and leads to the introduction of complex, in general, frequency dependent Lame parameters. The Love numbers (or combinations like h^c) become complex too, the imaginary part being related to damping. Whatever rheological model we take, either an extrapolation from the seismic band by introducing a shear modulus quality factor (e.g., Anderson and Minster, [1979]; Smith and Dahlen, [1981]) or a model of visco-elastic nature (e.g., Peltier et al., [1980]) as required for explaining long time scale geophysical phenomena: the major consequence is to decrease the NDFW frequency. This implies that anelasticity leads always to a frequency shift opposite to the observed one.

Adopting the upper bounds for mantle anelasticity from the study by Wahr and Bergen [1986], we get -20 percent for frequency shift and 4.4 percent for damping. We see that, even for the most dissipative anelastic models, the theoretical damping of the NDFW is still very weak in comparison with observation.

CMB Ellipticity

The change in the eigenfrequency due to variation of the core's dynamical ellipticity $\Delta\alpha^c$ is

$$\Delta\sigma_{ND} = -\Omega \frac{A}{A^m} \Delta\alpha^c. \qquad (9)$$

The change in dynamical ellipticity $\Delta\alpha^c \simeq (\Delta C^c - \Delta A^c)/A^c$ is essentially controlled by the shape of the CMB and the corresponding changes in the moments of inertia. For those one finds in a linear approach:

$$\Delta C^c = \rho^c \int_0^{2\pi}\int_0^\pi R^4(\theta) \sin^3\theta\, \Delta R(\theta,\lambda)\, d\theta d\lambda \qquad (10)$$

$$\Delta A^c = \rho^c \int_0^{2\pi}\int_0^\pi R^4(\theta)(\sin\theta - \sin^3\theta \cos^2\lambda)\Delta R(\theta,\lambda)d\theta d\lambda \qquad (11)$$

where ρ^c is the core density near the CMB, $R(\theta)$ describes the undisturbed CMB in hydrostatic equilibrium, and $\Delta R(\theta,\lambda)$ is a small deviation from it. ΔR can be specified in terms of a spherical harmonic

$$\Delta R(\theta) = C_2^0\, P_2^0(\cos\theta) \qquad (12)$$

to express an additional flattening of the core. When $\Delta\alpha^c_{EL}$ is the change of dynamical ellipticity that can explain the observed frequency shift $\Delta\sigma_{ND}$ completely, one finds with (10) to (12) a corresponding coefficient C_2^0 which then can be expressed as a change d of the polar radius.

The effect of mantle anelasticity lengthens the FCN period, increasing the discrepancy between theory and observation. The change of dynamical ellipticity, which explains even the extended frequency shift (compensating also the effect of mantle anelasticity) is $\Delta\alpha_{ANEL}$. Figure 3 shows for $\Delta\alpha_{EL}$ and $\Delta\alpha_{ANEL}$ the corresponding changes d in the polar radius of the core for two different densities at the CMB. That change turns out to be

$$-250m \geq d \geq -350m,$$

the minus sign indicating an increase of oblateness.

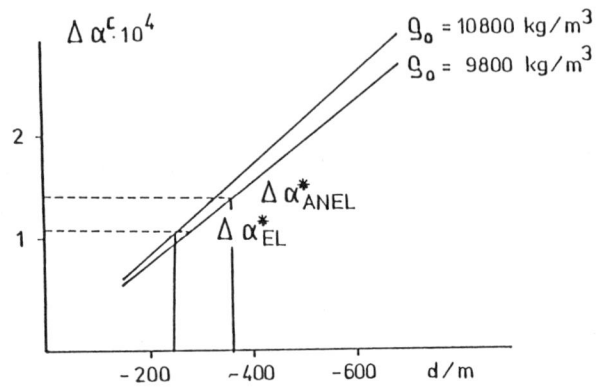

Fig. 3. Change in dynamical ellipticity versus change in the core's polar radius; $\Delta\alpha_{EL}$ and $\Delta\alpha_{ANEL}$ can explain the frequency shift for an elastic and anelastic Earth model, respectively.

CMB Topography

Morelli and Dziewonski [1987] provided a spherical harmonic expansion of the shape of the CMB inferred from PKP and PcP travel time residuals. We use their coefficients to calculate the corresponding changes in moments of inertia. By setting

$$\Delta R(\theta,\lambda) = \sum_{\ell=0}^{4} \sum_{m=0}^{\ell} (C_\ell^m \cos m\lambda + S_\ell^m \sin m\lambda) P_\ell^m(\cos\theta) \qquad (13)$$

we use our linear approach (10) and (11).

The deviation of the equatorial moment of inertia $\Delta A^c = \Delta A^c(\lambda_0)$ from the constant value usually assumed for a rotational symmetric Earth is controlled by the coefficients C_ℓ^2 and S_ℓ^2 only. This deviation turns out to be at most $\Delta A^c(\lambda_0)/A^c = 0.01$ percent where λ_0 indicates the position of the corresponding rotation axis in the equatorial plane. That confirms the assumption of an rotational symmetric Earth. The deviation ΔA^c and ΔC^c from the hydrostatic values is controlled by coefficients C_1^0, C_2^0, C_3^0, and C_4^0, respectively.

The resulting change in dynamical ellipticity $\Delta \alpha^c$ is essentially determined by the coefficient $C_2^0 = 0.14 \pm 0.24$ (Figure 4). This value produces a decrease of core flattening rather than an increase as required to explain the observations. Only due to their large error bars, Morelli and Dziewonski's [1987] results can be addressed as noncontradictory to ours. Using $C_2^0 = -0.10$ (corresponding to their lower bound) could explain 78 percent of the observed frequency shift.

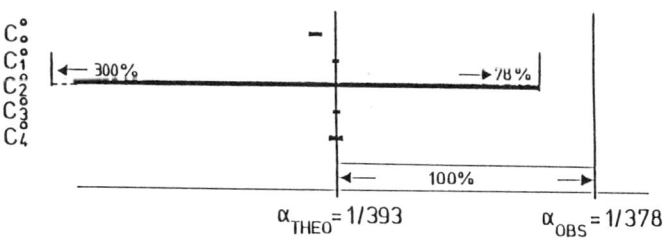

Fig. 4. Influence of coefficients of spherical harmonic expansion [Morelli and Dziewonski, 1987]; $\rho^c = 10^4$ kg/m^3.

Summary and Conclusions

The influences of the considered effects acting at the CMB are summarized in Figure 5, assuming commonly accepted values for involved parameters. We draw the following conclusions:

- Gravity-tide and VLBI observations indicate a decrease of FCN period by 30 days and a Q between 2800 and 8000.
- Mantle anelasticity causes a shift of NDFW frequency in the opposite direction [Wahr and Bergen, 1986] and enhances the discrepancy between observed and theoretical eigenfrequencies.
- The frequency shift is easily explained by an increase of CMB ellipticity by 200-400m (see also Gwinn et al. [1986]). Other effects do not contribute significantly to the shift of eigenfrequency, which is not certain for the oceans.
- Due to large error bars for C_2^0 in Morelli and Dziewonski's [1987] CMB model an increase of CMB ellipticity is not contradictory to seismological data. The strong sensitivity of NDFW frequency to C_2^0 coefficient can, in turn, provide an additional constraint for CMB models in future.
- The observed Q is much lower than reasonable models of mantle anelasticity can explain. A kinematic core viscosity of 3.3 m^2/s could explain the observation. This upper bound is close to the value derived by Toomre [1974] but six orders of magnitude larger than the value of Gans [1972]. With high probability additional damping mechanisms must be responsible (e.g., oceans, CMB-topography).

Acknowledgments. We like to thank Ursel Zimmermann, who drew the figures and helped with the calculations. The manuscript was typed by Alice Leykam. Financial support was provided by the DFG under grant number Wi 687/1-1 and INSU grant number ATP-Noyau 16-14. The computations were carried out at the Computer Center of Karlsruhe University.

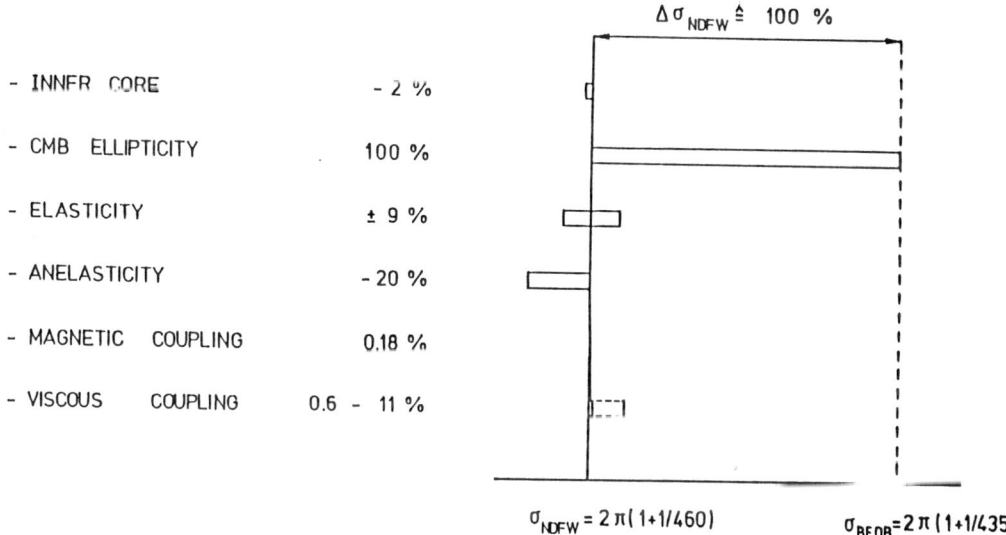

Fig. 5. Summary of discussed contributions.

References

Anderson, D.L. and J.B. Minster, The frequency dependence of Q in the Earth and implications for mantle rheology and Chandler wobble, Geophys. J.R. astr. Soc., 58, 431-440, 1979.

Dehant, V., Integration des equations differentielles aux deformations d'une terre ellipsoidale, inelastique, en rotation uniforme avec in noyau liquide, Thesis, Universite Catholique de Louvain, Belgique, 1986.

Eubanks, T.M., J.A. Steppe, and O.J. Sovers, An analysis and intercomparison of VLBI nutation estimates, Proc. of the Int. Conference on Earth Rotation and the Terrestrial Reference Frame, Ohio 326-340, 1986.

Gans, R.F., Viscosity of the Earth's core, J. Geophys. Res., 77, 360-366, 1972.

Gwinn, C.R., T.A. Herring, and I.I. Shapiro, Geodesy by radio interferometry: Studies of the forced nutations of the Earth, 2nd interpretation, J. Geophys. Res., 91, No B5, 4755-4765, 1986.

Herring, T.A., C.R. Gwinn, and I.I. Shapiro, Geodesy by radio interferometry: Studies of the forced nutations of the Earth, 1. Data Analysis, J. Geophys. Res., 91, 4745-4754, 1986.

Hinderer J., H. Legros, and M Amalvict, A search for Chandler and nearly diurnal free wobbles using Liouville equations, Geophys. J.R. astr. Soc., 71, 303-332, 1982.

Hinderer J., H. Legros, and M Amalvict, Tidal motions within the Earth's fluid core: resonance process and possible variations, Phys. Earth. Planet. Int., in press, 1987.

Loper, D.E., Torque balance and energy budget for the precessional driven dynamo, Phys. Earth Planet. Int., 11, 43-60, 1975.

Morelli, A. and A.M. Dziewonski, Topography of the core-mantle boundary and lateral homogeneity of the liquid core, Nature, 325, 678-683, 1987.

Neuberg, J., J. Hinderer, and W. Zürn, Stacking gravity tide observations in Central Europe for the retrieval of the complex eigenfrequency of the nearly diurnal free wobble, Geophys. J.R. astr. Soc., 91, 853-868, 1987.

Peltier, W.R., D.A. Yuen and P. Wu, Post glacial rebound and transient rheology, Geophys. Res. Let., 7, 733-736, 1980.

Rochester, M.G., The secular decrease of obliquity due to dissipative core-mantle coupling, Geophys. J.R. astr. Soc., 46, 109-126, 1976.

Sasao, T., S. Okubo, and M. Saito, A simple theory on the dynamical effects of a stratified fluid core upon nutational motion of the Earth, Proc IAU Symp. No.78 Nutation and the Earth's Rotation, Kiev 1977, eds Fedorov, E.P., Smith, M.L., and Bender, P.L., Reidel, Dordrecht, pp. 165-183, 1980.

Smith, M.L. and F.A. Dahlen, The period and Q of the Chandler wobble, Geophys. J.R. astr. Soc., 64, 223-281, 1981.

Toomre, A., On the coupling of the core and mantle during the 26000 yr precession, The Earth-Moon System, eds B.G. Mardsen and A.G.W. Cameron, Plenum Press, New York, pp. 33-45, 1966.

Toomre, A., On the 'nearly diurnal free wobble' of the Earth, Geophys. J.R. astr. Soc., 38, 335-348, 1974.

Wahr, J.M., Body tides of an elliptical, rotating, elastic and oceanless Earth, Geophys. J.R. astr. Soc., 64, 677-703, 1981.

Wahr, J.M. and Z. Bergen, The effects of mantle anelasticity on nutations, Earth tides, and tidal variations in rotation rate, Geophys. J.R. astr. Soc., 87, 633-668, 1986.

NUMERICAL SOLUTION FOR THE ROTATION OF A RIGID MODEL EARTH

Joachim Schastok, Michael Soffel and Hanns Ruder

Lehrstuhl für Theor. Astrophysik, Auf der Morgenstelle 12, D - 7400 Tübingen, FRG

Abstract. Using the ephemerides DE200 of the JET PROPULSION LABORATORY we integrated the Euler equations for a rigid model Earth taking into account torques exerted by the Moon and the Sun over a span of 40 years. Inserting appropriate initial conditions and removing precessional effects we were able to compare our results for the nutational motion directly with Kinoshita's analytical theory. Differences to Kinoshita's theory for the nutations in longitude and latitude are as large as 2 mas and 1 mas, respectively, with main frequencies of 18.6 and 9.3 years.

Astronomers usually represent the rotation of the Earth in a space - fixed equatorial system by employing four matrices in the sense that a vector a given w.r.t. to equator and equinox of a certain epoch is transformed to a corotating terrestrial system by

$$a' = \hat{W}\hat{S}\hat{N}\hat{P} a \qquad (1)$$

where $\hat{P}$, $\hat{N}$, $\hat{S}$ and $\hat{W}$ denote the matrices of precession, (forced) nutation, diurnal spin and polar wobble, respectively. Here, we are interested in the problem of forced nutations only.

The nutation matrix recommended by the International Astronomical Union was developed by *Wahr* [1981] on the basis of *Kinoshita's* [1977] analytical results for a completely rigid Earth. The model used by Wahr is that for an oceanless Earth with liquid outer and solid inner core, taking into account effects of elasticity as perturbations to the behaviour of a rigidly rotating Earth. Recently discrepancies of some mas between the theoretically computed nutations and those derived from VLBI observations have been detected (*Herring et al.* [1986]). Although these discrepancies arise mainly from insufficient modelling of the core - mantle coupling it seems desirable to improve the theoretical results for the nuta-

Copyright 1990 by
International Union of Geodesy and Geophysics
and American Geophysical Union.

tional motion of a rigid Earth with the aid of numerical methods. This has been done for the first time by *Kubo and Fukushima* [1987] who found differences of the order of 1 mas between their numerically derived values for the nutations in longitude and obliquity and those of Kinoshita's analytical theory. However, Kubo and Fukushima used a somewhat simplified model including e.g. lunar ephemerides from Brown's theory. We have improved this work for example by using the DE200 ephemerides from the Jet Propulsion Laboratory (see e.g. *Newhall et al.* [1983]) to evaluate torques exerted by the Moon and the Sun on the figure of the Earth.

The integration of the Euler - eqs. for the rotational motion of a rigid Earth over a span of about 40 years was performed w.r.t. to the ecliptical reference system of the epoch J2000.0. Initial conditions for two of the Euler angles (Θ and Φ related to the obliquity of the ecliptic and the direction to the vernal equinox at the initial epoch) and their time derivatives were taken from the theory of precession by *Lieske* [1977] and Kinoshita's nutational series. Effects from the geodetic precession were removed for our Newtonian treatment. The z - component of the Earth's angular velocity was taken to be 1 299 548. 204"/day. As with the ephemerides most of the dynamical parameters like the GM - values for Earth, Moon and Sun and multipole moments and equatorial radius of the Earth were taken from DE200. In addition the Earth's quadrupole moment and dynamical ellipticity were taken to be $1,771 \cdot 10^{-6}$ and 0. 003 273 952, respectively, in agreement with Kinoshita's theory. We computed the corresponding principal moments of inertia according to the theory for a rigid body.

To compare our results with those of Kinoshita's theory we first had to remove secular drifts due to precession in the angles Θ and Φ which yields the nutations w.r.t. to the ecliptical system of our fixed reference epoch. This result is transformed to a mean of - date system by the reduction formula given by Kubo and Fukushima with sufficient accuracy. A detailed analysis shows that the

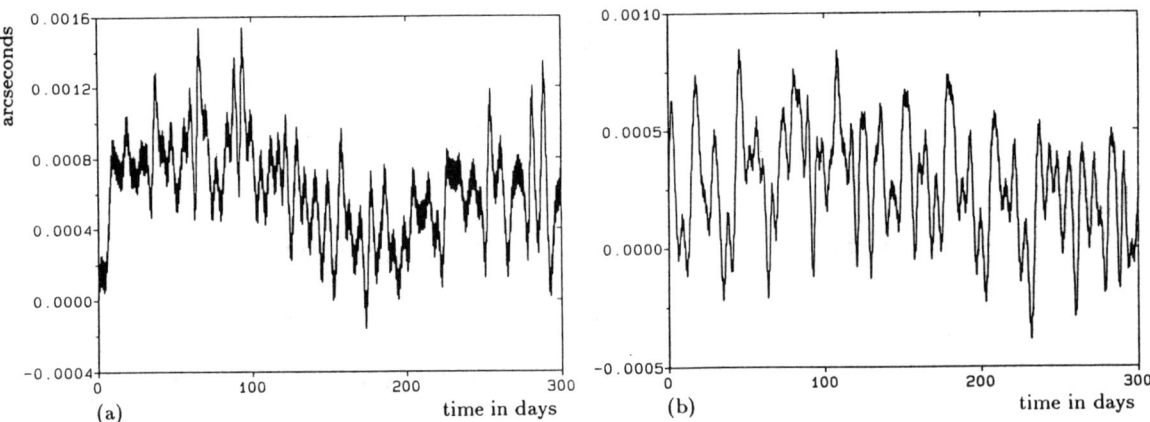

Fig. 1. Short - periodic differences between our numerical results and those of Kinoshita's analytical theory for nutation in longitude (a) and obliquity (b) of the Earth's figure axis.

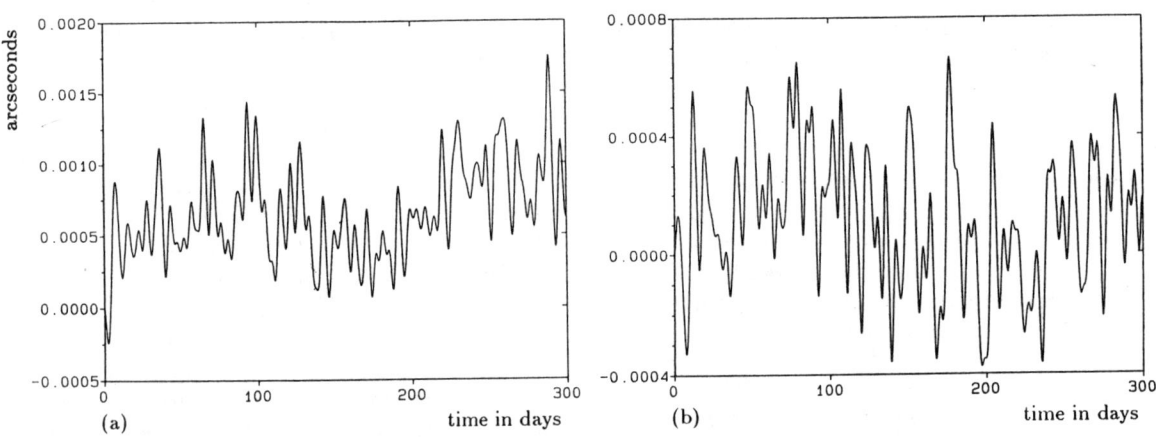

Fig. 2. Short - periodic differences between our numerical results and those of Kinoshita's analytical theory for nutation in longitude (a) and obliquity (b) of the Earth's angular momentum axis.

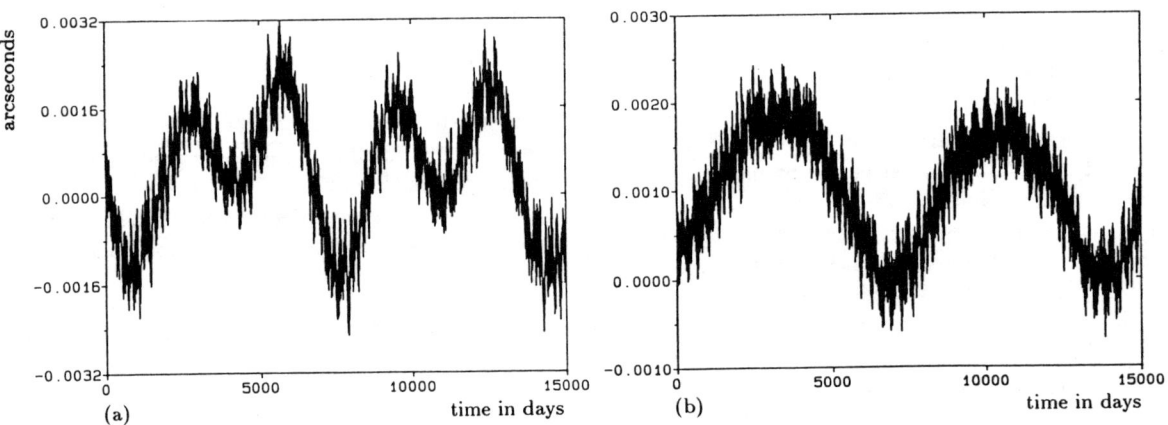

Fig. 3. Long - periodic differences between our numerical results and those of Kinoshita's analytical theory for nutation in longitude (a) and obliquity (b) of the Earth's figure axis.

free nutations are negligibly small for the initial conditions chosen by us. This can also be seen from the short - periodic differences between our and Kinoshita's results for the motion of the figure axis of the Earth and the Earth's angular momentum. Since free nutations do not affect the orientation of the angular momentum axis we conclude that the "noise" in the differences for nutations of the Earth's figure (Figs. 1a and b) when compared to the smooth curves for differences in the nutations of the angular momentum axis (Figs. 2a and b) is just caused by free nutations. In agreement to Kubo and Fukushima the dominant discrepancies between numerical and analytical results have periods of about 9 and 18 years and lie in a range of ~ 1 mas for for the nutations in obliquity and of ~ 2 mas for those in longitude (see Figs. 3a and b). In addition they show a phase shift w.r.t. the longitude of the Moon's ascending node and so to the arguments of the leading terms of the current nutational series.

A purely numerical derivation of an improved nutation series for a rigid Earth is presently in work.

References:

Herring, T.A., Gwinn, C.R., Shapiro, I.I., 1986, Journal of Geophysical Research Vol. 91 No. B5, 4745

Kinoshita, H., 1977, Celestial Mechanics 15, 277

Kubo, Y., Fukushima, T., 1987, in: G. Wilkins and A. Babcock (Eds.), The Earth's Rotation and Reference Frames for Geodesy and Geodynamics, Proceedings of IAU - Symp. No. 128, Reidel

Lieske, J.H., Lederle, T., Fricke, W., Morando, B., 1977, Astronomy and Astrophysics 58, 1

Newhall, X.X., Standish, E.M., Williams, J.G., 1983, Astronomy and Astrophysics 125, 150

Wahr, J.M., 1981, Geophysical Journal of the Royal Astronomical Society 64, 705

THE LONG PERIOD ELASTIC BEHAVIOR OF THE EARTH

Bernd Richter

Institut fuer Angewandte Geodaesie, Richard-Strauss-Allee 11, D-6000 Frankfurt 70, FRG

Abstract. A 5 year series of gravity observations with Superconducting Gravimeters (SCG) has been investigated. The last 2 years of these series have been registered with two SCG's, running side by side at the Bad Homburg Earth tide station. This experiment underlines the high quality of the data and gives informations about the accuracy of the instruments.

Different models have been used for the reduction of the air pressure influence, at which the dependence on tidal frequencies and regional air pressure distribution are taken into account.

The long term gravity variations can be influenced by the instrumental drift. It is shown that most of the instrumental drift of SCG TT40 is caused by variations of the temperature feedback system which can be reduced from the data. The reduced observation series are used to derive the gravity factors and phase delays of the long periodic tides. Besides the tidal frequencies a strong correlation with the polar motion observed by other techniques is given. Analyzing the data of the IRIS-project together with the gravity data set a gravity factor of the Chandler period is determined.

Introduction

The analysis of Earth tide observations is one tool to get information on the elastic behaviour of the Earth in different frequency domains. The major tides of the tidal spectrum appear in the semi-diurnal and diurnal band and the worldwide behaviour is well known [Melchior, 1983]. By contrast the information on long period gravity variations with frequencies between 14 and 435 days is poor. This is due to the fact that these variations are small and latitude dependent (5 cercent of the total tidal signal at 45°) and that the behavior of spring gravimeters is not predictable over a long term. The first suitable information on the long period frequency band arises from observations obtained with the Superconducting Gravimeter (SCG).

In the beginning of 1981 the SCG TT40 was installed at the Earth tide observation station Bad Homburg ($\Phi = 50.2285°N, \lambda = 8.6113°E, h = 190m$) established by the IfAG. The continuous registration of SCG TT40 had to be interrupted in May 1984, due to a problem arising from frozen air in the neck of the Dewar, so that the vibrations of the cold-head disturbed the observation. To solve this the instrument had to be warmed up, during which time additional modifications were made to the instrument. In May 1985 a second SCG, TT60, was installed at the same station. The first nine months were used for intensive studies of the instrumental behavior of the 2 gravimeters. The total schedule is given in Figure 1.

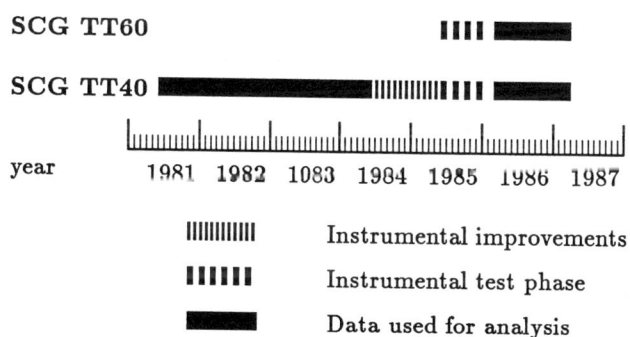

Fig. 1. Gravimeter schedule.

Apart from the Gravimeter signal (sampling interval 10 seconds) the air pressure, the temperature feedback voltage, the x-, y-tilt and the pillar temperature are registered at 2 minute intervals.

The calibration factor of SCG TT40 has been derived from a 9 month parallel registration with the La Coste Romberg Earth tide meter ET-15 from the Oceanographic Institute, Bidston, GB. The manufacturer's calibration of the gravimeter measuring screw of ET-15 has been checked on the well established gravity calibration line at Hanover, FRG. [Edge et al., 1986]. The calibration factor of SCG TT60 has been matched to that of the SCG TT40.

The data analysis is performed by the HYCON-MC method, which allows a simultaneous solution of tidal parameters, instrumental drift and the determination of regression factors for different input channels [Schueller, 1986].

Instrumental Drift Behaviour

By theory the SCG should be a drift-free instrument, but the mechanical imperfection of the vacuum-can and the lines, the reaction of the helium gas with the thin layer of lead covering the test mass as well as the aging of the electronic components are named as example candidates which can produce a drift in the electronic readout of the gravimeter. For the interpretation of the signal those apparent gravity variations resulting from instrumental drift have to be separated from 'real' gravity fluctuations.

Most of the drift of SCG TT40 is caused by a small leak in the vacuum-can. During the 3 year registration series the relevant data for performing a regression analysis of the drift characteristic could not be recorded. Different mathematical drift models (polynomial-, e-, cos- functions) have been fitted in a least square procedure [Richter, 1986]. They influence mainly the phase and not the amplitude relationship of the long period tidal parameters. The remaining analysis of the 3 year interval uses a cubic polynomial drift model which gives the best fit.

In 1985, parallel to the installation of the second SCG, the registration system were also extended to obtain the required data. Investigations show that the drift of SCG TT40 can be described by the physically realistic variation of the temperature feedback voltage. By a regression analysis the drift of SCG TT40 can be reduced to a level below $2 \cdot 10^{-8} ms^{-2}$/ year. This remaining drift, for which no physical model can be given at the moment, has been modelled by a polynomial approximation [Richter, 1987].

The SCG TT60 does not show any abnormal drift behaviour. The strong drift at the beginning dropped out within 3 months and is now at the same level as SCG TT40 ($2 \cdot 10^{-8} ms^{-2}$).

Long Period Tidal Parameter

The tidal parameters (i.e. gravimetric factor δ and phase lead κ were obtained by a harmonic least squares procedure (HYCON-MC), with the local barometric pressure, the x- and y-polarmotion coordinates and the drift model being used as additional input signals to the theoretical tides.

An extract of the result of the analysis of the 3 year observation data (series I) is given in table 1. The com-

TABLE 1. The long period tidal parameters determined from an uninterrupted 3 year data set.

No.	Tide	Amplitude $10^{-8} ms^{-2}$	δ	σ	κ [°]	σ [°]
1	Sa	0.365	7.4451	0.4338	-8.65	5.46
2	Ssa	2.300	1.0810	0.0474	-0.13	2.50
3	Msm	0.600	1.1845	0.0543	3.70	2.62
4	Mm	2.611	1.1274	0.0104	0.13	0.53
5	Msf	0.432	1.1718	0.0630	4.85	3.08
6	Mf	4.944	1.1462	0.0056	0.33	0.28
7	Mstm	0.180	1.1265	0.1638	5.64	8.35
8	Mtm	0.947	1.1305	0.0290	-1.53	1.47
9	Msqm	0.152	1.1084	0.0606	1.25	3.13

plete results including the diurnal and semi-diurnal frequencies have been published in Richter [1986] or Richter [1987]. The tidal parameters in Table 1 are not corrected for ocean loading effects. The result of the annual tide SA is far away from the theoretically anticipated value of 1.16. The reason is that besides the annual gravitational effect other effects (e.g., meteorological effects) occur with the same period and were accumulated in their phase relation, leading thereby to large uncertainties. At the moment there are no models available to separate the different annual influences.

Determination of the Elastic Earth Parameter in Relationship to the Polar Motion

The x- and y-polar coordinates describe the instantaneous position of the rotation axis with respect to the CIO. The change in the position of the rotational axis leads to a change in the centrifugal force at a fixed station on the Earth's surface. The mathematical background has been derived by several authors [Lambeck, 1973; Heitz, 1980, 1983; WAHR, 1985] so that only the final equation is cited here:

$$b^a_{3,z} = (\partial W_R/\partial \bar{r}) = -\frac{2\delta}{\bar{r}} W_R \quad (1)$$
$$\approx -\delta \bar{\omega}^2 \bar{r} \sin\Phi(n_1 \cos\lambda + n_2 \sin\lambda)$$

with W_R denoting the rotational potential, $\bar{r}$ the Earth's radius, δ the gravimetric factor, $\bar{\omega}$ the Earth's rotational speed and n_1, n_2 the pole coordinates in rad.

To determine a possible phase lag equation (1) is rewritten as:

$$b^a_{3,z} = n_1 k_0 \delta \cos\lambda + n_2 k_0 \delta \sin\lambda = n_1 U_1 + n_2 U_2 \quad (2)$$

with

$U_1 = k_0 \delta \cos\lambda$; $U_2 = k_0 \delta \sin\lambda$ and $k_0 = -3.36 \cdot 10^6 \sin 2\Phi$.

The raw pole-coordinates observed by the IRIS-network, published by the U.S. Naval Observatory are added to the harmonic analysis as additional input channels in the time domain. The simultaneous solution estimates the following δ- and κ-parameters of the polar motion:

$$\delta = 1.27 \pm 0.16 \text{ and } \lambda = 7.7° \pm 3.2°.$$

Taking the longitude of the Earth tide station into account no phase lag κ can be detected between the gravitational and the celestrially observed polar motion (see Figure 2). The same procedure has been repeated with the smoothed BIH-data [Richter and Zürn, 1987].

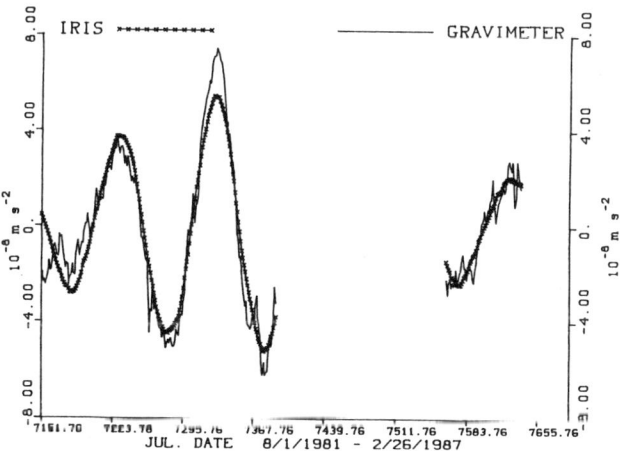

Fig. 2. The changes in the centrifugal force derived from IRIS data and SCG TT40 observations.

The criticism of this method is that with a 3 year data set one cannot separate the 435 day period from the annual tide. The following method is based on the assumption that the polar motion can be divided into two parts: a periodical (Chandler and annual period), and an aperiodical part. To determine the periodical components the 7 year IRIS data set has been analyzed with respect to the Chandler, annual, and semi-annual period. The result is:

Chandler period:	amplitude = 3.55	$\cdot 10^{-8} \text{ms}^{-2}$
SA (polar motion):	amplitude = 1.69	$\cdot 10^{-8} \text{ms}^{-2}$
SSA (polar motion):	amplitude = 0.11	$\cdot 10^{-8} \text{ms}^{-2}$

In the spectrum of the residuals no significant frequency is left which can interfere with tidal waves (see Figure 3). Now the determined periodic part of the polar motion multiplied by the theoretically anticipated δ-factor of 1.16 is subtracted from the observed gravimetric signal and the analysis of the gravity data is repeated with the residuals (non-periodic part of the IRIS data set analysis) as the input channel. The result of the δ-factor of the Chandler period derived from the 3 year data set is then:

$$\delta = 1.23 \pm 0.01.$$

A common analysis of series I and series II of SCG TT40 (together more than four years) leads to the following result:

$$\delta = 1.21 \pm 0.01.$$

The formal errors are somewhat underestimated, but it is demonstrated that the gravity data are also sensitive to the non-periodic variations of the polar motion. More data are necessary to stabilize this result.

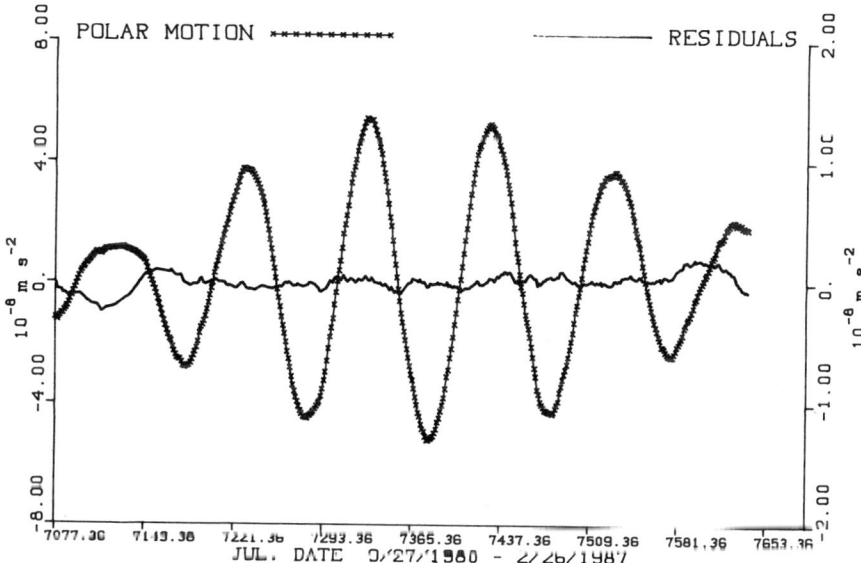

Fig. 3. The changes of the centrifugal force calculated from the IRIS data set and the residuals of the analysis of them.

Air Pressure Reduction

To be more sensitive for short period variations in the polar motion the noise level has to be reduced. In Figure 4 the residuals of the 1-year parallel registration of SCG TT40 and SCG TT60 (series II, series III) are plotted. It can be seen that the different drift behavior of the two instruments is well determined by the mathematical model used, but there is a lot of common energy in both data sets which needs to be described by better models. At the moment only the local air pressure is taken into account. Beside a frequency dependency [Warburton and Goodkind, 1978; Richter, 1987] the regional and global air pressure distribution influences the gravity signal by 10 - 20 cercent of the local one. To study these effects theoretical models have been developed by Rabbel and Zschau [1985] and Van Dam and Wahr [1987].

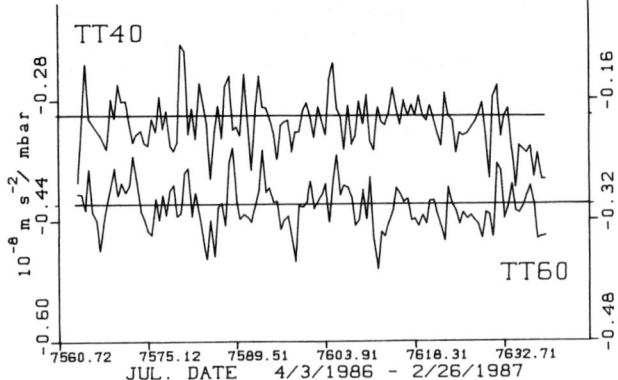

Fig. 5. Time dependent air pressure regression coefficients of SCG TT40 and SCG TT60.

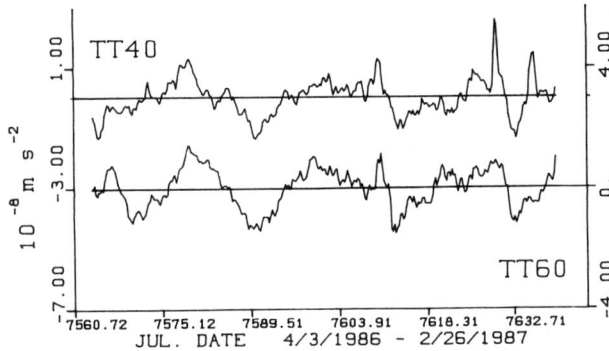

Fig. 4. Residuals of the analysis of parallelregistration of SCG TT40 and SCG TT60.

As a first approximation Rabbel and Zschau [1985] suggested a two coefficient solution which takes into account the mean pressure value over an area of 2000 km separated by an ocean and overland distribution. A check of this simple model with the 3 year data set was not very satisfactory. The meteorological data still have to be pre-processed to perform the same computations with the complete mathematical models suggested in both papers.

An empirical solution is derived from the 1-year parallel registration of the two SCG's. Related to the regression coefficient of the local air pressure the regional air pressure effect causes a time dependent variation. The gravimeter signal has been correlated with the local air pressure data over a time span of 2 and a half days after which the correlation interval has been shifted by 60 hours and analyzed again. The variations of the local air pressure regression coefficient are plotted in Figure 5 for both gravimeters. It can be seen that the variations are not random processes but are systematically common for both gravimeters. The result is in the range (± 10 cer-

cent) which is theoretically expected, but it has to be checked wheather the systematic behavior of both SCG's is mainly influenced by air pressure distribution or due to other physical processes.

Conclusions

The high sensitivity and long term stability of Superconducting Gravimeters (SCG) allows the observation of long term gravity variations such as those induced by Earth tides and polar motion. Studying the behavior of the instruments the drift of SCG TT40 can be reduced to the level of $1 - 3 \cdot 10^{-8} \mathrm{ms}^{-2}/$ year. The remaining drift is linear and can be described by simple mathematical models.

The experience with two SCG's running side by side demonstrates that the residuals of the analysis are not random. Therefore in order to exploit the full instrumental sensitivity the modelling of physical processes such as, e.g., air pressure have to be improved to account for both regional and global influences; the ocean tidal models have to be extended to include annual variations. The experience obtained with these improvements can be applied to the reduction models used in space techniques.

For the study of long period gravity variations longitudinal effects such as those introduced by the polar motion should be observed in parallel at stations separated by 90°.

References

Edge, R.J., T.F. Baker, and G. Jefferies: Improving the accuracy of tidal gravity measurements, <u>Proc. 10th Int. Symp. Earth Tides</u>, (ed. Vieira), Madrid, 213-221, 1986.

Heitz, S.: Grundlagen kinematischer und dynamischer Modelle der Geodäsie, Geophysik und Astronomie, Band 1 und 2, <u>Ferd. Dümmler's Verlag</u>, Bonn, 1980 and 1983.

Lambeck, K.: Temporal variations of rotational origin in the absolute value of gravity, Studia geophys. et geod., 17, 269-271, 1973.

Melchior, P.: The Tides of the Planet Earth, 2nd edition, Pergamon Press, Oxford, 1983.

Rabbel, W. and J. Zschau: Static deformations and gravity changes at the Earth's surface due to atmospheric loading. Geophysics, 56, 81-99, 1985.

Richter, B.: The spectrum of a registration with a superconducting gravimeter, Proc. 10th Int. Symp. Earth Tides, (ed. Vieira), Madrid, 131-139, 1986.

Richter, B.: Das supraleitende Gravimeter — Anwendung, Eichung und Ueberlegungen zur Weiterentwicklung —, Dt. Geod. Komm., Reihe C, Nr.329, Frankfurt, 1987.

Richter, B., and W. Zürn: Chandler effect and nearly diurnal free wobble as determined from observations with a superconducting gravimeter, Proc. IAG Symp. 128, The Earth's rotation and reference frame for geodesy and geodynamics, Coolfont, W. Virginia, Oct. 20. - 24. 1986, in press, 1987.

Schüller, K.: Simultaneous tidal and multi-channel input analysis as implemented in the HYCON method, Proc. 10th Int. Symp. Earth Tides, (ed. Vieira), Madrid, 515-520, 1986.

Van Dam, T.M., and J.M. Wahr: Deformations of the Earth's surface due to atmospheric loading: effects on gravity and baseline measurements, J. Geophys. Res., 92, B2, 1281-1286, 1987.

Wahr, J.M.: Deformation induced by polar motion, J. Geophys. Res. 90, 9363-9368, 1985.

Warburton, R.J., and J.M. Goodkind: The influence of barometric-pressure variations on gravity, Geophys. J. R. astr. Soc., 48, 281-292, 1978.

THE EARTH'S DIFFERENTIAL ROTATION;
HYDROSPHERIC CHANGES

Nils-Axel Mörner

Paleogeophysics & Geodynamics, Geological Institute
Stockholm University, S-106 91 Stockholm, Sweden

Abstract. The Earth experiences a differential rotation interchanging angular momentum between its various layers and sublayers. It is demonstrated that the hydrosphere plays a very active role in this interchange of angular momentum, and that it constantly gains and looses angular momentum in a feed back relation with the pulsation of the ocean currents systems (especially the currents responsible for the redistribution of mass between low and high latitudes).

Introduction

The Earth is a multi-layered system that experiences differential rotation, i.e. where angular momentum is interchanged between the different layers and sublayers. The intra-annual changes in the length of the day (LOD) are compensated by the atmosphere, or rather, vice versa. The long-term changes in LOD are generally thought to be compensated by and/or caused by core/mantle dislocations and differential rotation. The role of the hydrosphere has remained obscure and hard to model and calculate. Today, however, it is possible to demonstrate that the hydrosphere plays a very active role in the interchange of angular momentum.

The ENSO-Events

The El Nino/Southern Oscillation (ENSO) events represent interesting and important anomalies in the Earth's climatic, oceanographic and marine biological systems. They have, therefore, attracted a significant interest during the last decade, or so. In 1982/83, the Earth experienced a very strong ENSO event. This event has been studied in details and been given an impressive description within the World Climate Data Programme of WMO (WCDP, 1985). I will here discuss the interchange of angular momentum between the "solid" Earth (the LOD records) and the atmosphere and hydrosphere. The ENSO events seem to provide a mechanism that also operates on longer time scales causing signals/effects in the time reange of about 50-150 years (Mörner, 1984a, b, 1987a, b).

LOD and the Atmosphere

The annual and intra-annual LOD variations are balanced via the interchange of angular momentum between the "solid" Earth and the atmosphere (Barnes et al., 1983; Eubanks et al., 1983, 1984; Rosen and Salstein, 1983).

During the great 1982/83 ENSO event, there was an extra loss of angular momentum from the "solid" Earth record (LOD) which was transferred to the atmosphere, causing the westerly jet-streams to increase their velocity and to be displaced equatorwards.

In Fig. 1, we compare the 1980/81, 1981/82 and 1982/83 cycles. The 1982/83 cycle (ENSO) differ from the others in the occurence of a high peak in early 1983. The difference to the other cycles (non-ENSO) signifies the amount of angular momentum lost from the "solid" Earth to the atmosphere. It is a question of at least 0.3 milliseconds and maybe up to as much as 1.0 milliseconds. The effect on the jetstream is illustrated in Fig. 2.

LOD and the Hydrosphere

The 1982/83 ENSO event is linked to a significant loss in angular momentum (about 0.4 ms) from the "solid" Earth (Fig. 3B). This seems neither to have been compensated by the atmosphere nor by

Copyright 1990 by
International Union of Geodesy and Geophysics
and American Geophysical Union.

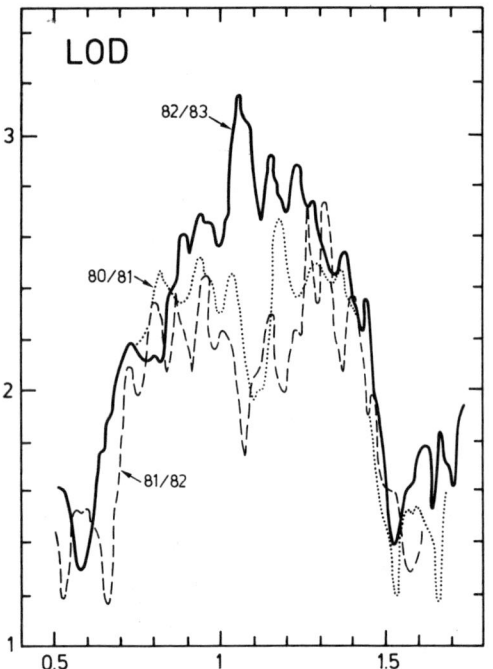

Fig. 1. LOD-changes for the three half-to-half year records 1980 to 1983. The differences between the two non-ENSO years and the 1982/83 ENSO year provide a rough estimate of the angular momentum interchanged between the "solid" Earth and the atmosphere. The early 1981 and 1982 lows are turned to an early 1983 peak (with an about 1.0 ms peak-to-bottom difference). Vertical scale in milliseconds.

the Earth's core. Instead, it seemed likely that it was compensated by the hydrosphere (Mörner, 1987a, b). Fig. 3 seems to provide conclusive evidence that this was really the case.

The volume of water in the upper layer of the tropical Pacific increased from 1981 up to mid 1982 (curve A). In mid 1982, the optimum sea surface temperature anomaly (curve D) was in the western equatorial Pacific (N-4). Angular momentum was transferred from the "solid" Earth, increasing the LOD which peaked in earliest 1983 (curve B). Obviously, angular momentum was transferred to the hydrosphere. The water volume (C) had started to decrease and the optimum sea surface temperature anomaly (D) was displaced to the eastern equatorial Pacific (N-3). During 1983, the water volume (C) rapidly fell to a minimum during the later half of the year. The optimum sea surface temperature anomaly (D) was displaced to the coastal zone (N-1-2), indicating that water started to be piled up along the west coasts of the American continents. Obviously, a continental "mountain torque" forced the hydrosphere to loose angular momentum which was transferred back to the "solid" Earth (B). Simultaneous rises and falls of the sea level have been documented all along the American west coasts for previous ENSO and non-ENSO years (Ensfield and Allen, 1980; Mörner, 1987b). This is illustrated in Fig. 4.

This indicates that the hydrosphere is much more susceptible to changes in the Earth's rate of rotation than generally assumed. Our data show that there, indeed, is an interchange of angular

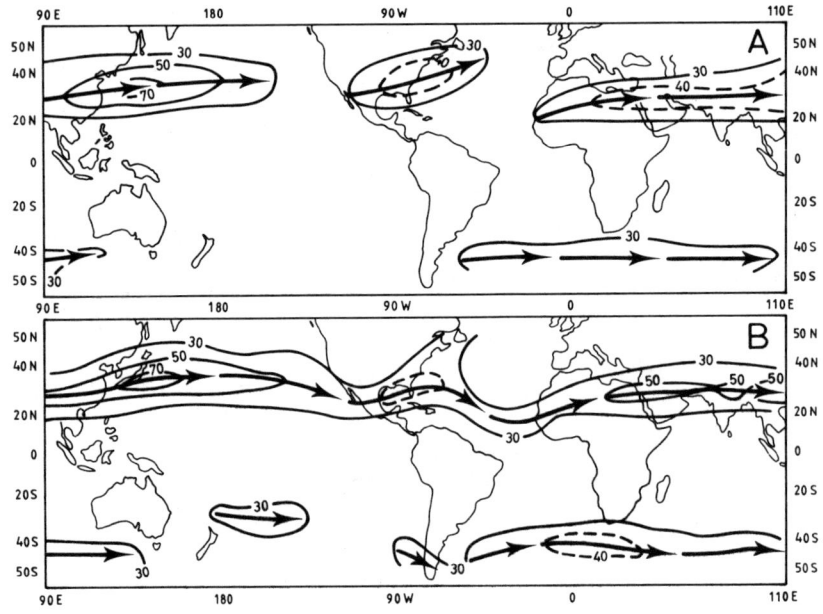

Fig. 2. Westerly jet-streams during normal years (A) and during the 1982/83 ENSO year (B). Wind velocity in m/s (WCDP, 1985).

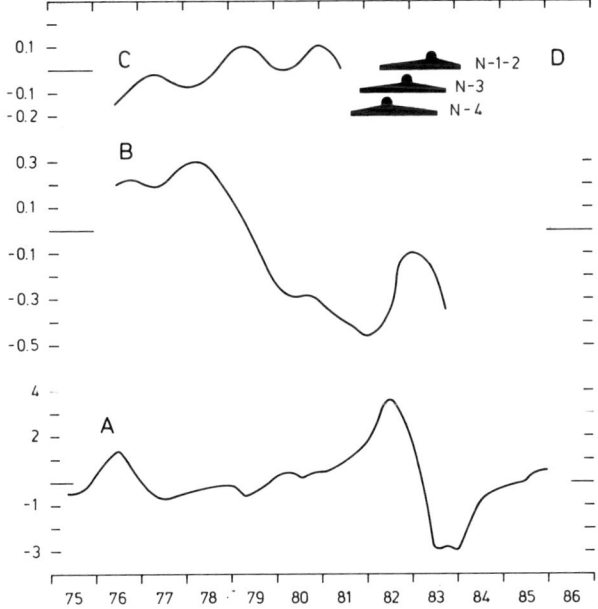

Fig. 3. Records indicating the transfer of angular momentum first from the "solid" Earth to the hydrosphere and then back to the Earth again in connection with the great 1982/83 ENSO event. A: water volume changes in the upper layer of the tropical Pacific in 10^{14} m^3 (WCDP, 1985) B: non tidal changes in LOD for the "solid" Earth in ms (Barnes et al., 1983; Eubanks et al., 1983; Rosen and Salstein, 1983). C: corresponding LOD changes for the atmosphere. D: time position of optimum sea surface temperature anomalies in equatorial Pacific within the four NINO areas; N-4 in the western part, N-3 in the eastern part and N-1-2 in the coastal part (WCDP, 1985).

momentum between the "solid" Earth and the hydrosphere during ENSO events.

Decadal Signals

The mechanism established for the interchange of angular momentum with the hydrosphere during ENSO-events is found also to apply for the decadal signals in the last centuries' LOD-records and for the 50-150 years' signals in the Holocene climatic-eustatic records (Mörner, 1984a, b, 1987a, b). This is of fundamental significance for our understanding of the Earth's dynamicity and its "global changes" (Mörner, 1988). Via the effects on the coastal upwelling, this model also explains variations in marine biological productivity and pre-industrial atmospheric CO_2 fluctuations (Mörner, 1987a, b).

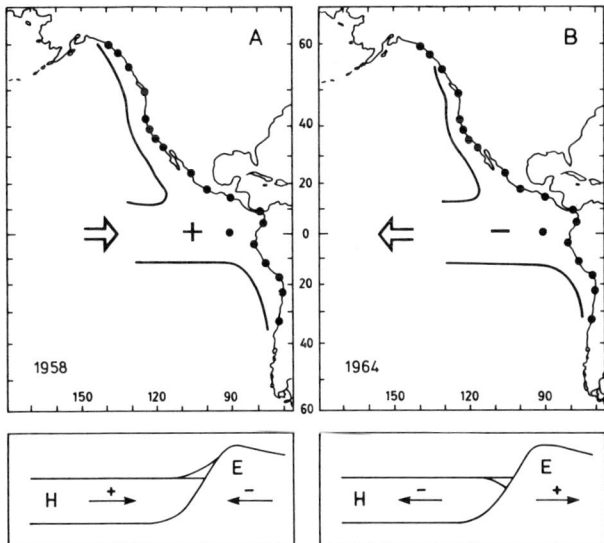

Fig. 4. Sea level changes along the American west coast in the ENSO year 1958 (A) and in the non-ENSO year 1964 (B) according to data by Enfield and Allen (1980). The mass redistributions are here (Mörner, 1987a, b) primarily claimed to be the cause of changes in the hydrospheric angular momentum (and not wind forces as generally assumed). Generalized models of the corresponding "solid" Earth deceleration and hydrospheric acceleration in 1958 (A) and of "solid" Earth acceleration and hydrospheric deceleration in 1964 (B) are given at the base (Mörner, 1987a, b).

The Last Centuries' LOD Records

The Earth's rate of rotation, or rather the lengthening of the day (LOD), has been continually measured since the beginning of the 17th century (Stephenson and Morrison, 1984). The non-tidal component of this record shows significant fluctuations between periods of acceleration and deceleration. This LOD record cannot be understood in terms of atmospheric variations, nor does core/mantle motions seem capable of explaining more than a part of the record (i.e. changes related to the 1840, 1910 and 1970 geomagnetic "jerks"). I have previously shown (Mörner, 1987b) that these LOD variations primarily seem to be compensated by the interchange of angular momentum between the "solid" Earth and the hydrosphere.

The Gulf Stream (like the Kuroshio Current) redistributes a significant amount of mass between lower and higher latitudes, and must hence be a powerful source for variations in the Earth's angular momentum budget (Mörner, 1984a, 1987b). Changes in the coastal water temperature, in the continental air temperature and precipitation and in the coastal sea level changes in northwestern Europe all reflect the pulsation of the Gulf Stream activity. These changes possess a very good correlation with the LOD variations (Fig. 5; see Mörner, 1987b, for further details) suggesting a causal feed back relation between the pulsation of the Gulf Stream activity and the Earth's rate of rotation (Mörner, 1984a, 1987b).

In the North Atlantic, a speeding up of the "solid" Earth (decreasing LOD) and loss of angular momentum (slowing-down) of the hydrosphere would lead to an intensified equatorial eastward current, a northward migration of water masses (i.e. the Gulf Stream) and a restriction of Arctic surface water influx. This would result in a warming of northwestern Europe and fits perfectly well for all major periods of LOD decrease (mantle speeding-up); viz. 1845-1872, 1903-1929 and after 1972 (also for the generally warm but poorly defined period 1695-1797). At the same time the main atmospheric jetstream would decrease and move northwards. A slowing down of the "solid" Earth (increasing LOD) and a gain of angular momentum (speeding-up) of the hydrosphere would lead to a decrease in the equatorial easterly current, a southward migration of the Gulf Stream and a southward penetration of Arctic surface water. At the same time the atmospheric westerly jetstream would increase and move southwards so that it would tend to block (i.e. retard) both the equatorial current and the Gulf Stream and may even lead to the loss of hot equatorial surface water to the southern hemisphere. All this would result in a cooling in northwestern Europe. This fits perfectly well with the periods of LOD-increase (mantle slowingdown); viz. prior to 1695 (representing the 2nd Little Ice Age), 1827-1845 and 1929-1972 (Fig. 5).

The Holocene Records

An analysis of the geographical validity of major short-term climatic changes and shifts during the last 20,000 years, or so (Mörner and Karlén, 1984) revealed (Mörner, 1984b) that these changes and shifts - ranging from a few parts of a centigrade up to 5, maybe 10 centigrades - in no case represented global rises and falls in temperature, but instead represented local and regional events that seemed to be of compensational type on a global scale; i.e. implying the redistribution of heat over the globe (and not global rises and falls in temperature). These events were found to have a duration in the order of 50-150 years. Redistributions of heat over the globe with such durations can only take place in the hydrosphere (Mörner, 1984b). This gave evidence of frequent short-term changes in the oceanic circulation system.

In combination, short-term circulation changes would lead to the distribution of water masses (seen in sea level changes) and of stored energy (seen in paleoclimatic changes). Such redistributions of mass would inevitably lead to corresponding changes in the Earth's rate of rotation and vice versa. This lead me to propose a "graviational-rotational-oceanographic" model for short-term paleoclimatic changes (Mörner, 1984a, b) and related sea level changes and rotational changes.

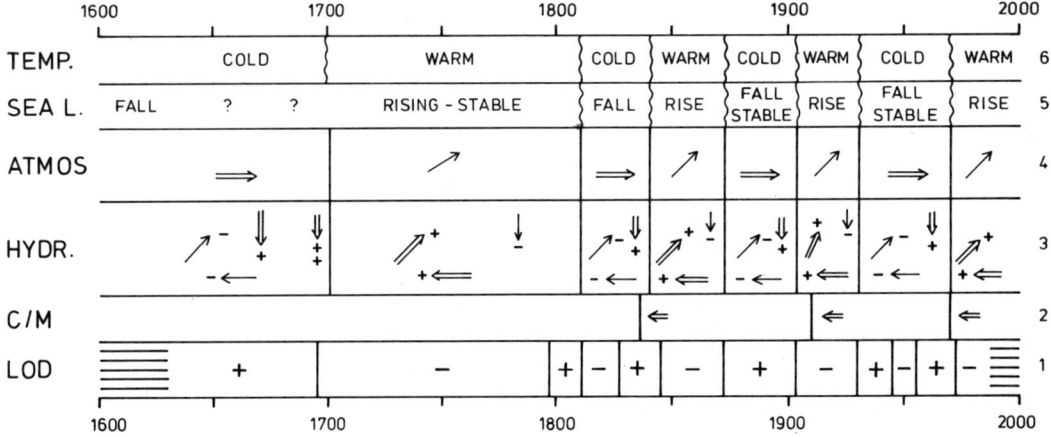

Fig. 5. Non-tidal LOD fluctuations (1). Corresponding core/mantle dislocations (2), hydrospheric ocean circulation changes in the North Atlantic (3) and atmospheric changes in the westerly jetstream over the orth Atlantic-European region (4), and causative eustatic sea level changes (5) and general temperature changes (6) in northwestern Europe.

The North Atlantic is an excellent test area for this model (Mörner, 1984a, b, 1987a, b). The Gulf Stream is shown to possess clear short-term pulsations that correlate with deep-sea indices, northwest European sea level changes and Scandinavian air temperature variations (Fig. 6). These changes occur with a frequency changing "cyclicity" (Mörner, 1973) consistent with a terrestrial feed back mechanism (instead of an extraterrestrial beat). Obviously, the Gulf Stream variability is driven by rotational changes at the same time as it leads to counteracting rotational changes (Mörner, 1984a). This mechanism works perfectly well for the Late Pleistocene and Holocene records. It also explains the correlations according to Taira (1981) between Europe and Japan when it concerns sea level and climate; both areas are affected by similar NE-wards ocean current systems (the Gulf Steam and the Kuroshio Current) as well as the general differences between northern and southern hemisphere eustatic-climatic records (because the two hemispheres have so drastically different circulation systems.

Ocean circulation changes will affect the coastal upwelling. The present theory of rapid interchange of angular momentum between the "solid" Earth and the hydrosphere, therefore, also provides a simple and logical explanation to the recorded variations in marine biological productivity and to the high-amplitude (30 ppm in 100-150 years) pre-industrial variations in atmospheric CO_2 contents (Mörner, 1987a).

Longer Term Records

The main task of this paper has been to demonstrate the effect of the hydrosphere in the interchange of angular momentum in association with the Earth's differential rotation. The ENSO events, the last centuries' records and the Holocene records demonstrate that the hydrosphere has taken a very active part in the interchange of angular momentum and pays an important role in a complex system of redistribution of mass (evidenced by sea level fluctuations) and energy (evidenced by climatic changes, biological productivity changes, etc.) in a feed back relation to the loss and gain of angular momentum.

Besides this, however, it should also be noted that this model also applies for longer term changes such as the main Holocene cycle, the Milankovitch variables and some long-term deep-sea events (Mörner, 1987b, Figs. 15-17).

Conclusions

The hydrosphere plays a very active role in the Earth's differential rotation, loosing and gaining angular momentum in a feed back relation to changes in the ocean circulation (current pulsations).

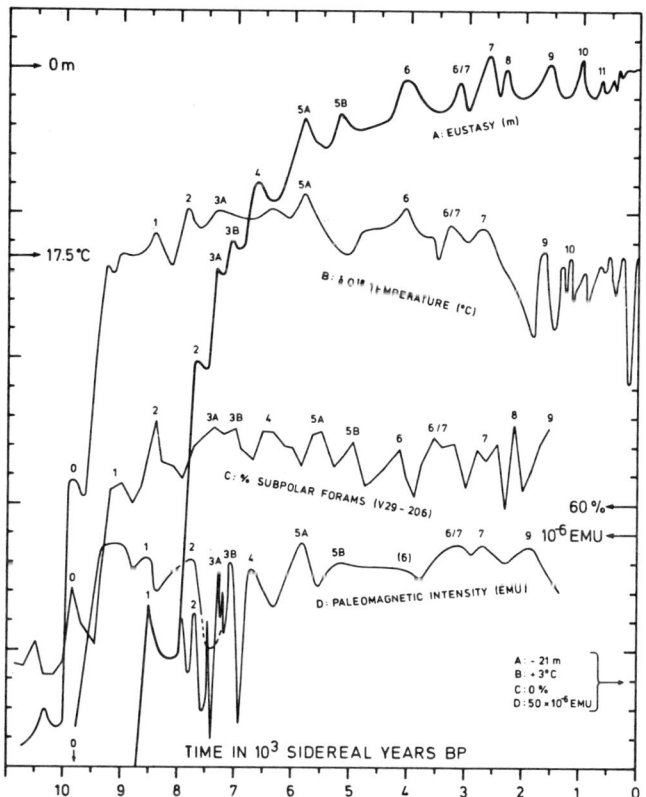

Fig. 6. Holocene short-term fluctuations (16 of them) in eustasy (A) in northwestern Europe, in continental temperature (B) on the island of Gotland in the Baltic, in percent subpolar forams (C) in the Denmark Strait between Iceland and Greenland sensitively registering the Gulf Stream pulsation, and in paleomagnetic intensity (D) in sediment cores from Southern Scandinavia. From Mörner (1984a, Fig. 6). Vertical scale in meters (A), centigrades (B), 10% (C) and 10×10^{-6} EMU (D). Geographical locations in Mörner, 1984a, Fig. 11.

References

Barnes, R.T.H., Hide, R., White A.A. and Wilson, C.A., 1983. Atmospheric angular momentum fluctuations correlated with length of the day changes and polar motion. Proc. Roy. Soc. London, Ser. A, 31 (1983);

Enfield, D.B. and Allen, J.S., 1980. On the structures and dynamics of monthly mean sea level anomalies along the Pacific coast of

North America, J. Phys. Oceanogr. vol. 10, p. 577-578.

Eubanks, T.M., Steppe, J.A., Dickey J.O. and Callahan, P.S., 1983. A spectral analysis of the Earth's angular momentum budget, JPL Geod. Geophys. Prepr., No. 102, p. 1-50.

Eubanks, T.M., Dickey J.O. and Steppe, J.A., 1984. The 1982-83 El Nino, The Southern Oscillation and changes in the length of the day. JPL Geod. Geophys. Prepr., No. 111, p. 111-19.

Mörner, N.-A. 1973. Climatic changes during the last 35,000 years as indicated by land, sea and air data. Boreas, 2, p. 33-53.

Mörner, N.-A., 1984a. Planetary, solar, atmospheric, hydrospheric and endogene processes as origin of climatic changes on the Earth. In Climatic Changes on a Yearly to Millenial Basis, (N.-A. Mörner & W. Karlén, Eds.), p. 487-507. Reidel Publ. Co.

Mörner, N.-A., 1984b. Climatic changes on a yearly to millenial basis. Concluding remarks. In Climatic Changes on a Yearly to Millenial Basis (N.-A. Mörner and W. Karlén, Eds.), p. 637-651. Reidel Publ. Co.

Mörner, N.-A., 1987a. Ocean circulation changes and redistribution of energy, mass and momentum on a yearly to centuary time-scale. In: "Long Term Changes in Marine Fish Populations" (T. Wyatt, Ed.), in press. Vigo (Spain).

Mörner, N.-A., 1987b. Terrestrial variations within given energy, mass and momentum budgets; Paleoclimate, sea level, paleomagnetism, differential rotation and geodynamics. In: Secular Solar and Geomagnetic Variations in the last 10,000 years (F.R. Stephenson and A.W. Wolfendale, Eds.), Reidel Publ., Co., in press.

Mörner, N.-A., 1988. Global changes. The lithosphere. Internal processes and Earth's dynamicity in view of Quaternary observational data. In: Proc. Global Changes Spec. Session, 12th INQUA Congr., Ottawa 1987, in press.

Rosen, R.D. and Salstein, D.A., 1983. Variations in atmospheric angular momentum on global and regional scales and the length of day. J. Geophys. Res., vol. 88, p. 5451-5470.

Stephenson, F.R. and Morrison, L.V., 1984. Long-term changes in the rotation of the Earth: 700 B.C. to A.D. 1980. Phil. Trans. Soc. Lond., vol. A313, p. 47-70.

Taira, K., 1981. Holocene tectonism in eastern Asia and geoid changes. Palaeogeogr. Palaeoclim. Palaeoecol., vol. 36, p. 75-85.

WCDP, 1985. The global climate system. Climatic System Monitoring (CSM) of the World Climate Data Program (WCDP), WMO Geneva, 52 pp.

THE INFLUENCE OF OCEAN AND SOLID EARTH PARAMETERS ON OCEANIC EIGENOSCILLATIONS, TIDES AND TIDAL DISSIPATION

Wilfried Zahel

Institut für Meereskunde, University of Hamburg, Germany

Abstract. A model of tidal waves and normal modes in a hemispherical ocean is given with the dependent variables expanded in terms of spherical harmonics and with the objective to investigate in particular the influence of ocean and solid Earth parameters on the quantities relevant to secular variations in Earth rotation. Comparisons with the results obtained by realistic models considering loading and self-attraction and being forced by the tidal potential solely, show that principal features of the tides in the North Pacific, total rates of energy dissipation and contents, and specific loading and self-attraction effects in the ocean also appear when the hemispherical model is applied. Computing the normal modes which determine tidal resonances and computing the response to Y_2^{-1} and Y_2^{-2} spherical harmonic tidal potentials over a broad range of periods yields essential effects of loading and self-attraction in the ocean within the complete range of realistic parameter values. For individual near-resonance tidal constituents the rate of tidal power, e.g., can be reduced or enhanced by more than a factor two, thus indicating the possibility of important consequences of solid Earth response for variations in Earth rotation.

Introduction

Computations having been performed with realistic ocean tide models have provided considerable insight into the response of the ocean to tidal forces, and valuable contributions to the understanding of the resonance behavior of the ocean have been made by numerically modelling the normal modes in a frictionless ocean with realistic topography on a rigid Earth by Platzman et al. [1981], Gotlib and Kagan [1982], and Gaviño [1984]. In particular when realistic tidal models were applied solely including homogeneous boundary conditions, thus allowing comparisons with coastal data and the investigation of the influence of various physical parameters without the constraint of prescribed tidal data, considerable differences between the results of the different model applications appeared. These differences and the deviations from observations might be attributed to the peculiarities of the numerical procedures used, to insufficient resolution of topography, and to varying parametrizations of physical processes. The dependence of the computed tidal regimes on solid Earth properties and the corresponding parametrizations were introduced when considering the loading and self-attraction effects (abbrev. LSA) in ocean tide models. The application of ocean tide models that consider these effects and include the tidal potential as the only forcing, see Estes [1977], Gordeev et al. [1977], Accad and Pekeris [1978] and Zahel [1978] yielded the common result that the main features of the tidal regimes were preserved when including the additional effects and that the total tidal dissipation was changed by an amount less than 10 percent. However, in some cases phase delays up to 45° were obtained for extended areas and in part also significant local modifications of the computed tide were found. Now, it has been noticed that the hemispherical model having been used by Longuet-Higgins and Pond [1970] for the computation of the frictionless normal modes and by Webb [1980] for the computation of the response to tidal potentials of the form of Y_2^{-1}- and Y_2^{-2} spherical harmonics with friction included can be generalized, taking into account the complete loading and self-attraction effects (see Zahel [1986]). This observation suggests investigating the influence of these effects on oceanic oscillations in conjunction with varying forcing periods, values of friction parameters and mean ocean depths.

This hemispherical model is based on the expansion of the dependent variables in terms of spherical harmonics. Thus the normal modes, the tidal regimes and the quantities being relevant in the realistic case to variations of the Earth's rotation can be computed with high precision excluding specific discretization effects and restrictions introduced when using finite differ-

Copyright 1990 by
International Union of Geodesy and Geophysics
and American Geophysical Union.

ence or finite element procedures. However, the simplification adopted by the hemispherical model, requires the tidal regimes in the schematic ocean to reflect principal features of the realistic tidal regimes in order to expect the results of the hemispherical model investigations to have a meaning for the real ocean tide and its computation. The comparison of the hemispherical M_2-tide, as computed by Zahel [1986], with the M_2-tide obtained by applying realistic ocean tide models, indicates fulfillment of this requirement in view of the tidal regimes in the Northern hemisphere resembling those in the North Pacific. Concerning the influence of loading and self-attraction on the M_2-tide, it appears that the hemispherical model yields, e.g., the most prominent effect obtained by Estes [1977], Accad and Pekeris [1978] and Zahel [1978] for the North Pacific, i.e., the phase delay at the Eastern boundary.

The Hemispherical Model

The Computation of Tides

The response of the ocean to tidal forcing on an oceanic scale is described by the integro-differential equations

$$\frac{\partial \underline{v}_h}{\partial t} + 2\underline{\Omega} \times \underline{v}_h + \underline{F}' = -g\nabla(\eta - \gamma_2 \bar{\eta} - \sum_n \alpha'_n \eta_n) \quad (1)$$

$$\frac{\partial \eta}{\partial t} + \nabla \cdot (H\underline{v}_h) = 0 \quad (2)$$

where

$$\sum_n \alpha'_n \eta_n = \iint_B \eta \frac{1}{4\pi} \sum_n \alpha'_n \sum_m \bar{P}_n^m(\sin\phi)\bar{P}_n^m(\sin\tilde{\phi})\cos(m(\tilde{\lambda}-\lambda)) d\tilde{\lambda} d\tilde{\phi} \cos\tilde{\phi}$$

and

$$\gamma_2 = 1 + k_2 - h_2 \qquad \alpha'_n = (1 + k'_n - h'_n)\alpha_n$$

The boundary conditions are given by $\underline{v}_h \cdot \underline{n} = 0$ and $\underline{v}_h \cdot \underline{m} = 0$ (if $\underline{F}'$ includes second order terms), where $\underline{n}$ and $\underline{m}$ denote unit vectors normal and tangential to the coastline, respectively.

These equations are the basis of the realistic ocean tide model applied by Zahel [1978] and [1980] to compute global diurnal and semi-diurnal tides and the corresponding complete energy balances. The equations which are the basis of the hemispherical model are obtained from (1) and (2) by referring to a hemispherical ocean centered at the equator with constant depth H, by using a linear bottom friction term, i.e., $\underline{F}' = R'/(\rho H)\underline{v}_h = R\underline{v}_h$, and by writing the horizontal velocity vector as

$$\underline{v}_h = \frac{\partial}{\partial t}(\nabla \Phi + \nabla \Psi \times \underline{z}) \quad (3)$$

with

$$\Phi = \sum_{r=1}^{\infty} p_r \Phi_r e^{-i\omega t} \qquad \Phi_r = a_n^m P_n^m(\sin\phi)\cos(m\lambda)$$

$$\Psi = \sum_{r=1}^{\infty} p_{-r} \Psi_r e^{-i\omega t} \qquad \Psi_r = a_n^m P_n^m(\sin\phi)\sin(m\lambda)$$

where the suffixes r are associated with pairs of suffixes (n,m) and where the validity of $\nabla \Phi_r \cdot \underline{n} = 0$ and $\Psi_r = 0$ in $\lambda = 0, \pi$ guarantees exact fulfillment of the boundary condition $\underline{v}_h \cdot \underline{n} = 0$. The equations for the coefficients p_r and p_{-r} are obtained by successively multiplying (1) with the functions $\nabla \Phi_r$ and $\nabla \Psi_r \times \underline{z}$ and integrating over the hemispherical domain. Considering the loading and self-attraction term requires some additional conversions and evaluations of integrals and finally leads without introducing any further simplification to the linear algebraic equations

$$-\omega^2 p_r - iR\omega p_r - 2i\omega\Omega\mu_r^{-1}\sum_{s=-\infty}^{\infty}\beta_{r,s}p_s + gH\mu_r(1-\alpha'_r/2)p_r$$

$$-\frac{1}{2}gH\sum_{\substack{r'=1\\r'\neq r}}^{\infty}\mu_{r'} q(r,r')p_{r'} = g\gamma_2 \bar{\eta}_r \quad (4)$$

$$-\omega^2 p_{-r} - iR\omega p_{-r} - 2i\omega\Omega\nu_r^{-1}\sum_{s=-\infty}^{\infty}\beta_{-r,s}p_s = 0 \quad (5)$$

where

$$q(r,r') = R_e^4 \bar{a}_{\bar{n}}^{\bar{m}} a_{n'}^{m'} \sum_n \alpha'_n \sum_m (2ma_n^m)^2 I\binom{m\ \bar{m}}{n\ \bar{n}} I\binom{m\ m'}{n\ n'} / ((m^2-\bar{m}^2)(m^2-m'^2))$$

$$I\binom{m\ m'}{n\ n'} = \int_{-1}^{+1} P_n^m(\mu) P_{n'}^{m'}(\mu) d\mu$$

and

$$\nu_r = \mu_r = n(n+1)/R_e^2 \qquad \bar{\eta}_r = \iint_B \Phi_r \bar{\eta} \, dA$$

With suitable factors a_n^m the systems $\{\Phi_r\}$ and $\{\Psi_r\}$ are orthonormal in the hemispherical domain. The $\beta_{r,s}$ denote the well known gyroscopic coefficients (see Webb [1980]) and $\bar{\eta}$ denotes the oceanic equilibrium tide on a rigid Earth. Equation (2) yields the expansion of the sea-surface elevation relative to the moving sea bottom

$$\eta = H \sum_{r=1}^{\infty} p_r \mu_r \Phi_r \quad (6)$$

When considering LSA the original factor 1 in the contributions of the pressure gradient term to the diagonal elements of (4) is replaced by the degree-dependent factors $\kappa_r = 1 - \alpha_r'/2$ and, moreover, off-diagonal elements appear in (4). As compared with the appearance of the latter elements, the changes of the diagonal elements mentioned clearly dominate the effects of LSA that will be referred to.

As described in detail by Longuet-Higgins and Pond [1970], the simplified eqs. (4) and (5) fall into two distinct systems, one determining solutions being symmetric, the other determining solutions being antisymmetric about the equator. The former yields semi-diurnal tides, the latter diurnal tides. In the same way the eqs. (4) and (5) describing the generalized problem fall into two distinct systems. Considering constituents up to degree and order 23 in the tidal computations, the infinite systems of equations turn into finite ones. Quantities describing the overall behavior of the ocean that are used for estimating the influence of parameters are the energy contents of the tidal regimes and the tidal power or tidal dissipation. When time averaged, these quantities can easily be evaluated once the coefficients p_r and p_{-r} have been calculated. The average potential energy contents E_p and the average kinetic energy contents E_k are given by

$$E_p = \langle \tfrac{1}{2}\rho g \iint \eta^2 dA \rangle + \langle \rho g \iint \eta \delta \, dA \rangle = \tfrac{1}{4}\rho g H^2 \sum_{r=1}^{\infty} \mu_r^2 p_r p_r^*$$
$$+ \tfrac{1}{2}\rho g H^2 \sum_{r=1}^{m} h_n' \alpha_n \mu_r^2 p_r p_r^* + \tfrac{1}{4}\rho g h_2 H \sum_{r=1}^{\infty} \mu_r (p_r^* \bar{\eta}_r + p_r \bar{\eta}_r^*) \quad (7)$$

$$E_k = \langle \tfrac{1}{2}\rho H \iint \underline{v}_h^2 dA \rangle = \tfrac{1}{4}\rho H \omega^2 \sum_{r=1}^{\infty} (\mu_r p_r p_r^* + \nu_r p_{-r} p_{-r}^*) \quad (8)$$

with the solid Earth deformation

$$\delta = h_2 \bar{\eta} + \sum_n h_n' \alpha_n \eta_n$$

The average work done by tidal forces and the moving sea bottom, being equal to the average tidal dissipation, is given by

$$\overline{W} = \langle \rho g \gamma_2 \iint \tfrac{\partial \eta}{\partial t} \bar{\eta} \, dA \rangle = \tfrac{1}{2}\rho g \omega H \gamma_2 \text{Im} \{ \sum_{r=1}^{\infty} p_r \mu_r \bar{\eta}_r^* \} \quad (9)$$

The tidal pressure work on the solid Earth is given by $h_2 \overline{W}/\gamma_2$ making approximately 89 percent of $\overline{W}$; see Schwiderski [1985] for a thorough discussion of the tidal energy budget with respect to variations of Earth's rotation.

The Computation of Normal Modes

Introducing the non-dimensional complex frequency $\bar{\omega} = \omega/(2\Omega)$ and the complex parameter $\tau = 1/(\epsilon\bar{\omega})$, with $\epsilon = 4\Omega^2 R_e^2/(gH)$, for calculating free oscillations, eqs. (4) and (5) can be written as

$$\bar{\omega}\mu_r p_r = \kappa_r \mu_r^2 R_e^2 \tau p_r - \tfrac{1}{2}R_e^2 \tau \mu_r \sum_{\substack{r'=1 \\ r' \neq r}}^{\infty} \mu_{r'} q(r,r') p_{r'}$$
$$- iR/(2\Omega)\mu_r p_r - i \sum_{s=-\infty}^{\infty} \beta_{r,s} p_s \quad (10)$$

$$\bar{\omega}\nu_r p_{-r} = - iR/(2\Omega)\nu_r p_{-r} - i \sum_{s=-\infty}^{\infty} \beta_{-r,s} p_s \quad (11)$$

With a constant real depth H prescribed, the truncated system (10), (11) represents a generalized linear algebraic eigenvalue problem of the form $A\underline{x} = \bar{\omega}B\underline{x}$, which is solved by calculating eigenvalues and eigenvectors of the matrix $L^{-1}A(L^{-1})^T$ being similar to $B^{-1}A$, where $B = LL^T$. At this τ is given as a complex number and varied such that the equation

$$4\Omega^2 R_e^2 /(gH) = \epsilon = 1/(\tau\bar{\omega}(\tau)) \quad (12)$$

is fulfilled where the value of the left hand side is prescribed. The iteration procedure for determining the roots of the nonlinear equation (12) turns out to rapidly converge in the $\bar{\omega}$ - interval in question. Taking into account constituents up to degree and order 23 the performance indices connected with the solution of (10) and (11) moreover prove to be less than 1. Therefore the computation of eigenfrequencies and eigenoscillations for the hemispherical ocean is regarded to have been achieved with high precision.

Symmetric and Antisymmetric Normal Modes

Normal modes in the hemispherical ocean have been computed by using the values $R = 9.26 \cdot 10^{-6} s^{-1}$, $H = 4420$ m and $h_2 = 0.612$, $k_2 = 0.302$. The values of loading Love numbers h_n' and k_n', based on a Guten-

TABLE 1. Computed Real Eigenperiods T and Imaginary Parts ω_I of Slowest Gravitational Normal Modes in a Hemispherical Ocean

LSA considered		LSA neglected	
T[s]	$\omega_I [10^{-6} s^{-1}]$	T[s]	$\omega_I [10^{-6} s^{-1}]$
Symmetric normal modes			
162,728	-4.2875	153,806	-4.2231
84,273	-4.4669	80,960	-4.4390
59,452	-5.1319	57,644	-5.1284
52,989	-6.1891	51,476	-6.1098
48,870	-5.5802	47,541	-5.5263
42,762	-5.0242	41,635	-5.0057
40,753	-5.2533	39,733	-5.2205
Antisymmetric normal modes			
110,937	-6.2428	107,145	-6.1237
74,672	-5.3368	71,992	-5.2776
54,628	-5.0022	52,999	-4.9707
43,821	-5.6154	42,778	-5.6286
41,553	-5.8475	40,561	-5.7437

LSA stands for loading and self-attraction. The ocean depth and the friction used are H = 4420 m and R = $9.26 \cdot 10^{-6}$ s^{-1}, respectively.

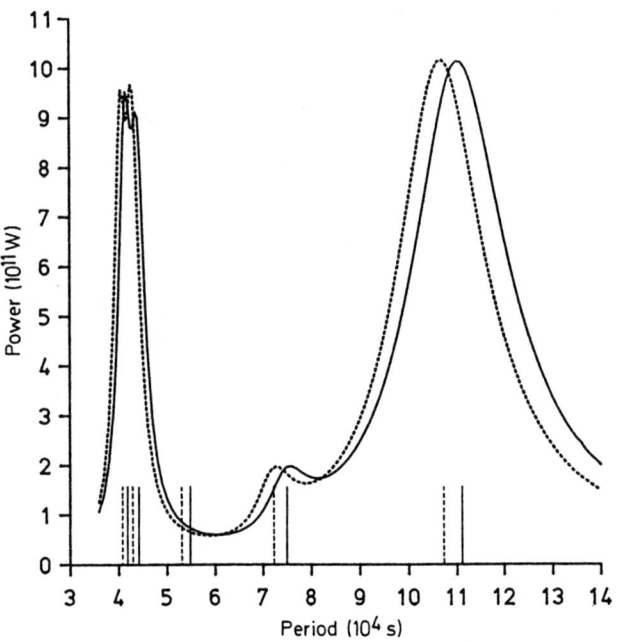

Fig. 1. Rate of work done by tidal forces and moving sea bottom. Model ocean with H = 4420 m and R = $9.26 \cdot 10^{-6}$ s^{-1} driven by Y_2^{-1}-tidal potential with K_1-amplitude and varying periods. Vertical lines mark eigenperiods of antisymmetric normal modes. Solid lines and dashed lines refer to computations with LSA considered and neglected, respectively.

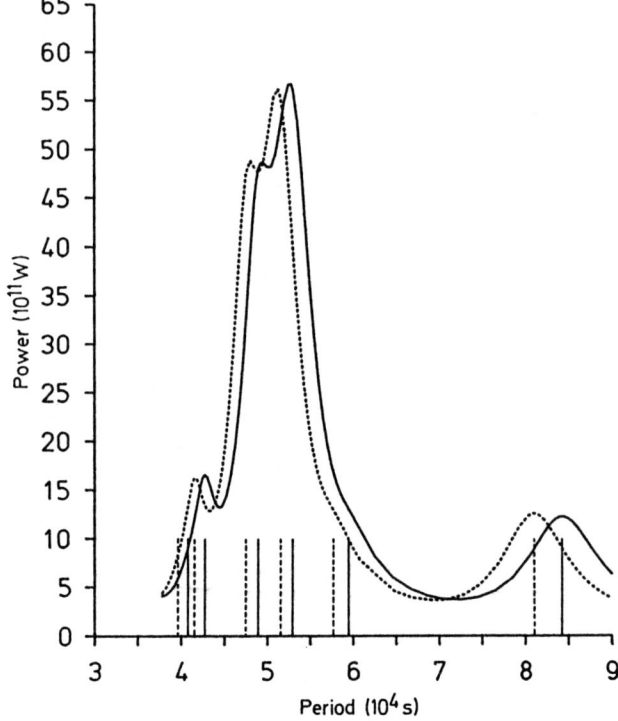

Fig. 2. Same as Figure 1 except the model ocean is driven by Y_2^{-2}-tidal potential with M_2-amplitude and vertical lines mark eigenperiods of symmetric normal modes.

TABLE 2. Computed Rates of Work $\overline{W}$ Done by Tidal Forces and Moving Sea Bottom of Energy Contents $\overline{E}$ and Resulting Q-Factor in a Hemispherical Ocean

Period	T_o	LSA considered $\overline{W}[10^{12}W]$	$\overline{E}[10^{16}J]$	Q	LSA neglected $\overline{W}[10^{12}W]$	$\overline{E}[10^{16}J]$	Q
K_{2m}	30	1.61	14.57	13.2	1.28	11.88	13.5
K_{2m}	60	2.06	39.55	28.0	0.81	15.77	28.6
M_2	30	1.33	11.81	12.5	1.70	15.02	12.4
M_2	60	0.79	14.57	26.0	1.04	18.78	25.5
$3N_2$	30	3.06	26.07	11.3	4.67	40.59	11.5
$3N_2$	60	2.66	46.03	22.9	7.37	131.47	23.6
K_1	30	0.20	1.56	5.8	0.22	1.72	5.7
K_1	60	0.11	1.71	11.4	0.12	1.90	11.2
O_1	30	0.31	2.38	5.2	0.38	2.99	5.2
O_1	60	0.19	2.85	10.2	0.25	3.87	10.4

The model is driven by Y_2^{-1} and Y_2^{-2} equilibrium tides with K_1-amplitude and M_2-amplitude, respectively. The ocean depth used is H = 4420 m. T_o denotes the frictional decay in hours, i.e., $R = 1/(T_o \cdot 3600)$ s^{-1}.

berg-Bullen Earth model, which was modified by including oceanic shield structures, have been chosen as used in Zahel [1978]. The depth used is typical of open ocean areas, and the value assigned to the friction parameter corresponds to an estimated approximate energy decay time of 30 hours (Webb [1980]), which belongs to the time interval regarded as realistic. Table 1 displays the computed eigenvalues, given by means of real eigenperiods and imaginary parts of eigenfrequencies, for the slowest gravitational modes including those which determine the resonant response to semi-diurnal and diurnal tidal forces. Obviously, considering loading and self-attraction leads in each case regarded to an increase of the real eigenperiods of the gravitational modes. This increase amounts to between 900 s and 3,800 s in the period interval 40,000 s to 111,000 s, and it becomes larger with larger eigenperiods. The corresponding eigenvalues have also been computed for a frictionless hemispherical ocean. The above described effect of LSA remains true when friction is neglected. The general increase of the real eigenperiods due to considering friction of the above magnitude proves to be independent of the consideration of LSA and to be far less than the described effect of LSA. This increase due to friction amounts to less than one tenth of the increase due to LSA and is even less than one twentieth for most of the normal modes regarded.

Thus, having in mind the representation of forced oscillations in terms of normal modes within a broad range of energy decay times, the tidal regimes are expected to be considerably influenced by the solid Earth parameters determining LSA.

Ocean response to Y_2^{-1} and Y_2^{-2} tidal potentials

Equations (4) and (5) have been solved including Y_2^{-1} and Y_2^{-2} equilibrium tides, respectively, with periods belonging to an interval containing the diurnal and semi-diurnal tidal bands, and with amplitudes of the K_1- and M_2-tide, respectively. These computations have been performed at an interval of 250 s using the same parameter values as in case of the calculation of normal modes. Figures 1 and 2 demonstrate the peak rates of work done by tidal forces and moving sea bottom to be due to the excitation of normal modes, where in the period range regarded three normal modes, two symmetric and one antisymmetric, do not even excite weak resonant responses because of spatial mismatch of the eigenfunctions concerned and the equilibrium tide. The resonance behavior displayed in Figures 1 and 2 could have been demonstrated equally by displaying the time averaged potential or kinetic energy contents in dependence of the forcing periods. Thus, the consideration of LSA in this model substantially affects the rates of tidal power and the related quantities as well as the rates of energy contents when the period of a tidal constituent is close to the eigenperiod of a tidally effective normal mode (key resonance). (See Figures 1 and 2 and Table 2.) Table 2 also shows that the rates that have been computed for the M_2-tide in the schematic ocean compare with those obtained by realistic models [Schwiderski, 1985]. This picture does not change in terms of relative magnitudes when the friction parameter is reduced within the range of realistic values, e.g., to a

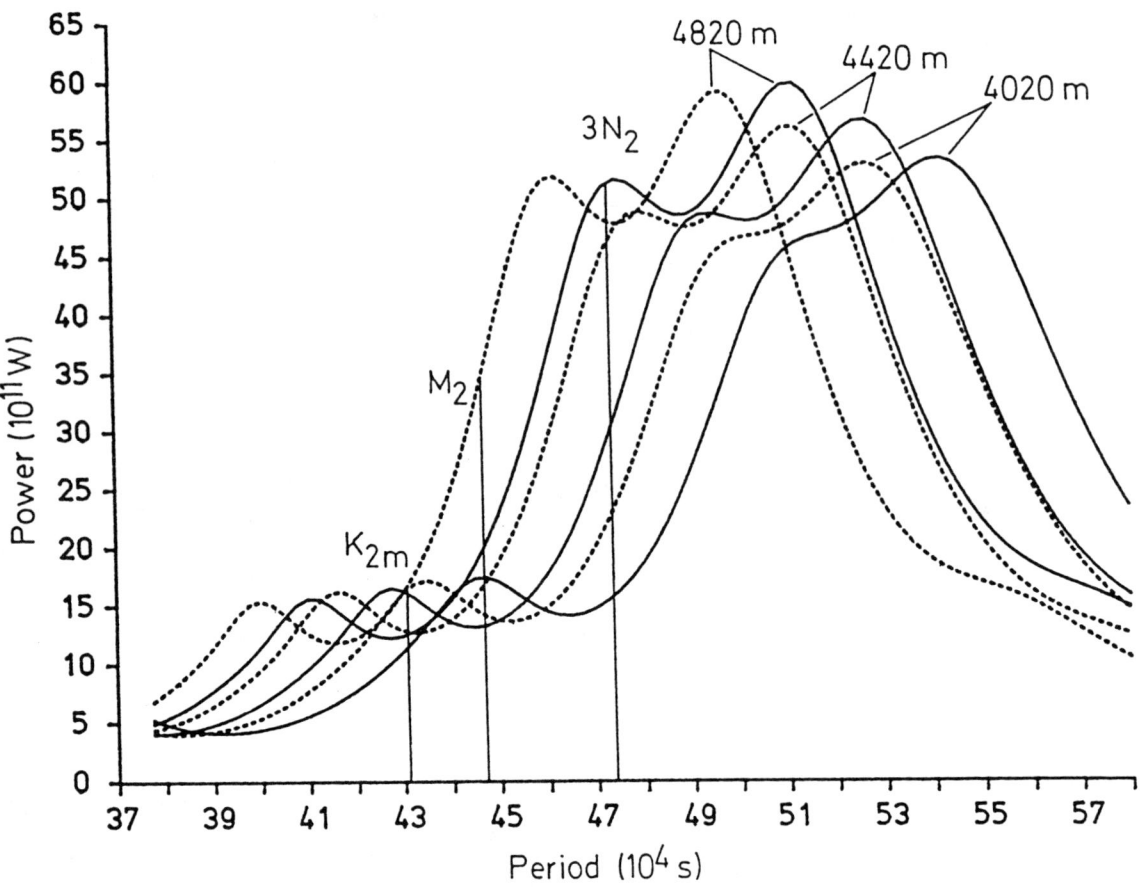

Fig. 3. Rate of work done by tidal forces and moving sea bottom. Model ocean with $R = 9.26 \cdot 10^{-6}$ s^{-1} driven by Y_2^{-2}-tidal potential with M_2-amplitude and varying periods using $H = 4020$ m, $H = 4420$ m and $H = 4820$ m, respectively. Vertical lines mark periods of lunar tidal constituents. Solid lines and dashed lines refer to computations with LSA considered and neglected, respectively.

value corresponding to a decay time of 60 hours. The described influence of LSA remains unaltered in the main, where independent of considering LSA or not reducing friction leads (see also Table 2) to lowering of the background level of tidal power, to an increase of energy contents and to a strengthening of resonances. Further, computations have been performed for two additional oceanic depths. The results again demonstrate the significant influence of LSA on the response to tidal forces and on normal modes. (See Figure 3.) In the near resonance situation of the semi-diurnal tides the consideration of LSA might change the rate of tidal power by up to 50 percent (decay time 30 h). This maximum rate of change is comparable with that caused by varying the oceanic depth. The situation is different in case of the diurnal tides because the resonance conditions are not fulfilled for the tidal band when H = 4420 m. (See Tables 1 and 2.) Nevertheless, independent of near resonance, diurnal as well as semi-diurnal tidal regimes are significantly affected by LSA (Figures 4 to 7), which is understandable in view of the modified contributions in the normal mode expansion. Similar phenomena due to LSA were found by Zahel [1978] when applying a realistic model. The near resonance of the $3N_2$-tide can easily be taken from the amplitude distribution of the sea surface, depicted in Figures 4 and 5.

When finally evaluating the influence of using different sets of realistic loading Love numbers [Melchior, 1983], it appears that the resulting small modifications of the influence of LSA on the rate of tidal power corresponds to the small changes in magnitude of the relevant coefficients α'_r.

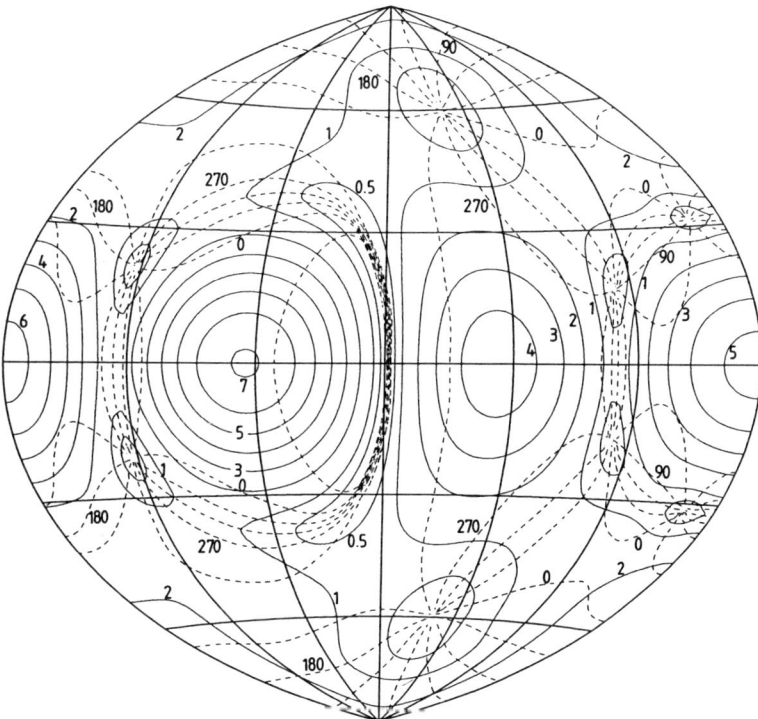

Fig. 4. $3N_2$-tide for a hemispherical ocean model with H = 4420 m and R = 9.26·10^{-6} s^{-1} when considering loading and self-attraction. Cotidal lines are dashed, with phases in degrees referred to meridian passage at the western boundary. Coamplitude lines are solid, with amplitudes in units of the maximum equilibrium tide amplitude.

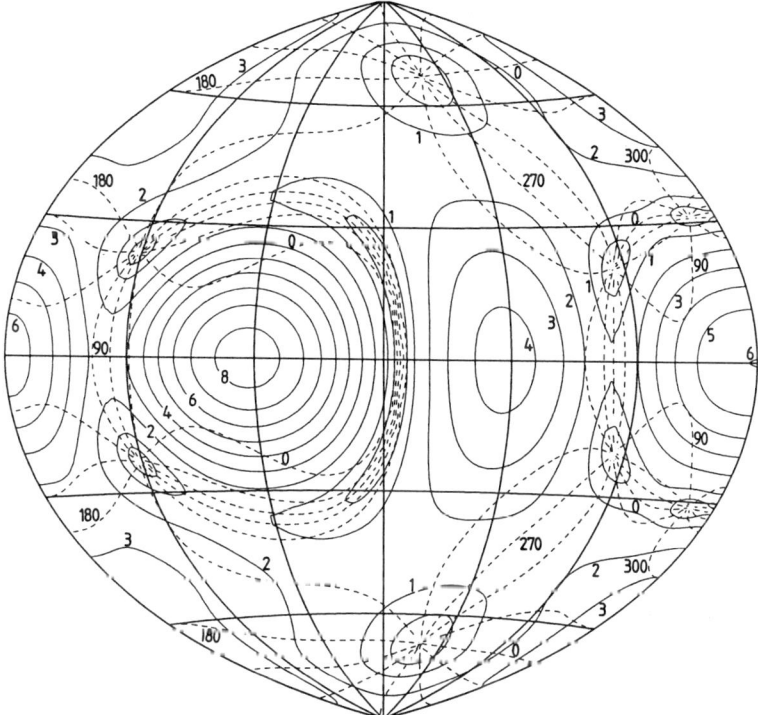

Fig. 5. Same as Figure 4 except for the model neglecting loading and self-attraction.

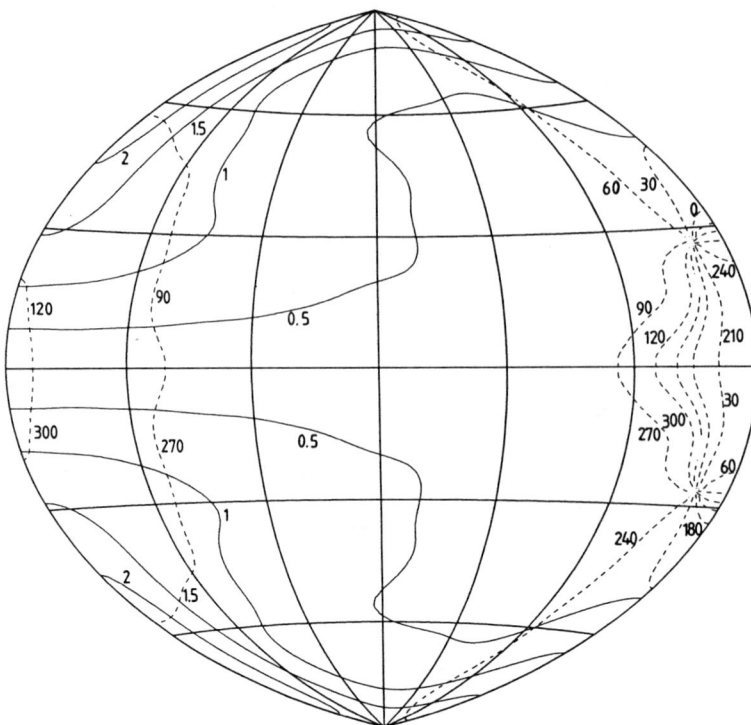

Fig. 6. K_1-tide for a hemispherical ocean model with H = 4420 m and R = 9.26·10^{-6} s^{-1} when considering loading and self-attraction. Cotidal lines are dashed, with phases in degrees referred to meridian passage at the western boundary. Coamplitude lines are solid, with amplitudes in units of the maximum equilibrium tide amplitude.

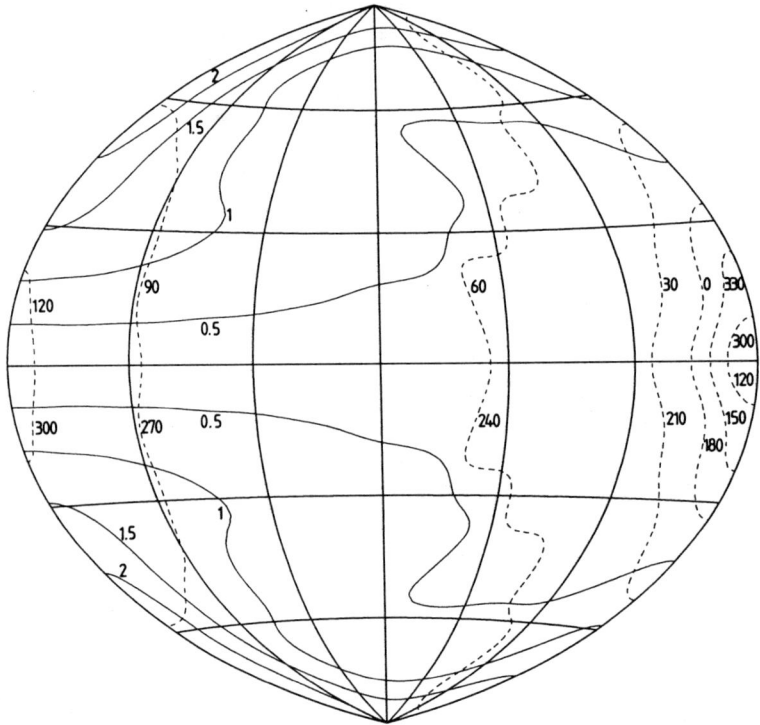

Fig. 7. Same as Figure 6 except for the model neglecting loading and self-attraction.

Conclusion

It has been demonstrated that applying a hemispherical model with analytical solution yields, when using realistic sets of ocean and solid Earth parameters, that the ocean response to tidal forces is significantly influenced by the loading and self-attraction effects and that this influence compares with that obtained by realistic models. These results indicate together with the recognized substantial influence on the normal modes that also in realistic ocean models key resonances might significantly be moved leading to changes in the rate of tidal power and in the corresponding total variation of the Moon's torque on the Earth, which even compare with the factor 2 to 2.5 obtained by Brosche and Hövel [1983] as a consequence of changing resonance conditions due to the continental drift occurring around the presence. In view of the substantially different modification the individual tidal constituents might experience, a careful consideration of loading and self-attraction effects appears necessary also when computing the ocean response to the complete tidal potential.

References

Accad, Y. and C. L. Pekeris, Solution of the tidal equations for the M_2 and S_2 tides in the world oceans from the knowledge of the tidal potential alone, Philos.Trans.R.Soc.London A290, 235-266, 1978.

Brosche, P. and W. Hövel, Tidal friction for times around the presence. In: Brosche, P. and Sündermann, J. (eds.), Tidal Friction and the Earth's Rotation II, Springer-Verlag, Berlin-Heidelberg, 175-189, 1983.

Estes, R. H., A computer software system for the generation of global ocean tide including self-gravitation and ocean loading effects, NASA/Goddard Space Flight Center, Report No. X-920-77-82, 1977.

Gaviño, J. H., On the calculation of resonance oscillations of a world ocean's finite difference model by means of the Lanczos method, Mitt.Inst.f.Meereskd.Univ.Hamburg 27, 1-78, 1984.

Gordeev, R. G., B. A. Kagan and E. V. Polyakov, The effects of loading and self-attraction on global ocean tides: the model and results of a numerical experiment, Journal of Physical Oceanography 7, 161-170, 1977.

Gotlib, V. Y. and B. A. Kagan, Numerical simulation of tides in the world ocean: 3. A solution to the spectral problem, Dt.hydrogr.Z. 35, 45-58, 1982

Longuet-Higgins, M. S. and G. S. Pond, The free oscillations of fluid on a hemisphere bounded by meridians of longitude, Phil.Trans.R.Soc. London A266, 193-223, 1970.

Melchior, P., The tides of the planet Earth. Pergamon, Oxford, 641 pp., 1983.

Platzman, G. W., G.A. Curtis, K. S. Hansen and R. D. Slater, Normal modes of the world ocean, Part II: Description of modes in the period range 8 to 80 hours, Journal of Physical Oceanography 11, 579-603, 1981.

Schwiderski, E. W., On tidal friction and the decelerations of the Earth's rotation and Moon's revolution, Marine Geodesy 9, 399-450, 1985.

Webb, D. J., Tides and tidal friction in a hemispherical ocean centered at the equator. Geophys.J.R.astr.Soc. 61, 573-600, 1980.

Zahel, W., The influence of solid earth deformations on semi-diurnal and diurnal oceanic tides. In: Brosche, P. and Sündermann, J. (eds.), Tidal friction and the Earth's rotation I, Springer-Verlag, Berlin-Heidelberg-New York, 98-124, 1978.

Zahel, W., Mathematical modelling of global interaction between ocean tides and earth tides, Phys.Earth.Planet.Inter. 21, 202-217, 1980.

Zahel, W., Astronomical tides. In: Sündermann, J., Oceanography, Landolt-Börnstein Vc, Springer-Verlag, Berlin-Heidelberg, 83-134, 1986.

SECULAR TIDAL AND NONTIDAL VARIATIONS IN THE EARTH'S ROTATION

Milan Burša

Astronomical Institute of the Czechoslovak Academy of Sciences
Praha, Czechoslovakia

Abstract. The angular momentum balance in the Earth-Moon-Sun system has been refined using the most recent value of the secular decrease in the Moon's mean motion detected by LLR as well as, the decrease in the second zonal geopotential harmonic detected by LAGEOS SLR. The secular decrease in the angular velocity of the Earth's rotation has been computed and compared with the observed one. The purely nontidal positive acceleration of the Earth's rotation has been determined as well as, its purely tidal deceleration. The last is of about 1/5 larger in magnitude of the total observed value.

The most recent value of the secular decrease in the Moon's mean motion n, detected by LLR [Dickey, personal communication, May 1987]

$$\frac{dn}{dt} = -(24.9'' \pm 1.0'') \text{ cy}^{-2}, \quad (1)$$

as well as, of the decrease in the second zonal geopotential harmonic J_2, detected by SLR [Rubincam, 1986]

$$\frac{dJ_2}{dt} = -(2.8 \pm 0.3) \times 10^{-9} \text{ cy}^{-1} \quad (2)$$

make it possible to refine the total, tidal and nontidal secular variations in the angular velocity ω of the Earth's rotation.

As regards the nontidal variation in ω, it may be computed directly from (2) under condition that $\delta A + \delta B + \delta C = 0$ (C > B > A are the principal moments of the Earth's inertia):

$$\frac{dC}{dt} \doteq -(4.5 \pm 0.5) \times 10^{29} \text{ kg m}^2 \text{ cy}^{-1}, \quad (3)$$

Copyright 1990 by
International Union of Geodesy and Geophysics
and American Geophysical Union.

$$\frac{dA}{dt} + \frac{dB}{dt} = (4.5 \pm 0.5) \times 10^{29} \text{ kg m}^2 \text{ cy}^{-1}, \quad (4)$$

$$\left(\frac{d\omega}{dt}\right)_{nontidal} = -\frac{\omega}{C}\frac{dC}{dt} \doteq \quad (5)$$

$$\doteq (1.29 \pm 0.14) \times 10^{-22} \text{ rad s}^{-2}.$$

The tidal secular decrease of the angular velocity $\left(\frac{d\omega}{dt}\right)_{tidal}$ may be computed on the basis of the observed secular decrease (1) in the Moon's mean motion which is of purely tidal origin. The basis for the solution is the angular momentum L of the Earth-Moon-Sun system which may considered as isolated, i.e.

$$L = \text{const.} \quad (6)$$

The tidal variation in the component of vector L directed along vector ω can be expressed as

$$\frac{dL}{dt} = \frac{M_\oplus M_☾}{M_\oplus + M_☾} \left[G(M_\oplus + M_☾) a_☾ (1 - e_☾^2) \right]^{1/2} \times \quad (7)$$

$$\times \cos i_☾ \left[\frac{1}{2} \frac{1}{a_☾} \frac{da_☾}{dt} - \frac{e_☾}{1-e_☾^2} \frac{de_☾}{dt} - \tan i_☾ \frac{di_☾}{dt} \right] +$$

$$+ \frac{1}{2} \frac{M_☉(M_\oplus + M_☾)}{M_☉ + M_\oplus + M_☾} \left[G(M_☉ + M_\oplus + M_☾) \times \right.$$

$$\left. \times a_B (1 - e_B^2) \right]^{1/2} \cos \epsilon_o \frac{1}{a_B} \frac{da_B}{dt} ;$$

43

M are the masses of the bodies, $G = 6\,673 \times 10^{-14}\, m^3\, s^{-2}\, kg^{-1}$ is the Newtonian gravitational constant, $a_{\mathrm{C}}, e_{\mathrm{C}}, i_{\mathrm{C}}$ and a_B, e_B, ε_o are the Keplerian orbital elements of the Moon's orbit and of the orbit of barycenter B of the Earth-Moon system respectively. From (1)

$$da_{\mathrm{C}}/dt = (3.68 \pm 0.15) \times m\, cy^{-1}; \quad (8)$$

the tidal variations in the remaining orbital elements are (for details, see [2]) as

$$di_{\mathrm{C}}/dt = -9.2\, rad \times 10^{-10}\, cy^{-1}, \quad (9)$$

$$de_{\mathrm{C}}/dt = 0.012 \times 10^{-7}\, cy^{-1}, \quad (10)$$

$$da_B/dt = (3.2 \pm 0.13) \times 10^{-4}\, m\, cy^{-1}. \quad (11)$$

With the use of (8) – (11), it is numerically

$$dL/dt = (5.1 \pm 0.3) \times 10^{35}\, kg\, m^2\, cy^{-2}. \quad (12)$$

Variation (12) should be compensated by the tidal variation in the angular momentum $C\omega$. Because the secular variation in C has no tidal contribution, it holds

$$\left(\frac{d\omega}{dt}\right)_{tidal} = -\frac{1}{C}\frac{dL}{dt} = -(6.3 \pm 0.5) \times 10^{-22}\, rad\, s^{-2} \quad (13)$$

The resulting tidal variation (13) incorporates two parts:

$$\left(\frac{d\omega}{dt}\right)_{tidal} = \left(\frac{d\omega}{dt}\right)_{tidal_{\mathrm{C}}} + \left(\frac{d\omega}{dt}\right)_{tidal_{\odot}}. \quad (14)$$

The first, generated by the Moon, equals

$$\left(\frac{d\omega}{dt}\right)_{tidal_{\mathrm{C}}} = -\frac{1}{C}\frac{M_{\oplus} M_{\mathrm{C}}}{M_{\oplus}+M_{\mathrm{C}}} \times \quad (15)$$

$$\times \left[G(M_{\oplus}+M_{\mathrm{C}})\, a_{\mathrm{C}}(1-e_{\mathrm{C}}^2)\right]^{1/2} \times$$

$$\times \cos i_{\mathrm{C}} \left[\frac{1}{2}\frac{1}{a_{\mathrm{C}}}\frac{da_{\mathrm{C}}}{dt} - \frac{e_{\mathrm{C}}}{1-e_{\mathrm{C}}^2}\frac{de_{\mathrm{C}}}{dt} - \tan i_{\mathrm{C}}\frac{di_{\mathrm{C}}}{dt}\right];$$

numerically

$$\left(\frac{d\omega}{dt}\right)_{tidal_{\mathrm{C}}} = -(5.29 \pm 0.3) \times 10^{-22}\, rad\, s^{-2}. \quad (16)$$

The second, generated by the Sun, is

$$\left(\frac{d\omega}{dt}\right)_{tidal_{\odot}} = -\frac{1}{2C}\frac{M_{\odot}(M_{\oplus}+M_{\mathrm{C}})}{M_{\odot}+M_{\oplus}+M_{\mathrm{C}}} \times \quad (17)$$

$$\times \left[G(M_{\odot}+M_{\oplus}+M_{\mathrm{C}})\, a_B(1-e_B^2)\right]^{1/2} \times$$

$$\times \cos \varepsilon_o\, \frac{1}{a_B}\frac{da_B}{dt};$$

numerically

$$\left(\frac{d\omega}{dt}\right)_{tidal_{\odot}} = -(1.04 \pm 0.05) \times 10^{-22}\, rad\, s^{-2}. \quad (18)$$

The ratio of (16) and (18) is

$$\left(\frac{d\omega}{dt}\right)_{tidal_{\mathrm{C}}} : \left(\frac{d\omega}{dt}\right)_{tidal_{\odot}} = 5.09. \quad (19)$$

The computed total decrease of the angular velocity of the Earth's rotation gives the sum of (13) and (5):

$$\frac{d\omega}{dt} = \left(\frac{d\omega}{dt}\right)_{tidal} + \left(\frac{d\omega}{dt}\right)_{nontidal} = \quad (20)$$

$$= -(5.04 \pm 0.4) \times 10^{-22}\, rad\, s^{-2}.$$

This computed value is very close to that obtained from observation directly. The difference does not exceed the standard error. That is why, variations (1) and (2) may be considered as realistic.

Variation (1) is of purely tidal origin. For (2), the dynamics of the Earth's interior should be responsible, rather than processes at the Earth's surface; the lower mantle contributes to J_2 by ~58%. E.g., the variation in the thickness $\delta\rho$ of layer

$$\delta\rho = a_o + a_2 P_2^{(o)}(\sin\phi) \quad (21)$$

at the core-mantle boundary as $da_2/dt = 17$ cm cy^{-1} may be responsible for the phenomenon:

$$\frac{dJ_2}{dt} = \frac{4}{5} \pi \frac{\bar{\sigma}}{M_\oplus a_o^2} \rho^{-4} \frac{da_2}{dt} ; \qquad (22)$$

$\bar{\sigma} = 9\,900$ kg m^{-3} is the density at the boundary, $\bar{\rho} = 3\,485$ km is the mean radius of the core, $a_o = 6\,378\,140$ m. Because of (3), there is no space for the expanding Earth hypothesis.

References

Rubincam, D.P., Personal communication, May 1986: in B. Chovitz, SSG 5-100 IAG Rep., 11, 1987.

Burša, M., Secular tidal and nontidal variations in the Earth's rotation, Studia geoph. et geod. 31, 1987, in print.

TIDAL DECELERATION OF THE EARTH

Peter Brosche

Observatorium Hoher List der Universitäts-Sternwarte Bonn,
D-5568 Daun, F.R. Germany

Abstract. The process under discussion results in a transport of angular momentum away from the rotating Earth into the orbit of the Moon and, to a smaller extent, into the Earth's orbit. This process is mediated by the oceans. Although not all of its parts are understood, it is clear that the resonance properties of the real oceans play a dominating role in determining the strength of the tidal interaction, that is, the average torque exerted on the Earth. In the long run, the geometry of the oceans varies due to continental drift; therefore the resonance properties vary accordingly. The tidal deceleration of the Earth is the only known mechanism which leads to large and monotonous changes in the Earth's rotation. Hence it has bearings for the climate, the life, the magnetic field and for the most early states of the Earth-Moon system.

In the study of Earth's rotation we are fascinated today but the most modern observational techniques. Because they have been available for only a few decades, specialists are now focusing on phenomena with short time scales. This is justified but one should not forget the long time phenomena. Among them, the only effect which is known to act monotonously in one direction for all geological epochs is the tidal friction process. It takes place mainly within the Earth-Moon system. Therefore it can be measured practically at two places: at the Earth and at the Moon (most measurements contain contributions from both sides but one is dominating). The moment of inertia of the lunar orbit is larger than that of the rotating Earth, hence the geometrical effects at the Moon are smaller. But because the orbital motion of the Moon as a whole is much simpler than the rotational motion of all the constituents of the Earth, the measurements of the lunar orbit are less problematic and are in fact the ones which we adopt today as the most precise ones. They are derived from:

Copyright 1990 by
International Union of Geodesy and Geophysics
and American Geophysical Union.

(a) Lunar occultations of stars embedded in a time scale delivered by Mercury transits (MTO).
(b) Lunar Laser Ranging data (LLR).

The results for the tidally caused change $\dot{n}$ in the mean motion of the Moon are

MTO -26 ± 2 [Morrison and Ward 1975]
LLR -25.3 ± 1.2 [Dickey et al. 1983]
or -24.9 ± 1.2 [Dickey 1986]
(units: seconds of arc per century^2)

Although it is now customary to quote the LLR value as the "best" one, I would attribute equal weight to both results. In view of the multitude of geophysical phenomena producing centimeter-variations in the Earth-Moon distance, we should be very happy that the outcomes from ~ 300 years and ~ 10 years agree so well.

The direct measurement of the tidal deceleration at the Earth itself is hampered by the many nontidal variations, especially within shorter time scales. Among the well established ones, the 'decadic' variations are the longest. In fact, they could be called centennial as well. Therefore only from averaging over millenia or longer periods, one could expect an information on the tidal effect. Indeed, the places of antique solar eclipses were the first empirical datum of our topic. Today, however, one prefers rather to use the torque observed at the lunar orbit for the Earth (with the reverse sign, of course) and to study the remaining nontidal variations in terms of geophysical phenomena [Stephenson and Morrison; 1982; Liu Ciyuan, 1987].

While our human observations consist in angles and distances, one observes growth rhythms in the outer skeletons of fossil and recent animals, which can provide the ratio of important periods: the length of the tropical year, the synodic month, and the solar day [Scrutton 1978]. These data are of greatest importance since they represent a time scale of about 10 percent of the age of the Earth. From what could be derived so far, the tidal deceleration was of the same order of magnitude as today, with some uncertain evidence for a period of smaller values around the Permian. However, not only the numbers are uncertain but

47

also the basic biological processes are not well understood. Studies of this processes in recent animals would be very valuable but are very rare (the latest study known to me is the work of Ohno [1985]). If the growth of stromatolites could be used, we had a time scale of 2 billion years [Vanyo, Hutchinson and Awramik, 1986]; unfortunately the problems related with this class seem even harder.

The value reported above for the tidal change in the lunar orbit corresponds to a torque of

$$L = 5 \cdot 10^{16} \text{ N m},$$

positive for that orbit and negative for the Earth. This means at the same time a dissipation $dE/dt = -4TW$ of the Earth's rotational energy, a value only one power of ten below the radioactive heat production from the interior of the Earth! It means likewise an increase of 4cm/year of the radius of the lunar orbit, just measurable with LLR. This increase does not correspond, however, to an increase of orbital energy equal to the loss of the Earth. Since angular momentum (which is preserved) is proportional to angular velocity but energy to its square, the latter is only marginally transferred to the Moon, but the greatest part must be dissipated into heat.

Here we come inevitably to the question as to how much do we understand theoretically the process, in addition to the application of conservation theorems. The picture of two retarded tidal bulges provides a principal insight into the cause of the tidal torque. An analytical treatment of this picture is only applicable to tides of the solid Earth, but we know that oceanic tides dominate the transport of angular momentum. Their temporal and spatial structure is much more complex and has to be treated on large computers in order to approximate the hydrodynamical equations with sufficient accuracy. It might be too pretentious but I will try to describe the outcome in a few words: the exchange of angular momentum between the oceans and the Moon can be computed with 10 or 20 percent accuracy and agrees with the observed value. The dissipation of energy is a small-scale phenomenon and it may be represented in a phenomenological manner but the constants have to be adjusted accordingly. Beyond these basic problems, the main difficulty with the oceans is their sensitivity to changes, in other words, their rich spectrum of eigenperiods which depends critically on the geometry of the oceans [Sündermann 1982, Gotlib and Kagan 1985]. That sensitivity leads already within 10 or 20 million years of present continental drift to a variation in the M_2-torque of a factor of 2 [Brosche and Hoevel, 1982].

This fact shows most clearly why so far we have been unable to answer the ultimate question of our topic: what is the status of the Earth-Moon system at its beginning? We have performed computations of tide models for several geological epochs but the paleobathymetry and the drift history are too uncertain to allow a real backwards integration of the Earth-Moon system for several 100 million years [Brosche and Sündermann 1984]. On the other side, it does not mean that any further efforts are useless at present. As the most important secular effect, the tidal deceleration is interesting also within shorter time scales. I want to mention two examples only:

(A) The rotation of the Earth is crucial for the existence of the magnetic field of the Earth. Correlations between both have been established beyond doubt for the decadic variations. Therefore it is not a far-fetched speculation to assume that also in the long run a causal nexus exists and that perhaps variations of the tidal deceleration can trigger the reversals of the magnetic field [Brosche, 1981].

(B) The climate must depend on the rotation of the Earth, simply because the atmosphere has also eigenperiods, and more intricately so because precession depends on the angular momentum of the Earth and precession influences among other parameters the climate [Berger, 1980].

References

Berger, A., The Milankovitch astronomical theory of paleoclimates: a modern review, Vistas in Astronomy, 24, 103-122, 1980.

Brosche, P., Geomagnetic reversals and tidal friction, Naturwissenschaften, 68, 139-140, 1981.

Brosche, P., and W. Hoevel, Tidal friction for times around the presence, in Tidal Friction and the Earth's Rotation II, (eds. P. Brosche and J. Sündermann) Berlin-Heidelberg-New York 175-189, 1982.

Brosche, P., and J. Sündermann, Tidal friction and dynamics of the earth-moon-system, in Landolt Börnstein, Neue Serie, Gruppe V, Geophysik und Weltraumforschung, (eds K. Fuchs and H. Soffel), Bd. 2a, 299-310, 1984.

Dickey, J. O., J. G. Williams, XX Newhall, and C. F. Yoder, Geophysical Application of Lunar Laser Ranging, Presentation XVIIIth Gen. Assembly IUGG Hamburg 1983.

Dickey, J. O., pers. comm. to M. Burša, 1986 (Final report of IAG - SSG 5.99 for XIXth Gen Assembly IUGG Vancouver 1987).

Gotlib, V. Yu., and B. A. Kagan, A reconstruction of the tides in the Paleocean: Results of a numerical simulation, Deutsche hydrogaph. Z., 38, 43-67, 1985.

Liu Ciyuan, Preprint Shaanxi Obs. to be published in Acta Astrophysica Sinica 1987.

Morrison, L. V., and C. G. Ward, An analysis of the transits of Mercury, Mon. Not. R. Astr. Soc., 173, 183-206, 1975.

Ohno, R., Experimentelle Analysen zur Rhythmik des Schalenwachstums einiger Bivalven und ihre paläobiologische Bedeutung, Paleontographica, Abt. A, 189, 63-123, 1985.

Scrutton, C. R., Periodic growth features in

fossil organisms and the length of the day and month, in *Tidal Friction and the Earth's Rotation*, (eds. P. Brosche and J. Sündermann) Berlin-Heidelberg-New York, 154-196, 1978.

Stephenson, F. R., and L. V. Morrison, History of the Earth's rotation since 700 B.C., in *Tidal Friction and the Earth's Rotation II*, (eds. P. Brosche and J. Sündermann) Berlin-Heidelberg-New York, 29-50, 1982.

Sündermann, J., The resonance behaviour of the world ocean, in *Tidal Friction and the Earth's Rotation II*, (eds. P. Brosche and J. Sündermann) Berlin-Heidelberg-New York, 165-174, 1982.

Vanyo, J. P., R. A. Hutchinson, and S. M. Awramik, Heliotropism in microbial stromatolitic growths at Yellowstone National Park: Geophysical inferences, *EOS*, April 1, 153-156, 1986.

EFFECTS OF THE TIDAL DISSIPATION ON THE MOON'S ORBIT AND THE EARTH'S ROTATION

M. Ooe, H. Sasaki

International Latitude Observatory of Mizusawa, Mizusawa-shi, Iwate-ken, 023 Japan

and

H. Kinoshita

Tokyo Astronomical Observatory, Mitaka-shi, Tokyo-to, 181 Japan

Abstract. Exact expressions for tidal perturbations of the Moon's orbit and the Earth's rotation are given. Then M_2 tidal models of present-day oceans and a permian ocean are computed by solving discrete tidal equations. These models are used to estimate the process of the dynamical evolution of the Earth-Moon system. Obtained results suggest that the Gerstenkorn date is not so far from 6 billion years ago.

Introduction

Tidal dissipation is an important factor to solve the evolution of the Earth-Moon system. This problem was studied by many authors with different approaches. MacDonald in his early work [MacDonald, 1964] used a constant tidal lag to estimate the historical effect of tides and excluded the solar tides. Goldreich [1966] took the Sun into consideration but ignored the eccentricity of the Moon's orbit. Mignard [1981] took the Sun into consideration but used a constant tidal lag. Webb [1980, 1982] computed a time dependent tidal lag for a hemispherical ocean by changing angular velocities in a tidal simulation. He found the Gerstenkorn event occurred 3.9 billion years ago, pointing out the importance of the effects of the solid tide. The eccentricity and the inclination, however, were ignored. Hansen [1982] used a time dependent tidal lag which he computed for ocean models with idealized continents. The eccentricity and the effects of the solar tides, however, were ignored.

These works clarified astonishing features of the tidal dissipation. These, however, leave some ambiguities in estimations which compel us to further implementations. Specially, the tidal lag is sensitive to the distribution of the continents. We investigate the effects of the tidal dissipation with more realistic continents by solving the tidal equations for the present day oceans and for a perimian ocean. We also allow for the eccentricity and the inclination in the estimation of the tidal evolution of the Earth-Moon system.

Tidal Perturbations of Lunar Orbit and Earth's Rotation

We ignore the effect of the Sun. In this case the sum of the angular momentum vectors of the Earth's spin and the revolution of the Moon is conserved. If we take an invariant plane, which is perpendicular to the total angular momentum vector, as a reference plane, the angular momentum conservation is expressed by

$$\{Mm_{\mathrm{C}}/(M+m_{\mathrm{C}})\}n_{\mathrm{C}} a_{\mathrm{C}}^2 (1-e_{\mathrm{C}}^2)^{\frac{1}{2}} \cos i^* + C\Omega \cos i_{eq} = const. \quad (1)$$

where i^* and i_{eq} are the inclinations of the lunar orbit and the Earth's equator to the invariant plane, respectively. n_{C} is the mean motion, a_{C} the semi-major axis and e_{C} the eccentricity of the Moon. M and m_{C} are masses of the Earth and the Moon, respectively. C and Ω are the principal moment of inertia and the angular velocity of the Earth, respectively. The inclination of the lunar orbit to the equator is given by $i_{\mathrm{C}} = i^* + i_{eq}$.

We use the equatorial coordinate system (r, ψ, λ), which is moving with respect to the invariant plane, to express the tidal potential and equation of motions. Lambeck [1977] developed the tidal potential at any position outside of the Earth as a function of orbital elements of the Moon as follows:

Copyright 1990 by
International Union of Geodesy and Geophysics
and American Geophysical Union.

$$\Delta U(r) = (Gm_{\mathrm{C}}/a_{\mathrm{C}}) \sum_{lm} (R/r)^{l+1} (R/a_{\mathrm{C}})^{l} k_{l} (2-\delta_{0m})$$

$$(l-m)!/(l+m)! \; P_{lm}(\sin\varphi) \sum_{p} F_{lmp}(i_{\mathrm{C}}) \sum_{q} G_{lpq}(e_{\mathrm{C}})$$

$$\begin{bmatrix} \cos \\ \sin \end{bmatrix} \begin{matrix} l-m \text{ even} \\ l-m \text{ odd} \end{matrix} (v_{lmpq} + E_{lmpq} - m\lambda) \quad (2)$$

with $v_{lmpq} = (l-2p)\omega_{\mathrm{C}} + (l-2p+q)M'_{\mathrm{C}} + m(\Omega_{\mathrm{C}} - \theta) + E_{lmpq}$ where R is the Earth's radius, $F_{lmp}(i)$ and $G_{lpq}(e)$ are polynomials in $\sin i_{\mathrm{C}}$ and e_{C}, respectively [Kaula, 1966], θ is the spin angle of the Earth, and E_{lmpq} is the phase lag of tidal response.

If the position (r, φ, λ) refers to the Moon, it is expressed in terms of Kepler elements $a_{\mathrm{C}}, e_{\mathrm{C}}, i_{\mathrm{C}}, \omega_{\mathrm{C}}, \Omega_{\mathrm{C}}, M'_{\mathrm{C}}$. The secular perturbations of the orbit are given by choosing parameters as periodic terms to be zero in the Lagrange planetary equations of motion.

$$(di/dt)_{lmpq} = K_{lm}\{(1-2p)\cos i_{\mathrm{C}} - m\}/\{a_{\mathrm{C}}(1-e_{\mathrm{C}}^2)^{\frac{1}{2}}\sin i_{\mathrm{C}}\}$$
$$\sin(E_{lmpq}) \quad (3a)$$

$$(da/dt)_{lmpq} = 2K_{lm}(l-2p+q)\sin E_{lmpq} \quad (3b)$$

$$(de/dt)_{lmpq} = K_{lm}\{(1-e_{\mathrm{C}}^2)^{\frac{1}{2}}/(a_{\mathrm{C}}e_{\mathrm{C}})\}$$
$$\{(1-e_{\mathrm{C}}^2)^{\frac{1}{2}}(l-2p+q)-(l-2p)\}\sin(E_{lmpq}) \quad (3c)$$

with

$$K_{lmpq} = (M+m_{\mathrm{C}})/M \; Gm_{\mathrm{C}}k_{l}/\{G(M+m_{\mathrm{C}})/a_{\mathrm{C}}\}^{\frac{1}{2}}$$
$$\{F_{lmp}(i_{\mathrm{C}})G_{lpq}(e_{\mathrm{C}})\}^{2}$$
$$(R/a_{\mathrm{C}})^{2l+1}\{(l-m)!/(l+m)!\}(2-\delta_{0m}).$$

(A factor $(M+m_{\mathrm{C}})/M$ is different from Lambeck's.) The time derivertive of the mean motion is given by

$$(dn/dt)_{lmpq} = -3/2(n_{\mathrm{C}}/a_{\mathrm{C}})(da/dt)_{lmpq}. \quad (4)$$

The variation of the angular momentum of the Earth's spin is given by the Euler equation ignoring the polar motion as follows:

$$d(C\Omega)/dt = N_3 \quad (5)$$

where Ω is $d\theta/dt$ and N_3 the tidal torque around the Earth's figure axis. N_3 is given by

$$N_3 = \partial(m \; \Delta U)/\partial\lambda. \quad (6)$$

The principal moment of inertia C depends on the velocity of Earth's spin. After Munk and MacDonald [1960], it is $C = C_0 + \tau\Omega^2$ with $C_0 = 0.32987MR$ and $\tau = 2k_s R^5/(9G)$, where k_s is the secular Love number. Taking (2) into (6), we have the secular acceralation of the Earth's spin as

$$(d\Omega/dt)_{\mathrm{C}\mathrm{C}} = -\{na/(C_0 + 3\tau\Omega^2)\}\{Mm_{\mathrm{C}}/(M+m_{\mathrm{C}})\}$$
$$mK_{lmpq}\sin E_{lmpq}. \quad (7)$$

The phase lag of the solid tide is given by $\sin E_{lmpq} = 1/Q$ where Q is the quality factor. We use 150 for Q which is inferred from analyses of the Chandler wobble and the solid tide. The equivalent phase lag of the ocean tide is given by the following expressions.

Let D^+_{lm} and ε^+_{lm} be the amplitude and the phase lag of the ocean tide in spherical harmonics at an arbitary epoch T. Unless the configuration of the ocean is changed, the phase admittance is constant during the orbit evolution and the tidal amplitude is proportional with the tidal potential on the Earth's surface. Hence D^+_{lm} is propotional with $(1/a_{\mathrm{C}})^{l+1} \sum_p F_{lmp}(i_{\mathrm{C}}) \sum_q G_{lpq}(e_{\mathrm{C}})$. Taking this effect into consideration, we can express the equivalent phase lag of the ocean tide to the solid tide as

$$\sin \tilde{E}_{lmpq} = X_{lmpq0} \; D^+_{lm0} \begin{bmatrix} \sin \\ \cos \end{bmatrix} \begin{matrix} l-m \text{ even} \\ l-m \text{ odd} \end{matrix} \varepsilon^+_{lm0} \quad (8)$$

with

$$X_{lmpq0} = (3M/m_{\mathrm{C}})(\rho_w/\rho)(1+k'_l)/k_l$$
$$\{1/(2l+1)\}(a_{\mathrm{C}0}/R)^{l+1}(1/R)$$
$$(l+m)!/\{(l-m)!(2-\delta_{0m})F_{lmp}(i_{\mathrm{C}0})G_{lpq}(e_{\mathrm{C}0})\} \quad (9)$$

where a suffix '0' represents values at some fixed epoch T_0. This means that the equivalent phase lag at time T is expressed by the tidal response and orbital elements of the 'fixed epoch' if the ocean are not changed.

Ocean Models and Tidal Simulations

We shall study the ocean tide using the tidal equations which is modified for the elastic deformation of the ocean bottom due to loading. With a linear friction term, the momentum equation is

$$\partial U/\partial t = gH/(R\cos\varphi)\partial(\alpha\eta - \beta\zeta)/\partial\lambda - BU/H + 2\Omega V \sin\varphi, \quad (10)$$

$$\partial V/\partial t = gH/R \; \partial(\alpha\eta - \beta\zeta)/\partial\varphi - BV/H - 2\Omega U \sin\varphi \quad (11)$$

and the continuity equation

$$\partial\zeta/\partial t = -1/(R\cos\varphi)\{-\partial(V\cos\varphi)/\partial\varphi - \partial U/\partial\lambda\} \quad (12)$$

where η and ζ are the forcing equilibruim tide and the ocean tide observed, respectively. U and V are east and north velocities which are integrated from the surface to the bottom of the ocean. B is the coefficient of bottom stress, which is expressed by $B = b\cos\varphi$. α is given by

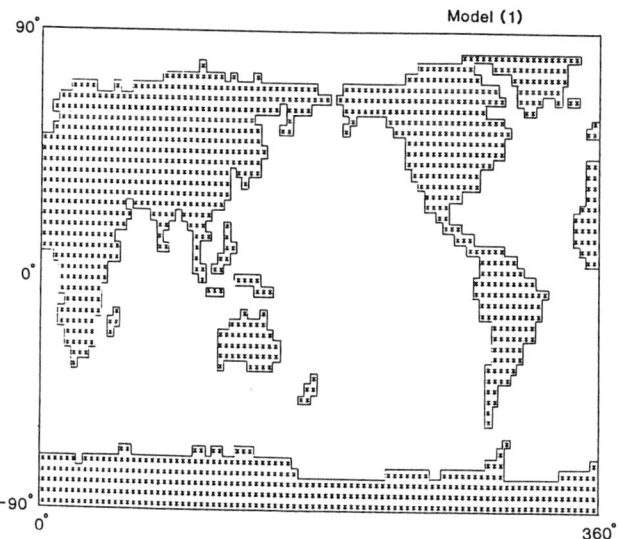

Fig. 1. Grid configuration of present day oceans, Model(1).

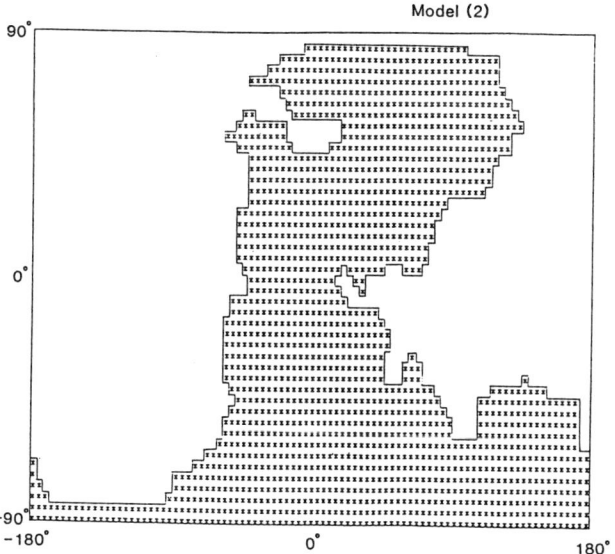

Fig. 2. Grid configuration of a permian ocean, Model(2).

$\alpha = 1 + k_2 - h_2$ and β is the coefficient for yielding of the ocean bottom.

Discrete tidal equations are developed corresponding to the above equations. A $4° \times 4°$ grid system is used. We design a finite-difference scheme fully staggered following Schwiderski [1980]. Analytic solutions are adopted at arctic regions. We assume no-flow across shorelines and free-slip along shorelines. We set α to 0.69 and β to 0.9. A time step is set to 90 or 180 seconds.

Discrete equations are solved for two ocean models. One is for present day oceans and the other is for a parmian ocean, which are called Model(1) and Model(2), respectively. Present day oceans are approximated with an uniform depth of 4400m. Figure 1 shows grid configuration of this model. A model for the permian ocean is shown in Figure 2. The decay time of these oceans is taken to be 30 hours [Webb, 1980] and also b is set to 0.0407m/s. The equations are solved for a sectorial wave at various frequencies, using the forcing equibrium tide of 0.1 meter in amplitude. Iteration of integrals almost converges after 8 to 16 periods of tidal wave.

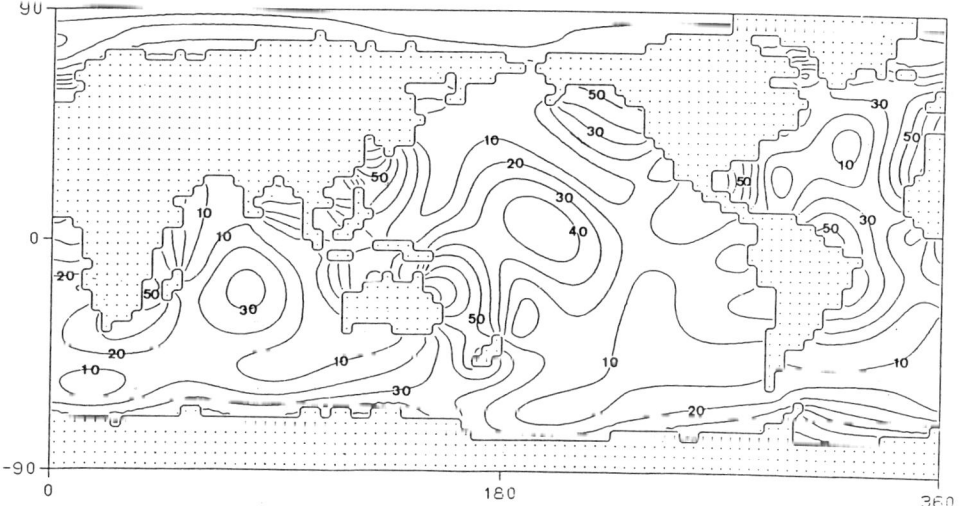

Fig. 3. Co-amplitude map of M_2 tide for Model(1).

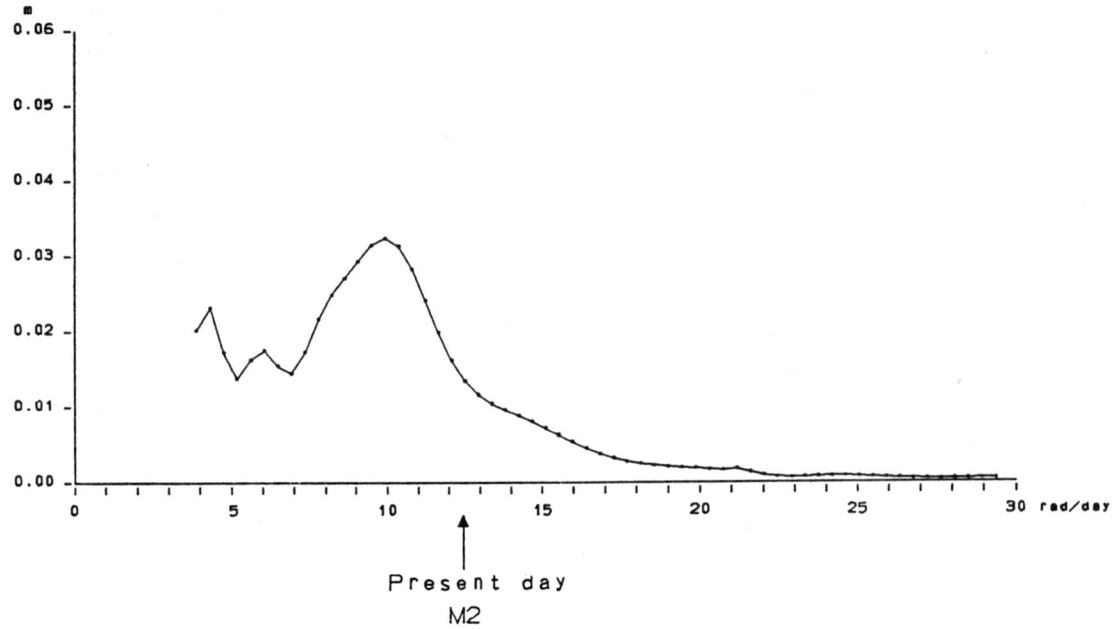

Fig. 4. Amplitude of spherical harmonics of Model(1) $D^+_{lm0} \sin \varepsilon^+_{lm0}$ of the sectoial wave is illustrated.

Figure 3 shows co-amplitudes map for the present-day M_2 tide (1.40519×10^{-4} rad/sec). Numerical results allow us to estimate spherical harmonics of the degree l and the order m at various frequencies. Figures 4 and 5 illustrate values of $D^+_{lm0} \sin \varepsilon^+_{lm0}$ in the unit of meter for Model(1) and Model(2), respectively. Resonance of the ocean appears at the frequency around 10 rad/day. As the amplitude of the forcing equilibrium M_2 tide is 0.242 m, we have to multiply the obtained amplitude by 2.42. Using this correction, Table 1 gives amplitudes and phase lags of spherical components at the frequency of the present-day M_2 tide.

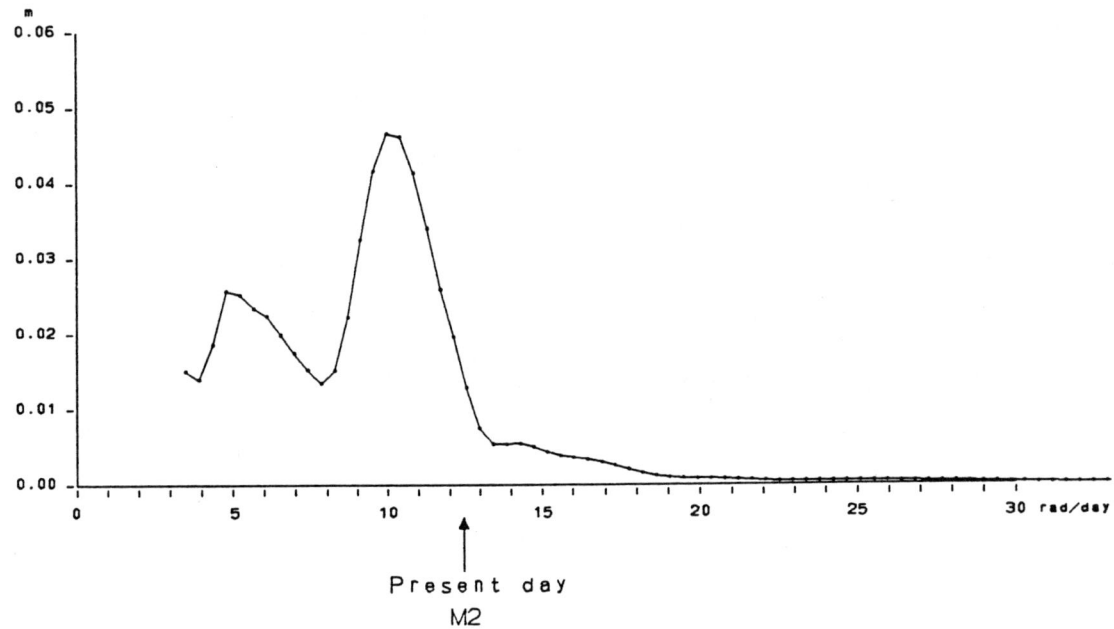

Fig. 5. Amplitude of spherical harmonics of Model(2). Definition is similar to Fig. 5.

TABLE 1. Amplitudes and Phase Lags of Spherical Components of the Ocean Tide. $\tilde{E}_{2200}$ is the equivalent phase lag.

M_2 tide	D^+_{220}	ϵ^+_{220}	$D^+_{220} \sin \epsilon^+_{220}$	$\tilde{E}_{2200}$
Hendershott(1972)	3.61 cm	105°	3.48 cm	6.4°
Zahel(1976)	4.66	110	4.38	
Model(1)	3.93	134	2.85	5.4°
Model(2)	7.17	138.5	4.75	8.9°

Variation of the Lunar Orbit and the Earth's Spin

Taking obtained amplitudes and phase lags of spherical harmonics into (8), we can estimate the equivalent phase lag. Then, we compute the perturbation of the orbit and the acceleration of the Earth's spin. Table 2 shows $\tilde{E}_{lmpq}$, da/dt, dn/dt and $d\Omega/dt$ obtained for M_2 tidal constituent (lmpq=2200), which suggests our results for the present-day oceans are almost consistent with astronomical observations. This encourages us to integrate the tidal purturbations up to the early stage of the Earth-Moon system. Equations (3a), (3b), (3c) and (7) are solved by giving the tidal phase lag at time T. The angular velocity of the tidal wave itself is a function of time given by

$$\sigma(T) = m\{\partial\theta/\partial t - \partial\Omega/\partial t\} - (l-2p)\partial\omega_{\mathbb{Q}}/\partial t - (l-2p+q)\partial M'_{\mathbb{Q}}/\partial t. \quad (13)$$

TABLE 2. Comparison of $\dot{a}$, $\dot{n}$ and $\dot{\Omega}$ obtained for the present models and those given by astronomical observations, etc.

	$\tilde{E}_{2200}$	$\dot{a}$ cm/y	$\dot{n}$ $10^{-23}/s^2$	$\dot{\Omega}$ $10^{-22}/s^2$
Astronomical			-1.36 ± 0.13	-7.2 ± 0.7
Lambeck				
M_2	6.4	4.1	-1.34	-4.7
Total Tide		4.5	-1.49	-6.4
Model(1)				
a. Ocean	5.4	3.6	-1.16	-5.2
b. Ocean+Solid	5.8	3.8	-1.25	-5.5
Model(2)				
a. Ocean	8.9	5.9	-1.94	-8.6
b. Ocean+Solid	9.3	6.2	-2.02	-8.9
SSG 5-99 (Bursa '87)				
LLR		3.68	-1.21	
Tidal				-6.4 ⎫ -5.1
Nontidal				1.3 ⎭

Hence we can easily estimate the time-dependent phase lag of a tidal constituent by simply shifting the angular velocity in Figures 4 and 5.

Figures 6a, 6b, 7a and 7b show the rsults obtained. Figure 6a shows effects of the present day ocean on the tidal evolution, in which we assumed oceans have not changed any configurations since the early stage of the Earth. In Figure 7a, we assumed the oceans took the same configuration with the permian ocean through the evolution of the Earth. In Figures 6b and 7b, we added the effects of the solid tide. In these figures, variations of the distance between the Earth and the Moon and the length of day are illustrated. These show the importance of the solid tide in the early stage of the evolution. Because about 80 percent of the tidal dissipation is expected to be due to M_2 tidal constituent, these models suggest the Gerstenkorn date is not far from 4.2 to 6.0 billion years ago. The length of day comes to 5.5 hours at the Roche limit.

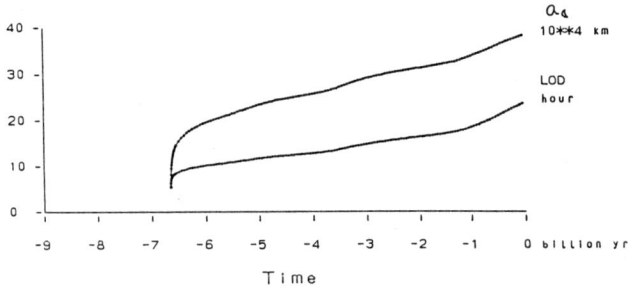

Fig. 6a. Tidal effects of Model(1) on the semimajor axis of the lunar orbit and the length of day.

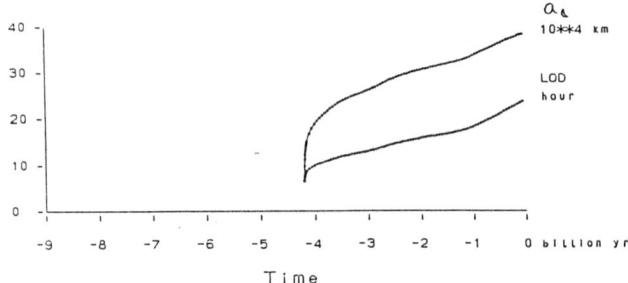

Fig. 6b. Tidal effect of Model(1) with the solid tide. The effect of the solid tide is added to the ocean tide.

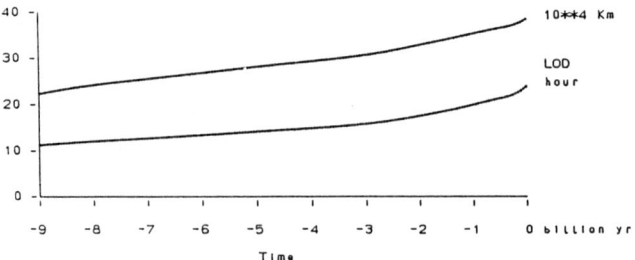

Fig. 7a. Tidal effects of Model(2) on the semimajor axis of the lunar orbit and the length of day.

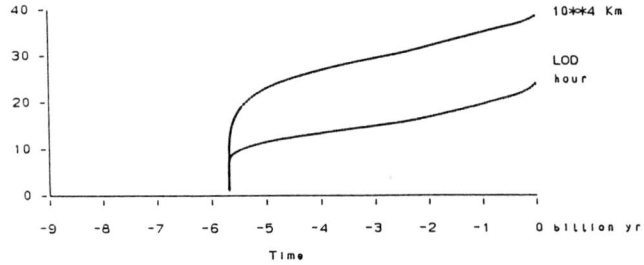

Fig. 7b. Tidal effect of Model(2) with the solid tide.

Concluding Remarks

In this paper, we ignored other tidal constituents than M_2. This effect, however, is not so serious. The more important factors are the configuration of the ancient oceans and the dissipation in the solid Earth. 150 for Q of the solid tide seems reasonable as compared with Q of Mars, 95, [Ooe, 1986]. We have to improve the model of the ancient oceans as examined by Sundermann and Brosche [1978]. Though the ocean models are crude, our computations show remarkable effects of time-dependent ocean tide on the evolution of the Earth-Moon system. Obtained results also suggest the Gerstenkorn date is not so far from 6.0 billion years ago. We did not solve variations of the inclinations of the lunar orbit and the Earth's equator to the invariant plane and effects of solar tides and the tide on the Moon. These shall be subjects of extended studies.

References

Bursa, M., Final Report of IAG Special Study Group 5-99: Tidal Friction and the Earth's otation, 19th General Assembly of IUGG, IAG, August 1987.

Goldreich, P., History of the Lunar Orbit, Rev. Geophys. 4, 411-439, 1966.

Hansen, K. S., Secular Effects of Oceanic Tidal Dissipation on the Moon's Orbit and the Earth's Rotation, Reviews of Geophys. and Space Phys., 20, 457-480, 1982.

Kaula, W. M., Theory of Satellite Geodesy, Blaisdell Publ. Company, 1966.

Lambeck, K., Tidal Dissipation in the Oceans: Astronomical, Geophysical and Oceanographic Consequences, Phi. Trans. R. Soc. London, 287, 545-594, 1977.

MacDonald, G. J. F., Tidal Friction Global Ocean Tides, Rev. Geophys. and Space Phys., 2, 467-541, 1964.

Mignard, F., The Evolution of the Lunar Orbit Revisited, III, The Moon and the Planets, 24, 189-207, 1981.

Munk, W. H. and MacDonald, G. J. F., 1960, The Rotation of the Earth, Cambridge at the Univ. Press.

Ooe, M., Tidal Effects and Q of the Solar System, Proceeding of the 9th ISAS Lunar and Planetary Symposium Institute of Space and Astronautical Science, Tokyo, 112-115, 1986.

Schwiderski, E. W., On Charting Global Ocean Tides, Rev. Geophys. and Space. Phys., 18, 243-268, 1980.

Sundermann, J. and Brosche, P., The Numerical Computation of Tidal Friction for Present and Ancient Oceans, in Tidal Friction and the Earth's Rotation Ed., P. Brosche, Springer-Verlag, 1978.

Webb, D. J., Tides and Tidal Friction in a Hemispherical Ocean centered at the Equator, Geophys. J. R. astr. Soc., 61, 573-600, 1980.

Webb, D. J., Tides and Evolution of the Earth-Moon System, Geophys. J. R. astr. Soc., 70, 261-271, 1982.

THE POLE TIDE IN DEEP OCEANS

S. R. Dickman

Dept. of Geological Sciences, State University of New York
Binghamton, New York 13901 U. S. A.

Introduction

My talk today will consist of three parts: I will spend some time developing the fluid dynamical theory of the pole tide--the oceanic response to the Chandler wobble--which I've been studying for the past few years; I will present a variety of results from my theory; and I will briefly discuss the implications of my results, for mantle anelasticity and for low frequency ocean dynamics.

Table 1 offers one reason why it is important to study the fluid dynamics of the pole tide. The most sophisticated calculations of the effect of a static pole tide on the Chandler wobble employ static oceans which are non-global (i.e. continents are present), loading and self-gravitating, and which overlie an elastic earth containing a fluid core; such a static tide would contribute about 30.6 days to the observed Chandler period of 433.2 days. By subtraction, the wobble period of an oceanless earth would be 402.6 days. In contrast, independent calculations of the wobble period of an elastic oceanless earth, by Smith & Dahlen [1981], yield a period of 396.9 days. The discrepancy of ∿ 6 days, as pointed out by Smith and Dahlen, could be due either to mantle anelasticity or to ocean dynamics, neither of which is included in those estimates.

Table 2 suggests another reason why the study of pole tide fluid dynamics is necessary. Pole tide amplitude estimates from actual tide data, accounting for noise level and statistical uncertainties, are compared here to static predictions based on a loading self-gravitating equilibrium tide forced by a Chandler wobble of the observed amplitude (analysis by D. Steinberg). The comparison at a variety of ports suggests that the actual pole tide may significantly exceed equilibrium, even in mid-ocean. We see for example that at Honolulu

Copyright 1990 by
International Union of Geodesy and Geophysics
and American Geophysical Union.

TABLE 1. Evolution of
Equilibrium Pole Tide Theory

Model[1]	Oceanic Contribution[2]
I. Global O Rigid M, C	62.4 days
II. Global O Elastic M, C	44.5
III. Non-global O Elastic M, C	31.0
IV. Non-global O Elastic M, Fluid C	34.6
V. Non-global, loading, self-gravitating O Elastic M, Fluid C	30.6

[1] Abbreviations are O=oceans, M=mantle, C=core.
[2] Contribution to observed Chandler period of 433.2 days.

(with a long span of good quality data) the pole tide may be 60%-70% larger than static. At Bermuda the tide may be enhanced by a factor of 3, although the data there is both shorter in duration and higher in noise level, thus more uncertain. With the help of dynamical pole tide theory, it will be possible to assess the reliability of results such as these.

Theoretical Basics

Any fluid dynamic tide theory begins with the Laplace tidal equations; with linearized bottom friction included, these are presented in Figure 1. The first two equations give the accelerations of the fluid (u_θ, u_λ are the meridional, azimuthal velocities) as a result of the Coriolis force; bottom friction with drag coefficient P; the tidal force, which is the

TABLE 2. Actual Versus Static Pole Tide

Location[1]	Noise Level	Actual Tide[2]	Static Tide[3]
Esbjerg (1900-1965)	0.86 cm	2.74 ± 0.60 cm	0.41 cm
Aberdeen (1916-1965)	0.53	0.90 ± 0.37	0.39
Balboa (1908-1969)	0.86	0.26 ± 0.61	0.12
Recife (1949-1968)	0.19	0.84 ± 0.14	0.13
Wash. DC (1932-1975)	0.34	1.03 ± 0.24	0.42
Ketchikan (1919-1972)	0.56	1.11 ± 0.40	0.36
Sydney (1900-1977)	0.17	0.65 ± 0.12	0.41
Midway (1947-1972)	1.20	below noise	0.44
Honolulu (1905-1975)	0.10	0.52 ± 0.07	0.30
Bermuda (1944-1972)	0.60	1.67 ± 0.42	0.49
Santa Cruz (1958-1974)	0.51	0.61 ± 0.36	0.53
Kwajalein (1947-1975)	0.93	below noise	0.18

[1] Duration of data set in years given in parentheses.
[2] Tide amplitudes found from unsmoothed Fourier spectra; error bars based on noise level. Monthly data from Permanent Service for Mean Sea Level [Lennon, 1978].
[3] Amplitude of loading, self-gravitating static tide based on Chandler wobble time series of same time span as tide data. Analysis from Dickman & Steinberg [1986].

gradient of the tide potential V; and the accompanying pressure forces, expressed in terms of the height T of the sea surface. The third equation is "continuity", and expresses the fact that any local horizontal outflow or divergence of tidewater must be accompanied by a change in the sea surface elevation.

In the case of the pole tide, the tidal potential is the change in centrifugal potential accompanying the Chandler wobble:

$$V = -\sin\theta\cos\theta \, \Omega^2 a^2 [\underline{m}\exp(-i\lambda) + \underline{m}^*\exp(i\lambda)]/2$$

V depends linearly on the wobble amplitude $\underline{m}$ ($\underline{m} = m_x + im_y$). In the case of global oceans, which we will discuss first for simplicity, the wobble has a prograde circular time dependence ($\underline{m} = M_p\exp(i\sigma t)$); the tidal potential, when written as a complex variable, $V = \text{Re}(\underline{V})$, can then also be

TIDAL EQUATIONS

$$\dot{u}_\theta - fu_\lambda = -\frac{g}{a}\frac{\partial}{\partial\theta}\left[T - \frac{V}{g}\right] - Pu_\theta$$

$$\dot{u}_\lambda + fu_\theta = -\frac{g}{a\sin\theta}\frac{\partial}{\partial\lambda}\left[T - \frac{V}{g}\right] - Pu_\lambda$$

$$\dot{T} = -\frac{1}{a\sin\theta}\left\{\frac{\partial}{\partial\theta}[hu_\theta\sin\theta] + \frac{\partial}{\partial\lambda}[hu_\lambda]\right\}$$

Fig. 1. Laplace tide equations with bottom friction included. Tidal variables are the horizontal velocities u_θ, u_λ and tide height T, functions of colatitude θ, longitude λ, and time. Dot denotes differentiation with respect to time. V is the imposed tidal potential; $f = 2\Omega\cos\theta$ is the Coriolis parameter; Ω is the Earth's mean angular velocity; a and g are surface radius and gravity; P is the bottom drag coefficient; and h is ocean depth. The first two of the equations are equations of motion, while the third represents conservation of mass.

described with a prograde circular time dependence, $\underline{V} = \hat{V}\exp(i\sigma t)$. It then follows that all the tidal quantities, expressed as complex variables, can be described with the same time dependence, i.e.

$T = \text{Re}(\underline{T})$ $u_\theta = \text{Re}(\underline{u}_\theta)$ $u_\lambda = \text{Re}(\underline{u}_\lambda)$

$\underline{T} = \hat{T}\exp(i\sigma t)$ $\underline{u}_\theta = \hat{u}_\theta\exp(i\sigma t)$ $\underline{u}_\lambda = \hat{u}_\lambda\exp(i\sigma t)$

The approach I have taken is to expand the time-independent portion of the tide height, $\hat{T}$, in spherical harmonics; see Figure 2. One reason for taking this approach is that the pole tide products of inertia $\underline{c}$, which are primarily responsible for the tidal effects on wobble, can (in the global case) be expressed simply in terms of one single tide height coefficient, $\hat{T}_2^{-1}$.

The result of using a spherical harmonic approach with complex variables is that it is possible to reduce the tide equations to a single, non-differential matrix equation,

$$\underline{B}\cdot\vec{\hat{T}} = \underline{M}_p\vec{b} \quad .$$

By inversion, this equation can be solved for the collection $\vec{\hat{T}}$ of unknown tide height coefficients $\hat{T}_2^{-1}$, $\hat{T}_4^{-1}$, $\hat{T}_6^{-1}$, ... ; for the simple case shown here of global oceans of uniform depth, i.e. constant h, the only non-zero coefficients are those of even degree, and order (-1). I will shortly discuss the significance of the fact that the pole tide height depends linearly on wobble amplitude.

The effects of the pole tide on wobble are determined from conservation of angular momentum,

A SPHERICAL HARMONIC APPROACH

$$\hat{\underline{T}} = \sum_{\ell} \sum_{n} T_{\ell}^{n} Y_{\ell}^{n}(\mu,\lambda)$$

$$\mu = \cos\theta$$

$$\hat{\underline{c}} = -\rho_w a^4 \int \hat{\underline{T}} \cos\lambda \sin\theta \cos\theta \, ds = -\frac{1}{2}\sqrt{\frac{8\pi}{15}} \rho_w a^4 \underline{T}_2^{-1}$$

Fig. 2. The approach followed in this paper is to expand the time-independent portion of the tide height, $\hat{\underline{T}}$, in spherical harmonics. Y_{ℓ}^{n} is the fully normalized complex harmonic function of degree ℓ, order n; the summations are over integers such that $\ell \geq 0$, $|n| \leq \ell$. One reason for this approach is that it is possible to express the products of inertia $\underline{c}$ of the pole tide in terms of a single harmonic coefficient. ρ_w is the sea water density.

i.e. the Liouville equation; see Figure 3. An oceanless earth is capable of prograde circular wobble with frequency σ_e; the addition of oceans modifies the polar motion's period, decay rate, and ellipticity, through the pole tide products of inertia $\underline{c}$ and relative angular momentum $\underline{\ell}$. (The formulation shown here accounts for the fluid core's near-lack of participation in wobble; $\kappa \approx 1$).

Variable Chandler frequency?

The hypothesis that the Chandler wobble period varies with time has appeared sporadically ever since Chandler's discovery of the wobble. As usually stated, the idea is that the wobble period depends on the wobble amplitude, which has varied over the century. The mechanism for achieving a variable period is often taken to be a dynamic oceanic response to wobble—a bigger wobble amplitude would cause a bigger pole tide, thus a bigger contribution to the Chandler period, etc. As outlined in Figure 4, however, as long as the tide height depends linearly on wobble amplitude, the dynamic pole tide effect on wobble is constant. If T is linearly proportional to the wobble amplitude, then so are the tidal products of inertia $\underline{c}$; substitution into the Liouville equation then leads to a shift in wobble frequency involving the <u>constant</u> of proportionality.

VARIABLE CHANDLER PERIOD BY DYNAMIC OCEANIC RESPONSE ?

$$\underline{T} \sim \underline{m} \quad \Rightarrow \quad \underline{c} \sim \underline{m}$$

IF $\quad \underline{c} = C\underline{m}$

THEN

$$\dot{\underline{m}} - i\sigma_e \underline{m} = -iC'\underline{c} = -iC''\underline{m}$$

SO

$$\dot{\underline{m}} - i(\sigma_e - C'')\underline{m} = 0$$

Fig. 4. If the pole tide height and thus products of inertia are linearly proportional to the wobble amplitude, then the tide can reduce the wobble frequency only by a constant amount. Substitution of $\underline{c}$ into the Liouville equation (see Figure 3) yields an equation for prograde circular motion with frequency $\sigma_e - C''$, different from σ_e by the constant C''.

Non-global Oceans

When the oceans are non-global, their asymmetry with respect to the rotation axis forces the pole path to be elliptical,

$$\underline{m} = \underline{M}_P \exp(i\underline{\sigma}t) + \underline{M}_R \exp(-i\underline{\sigma}^* t),$$

so that the wobble contains retrograde circular as well as prograde circular components. It is, however, possible to write

WOBBLE EQUATION

OCEANLESS EARTH: $\quad \dot{\underline{m}} - i\sigma_e \underline{m} = 0$

EARTH WITH OCEANS: $\quad \dot{\underline{m}} - i\sigma_e \underline{m} = -i\frac{\Omega}{A_m + (1-\kappa)A_c}\{\underline{c} + \underline{\ell}/\Omega\}$

Fig. 3. Liouville equation. The free wobble frequency of an oceanless earth is σ_e. The products of inertia $\underline{c}$ and relative angular momentum $\underline{\ell}$ of the pole tide can modify the wobble frequency and path. With $\kappa \approx 1$ as a result of core fluidity, the mantle's inertia A_m but not the core's inertia A_c scales $\underline{c}$ and $\underline{\ell}$, in effect amplifying the oceanic effect on wobble.

TABLE 3. Preliminary Results for Non-Global Oceans[1]

Model[2]	P	σ Period	σ Damping Time	ε	$\underline{T}_O^O$
unconstrained					
(2, 2)	1.5×10^{-5} sec^{-1}	460 days,	170 years	0.000	(9, -2) mm
	1.5×10^{-6}	459 ,	-159	0.001	(10, -9)
	1.5×10^{-7}	481 ,	26	0.040	(5, -46)
	1.5×10^{-8}	469 ,	6	0.034	(-8, -37)
	1.5×10^{-9}	468 ,	6	0.033	(-9, -35)
	0	468 ,	6	0.033	(-9, -35)
(10, 5)	1.5×10^{-5} sec^{-1}	461 days,	299 years	0.001	(43, 6) mm
	1.5×10^{-8}	461 ,	-68	0.002	(-2, -6)
constrained					
(10, 5)	1.5×10^{-4} sec^{-1}	461 days,	150 years	0.000	(0, 0) mm
	1.5×10^{-5}	460 ,	86	0.001	(-1, -9)
	1.5×10^{-6}	460 ,	-17000	0.001	(2, 5)
	1.5×10^{-7}	459 ,	79	0.002	(-3, -8)
	1.5×10^{-8}	458 ,	78	0.002	(-8, -7)
	1.5×10^{-9}	458 ,	84	0.002	(-8, -7)

[1] Results adjusted by specifying σ_e = 2 /383.3 rad/day so that the tide effects would include a Chandler period of 433.2 days, infinite damping time, and a path ellipticity of 0.017 if the tide were static.

[2] Unconstrained solutions involve no boundary conditions; constrained solutions are based on a small number of boundary condition equations. Numbers in parentheses refer to maximum harmonic degrees of the ocean function employed (2 or 10) and tide height determined (2 or 5).

$$V = - \sin\theta\cos\theta \, \Omega^2 a^2 \, \mathrm{Re}[\underline{M} \exp(i\underline{\sigma}t)]$$

where $\underline{M} = \underline{M}_P \exp(-i\lambda) + \underline{M}_R^* \exp(i\lambda)$;

thus if complex variables are employed we can still describe the tidal potential and all other tidal variables using a prograde circular time dependence.

With non-global oceans the ocean depth will vary from place to place; such variability complicates the tide equations immensely. If the depth function is expanded in spherical harmonics,

$$h = \Sigma \, \underline{h}_j^s Y_j^s \,,$$

it is nevertheless possible to reduce the equations to a single, non-differential matrix equation:

$$\mathbf{B} \cdot \vec{T} = \underline{M}_P \vec{b}_P + \underline{M}_R^* \vec{b}_R \,.$$

This time the entire collection of unknown tide height coefficients, $\vec{T} = (\underline{T}_O^O, \underline{T}_1^{-1}, \underline{T}_1^O, \underline{T}_1^1, \underline{T}_2^{-2}, \ldots)$, is involved; the matrix $\mathbf{B}$ and vectors $\vec{b}_P$, $\vec{b}_R$ are more complicated as well.

In non-global oceans the tidal currents must satisfy boundary conditions, such as the "no-flow" condition

$$\vec{u} \cdot \hat{n} = 0 \,,$$

at every point of coastline. Using the ocean function O, defined as zero on land and unity over the oceans, it is possible to restate the boundary conditions as the global constraint

$$\vec{u} \cdot \vec{\nabla} O = 0 \,.$$

The constraint can then be reduced to a non-differential matrix equation,

$$\mathbf{B}' \cdot \vec{T} = \underline{M}_P \vec{b}_P' + \underline{M}_R^* \vec{b}_R' \,,$$

exhibiting a similar form to the tide equation.

Table 3 summarizes the preliminary results I obtained last year. These results are adjusted so that the effects would be a period of 433.2 days, no damping (infinite decay time), and an ellipticity of 0.017 if the tide were static. As you can see, for a wide range of bottom drag, the oceanic products of inertia lengthen the wobble by ~ twice the static amount, causing also appreciable damping (under some conditions, negative damping -- amplified wobble)!

Refinements

One problem with these results is that the tidal solutions evidently failed to conserve mass globally, even though they were based on continuity; that is, the degree-zero order-zero tide coefficient $\underline{T}^o_o$ is very definitely non-zero in these solutions, though one would expect it to vanish if tidal mass is conserved globally. There are a number of approaches one may take to enforce mass conservation, such as adding specific, artificial constraints which forcibly lead to reduced $\underline{T}^o_o$. The approach I chose was simply to increase the number of boundary condition equations used in the solution--a desirable goal in any case. We then have significantly more equations than unknowns (an overdetermined problem); using a generalized inverse, i.e. least-squares, allows the solutions to be obtained. We find, for example, that $\underline{T}^o_o = -(0.4, 0.07)$ mm for $P = 1.5 \times 10^{-5}$ sec^{-1} and $\underline{T}^o_o = -(0.4, 0.9)$ for $P = 1.5 \times 10^{-8}$. These are representative, and show that in the overconstrained solutions the $\underline{T}^o_o$ are diminished to a satisfactory extent, by an order of magnitude or more. The overconstrained approach was followed in the remainder of my work.

The major problem with the preliminary solutions is that the tide height coefficients, when combined spherical-harmonically, yield a pole tide which can be as large in mid-continent as in mid-ocean. This is illustrated schematically in Figure 5, suggesting that the solution has been obtained to the tide equations within the oceans, and on land as if it were water-covered; but the solution is "crimped" or constrained at boundaries, because of the boundary conditions, so that there is no communication between the two regions of solution.

Such a situation was at first disturbing, but I think I can convince you that on second thought it is perfectly reasonable. Refer again to the original Laplace tide equations (plus friction), Figure 1; consider now a tide potential which is steady, i.e. constant over time. One solution to these equations is clearly the static one, with zero velocities and the traditional equilibrium tide height,

$$u_\theta = 0 = u_\lambda \quad , \quad T = V/g \quad .$$

Note, however, that this tide height, too, fails to vanish on land; the tide potential V doesn't vanish on land because the tidal force exists globally.

In traditional static theory, zero tide height on land is ensured by multiplying the tide solution times the ocean function,

$$T = O V/g \quad ,$$

since O is zero on land and unity over the oceans. Computationally it is easy to truncate our dynamic solution as well; each harmonic component of the truncated tide height is simply a linear combination of ocean function coefficients and the untruncated tide height coefficients:

$$[\underline{O T}]^n_\ell = \Sigma \ \Sigma \ A^{nsq}_{\ell j p} \underline{O}^s_j \underline{T}^q_p \quad .$$

The effects of the truncated, overconstrained pole tide on wobble are summarized in Table 4.

TABLE 4. Pole Tide Effects for Non-Global Oceans[1]

P	σ		ε
	Period	Damping Time	
1.5×10^{-5} sec^{-1}	434.8 days	599 years	0.0150
1.5×10^{-6}	434.9	506	0.0146
1.5×10^{-7}	434.9	466	0.0145
1.5×10^{-8}	434.9	463	0.0145
0	434.9	463	0.0145

[1] Adjusted using $2\pi/\sigma_e = 383.3$ days so that the effects would be a Chandler period of 433.2 days, infinite damping time, and an ellipticity of 0.017 if the tide were static. These solutions are overconstrained and truncated, as described in the text.

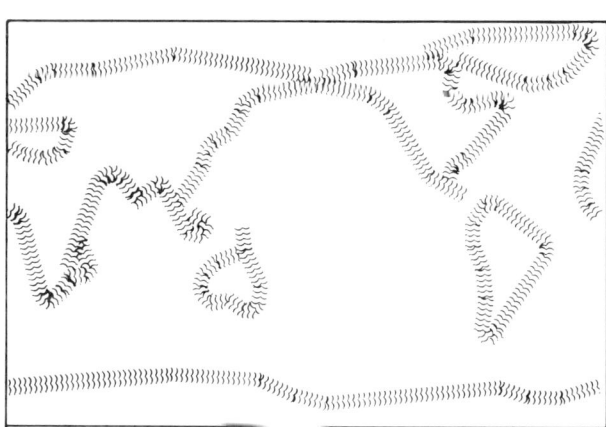

Fig. 5. Schematic map illustrating the continent - ocean distribution. As suggested in the text, solutions to the Laplace tide equations are inherently global; tide solutions will not vanish on land because the tidal force exists globally. However, boundary conditions constraining the flow at coastlines will inhibit communication between the land and ocean regions of solution.

As before, these are adjusted so that the results would be a wobble period of 433.2 days, no damping (infinite decay time), and a path ellipticity of 0.017 if the tide were static. It is now apparent that, for the entire range of bottom friction considered, the pole tide is actually very close to equilibrium. The dynamic tide contributes only ∼ 1.5 days more to the Chandler period than a static tide would, and dissipates wobble energy at a very mild rate.

The next step in refining the theory is to account for mantle elasticity, and for the self-gravitation and loading of the oceans. As indicated in Figure 6, the ability of the mantle to deform elastically in response to wobble causes the tide potential to be scaled by the expected Love number factor $(1+k_2-h_2)$. With our spherical harmonic approach, it is straightforward to account reasonably exactly for oceanic loading and self-gravitation; the result is that each harmonic degree of the tide height is scaled by a factor (β_ℓ) which depends on the load Love numbers of the same degree and on the oceanic density (ρ_w). It should be noted that numerical tide analysts usually account for loading and self-gravitation by scaling the tide height coefficients uniformly (same factor β); also, that factor is generally taken to be ∼ 0.9, whereas a more typical β_ℓ might be ∼ 0.8.

LOADING, SELF-GRAVITATING TIDE

$[\hat{T} - \frac{\hat{V}}{g}]$ becomes

(approx.) $[\beta\hat{T} - \alpha\frac{\hat{V}}{g}]$, $\alpha = 1 + k_2 - h_2$
$\beta \approx 0.9$

("exact") $[\Sigma \beta_\ell \hat{T}_\ell^n Y_\ell^n - \alpha\frac{\hat{V}}{g}]$, $\beta_\ell = 1 - 3\frac{\rho_w}{\bar{\rho}}\frac{1+k'_\ell-h'_\ell}{2\ell+1}$

Fig. 6. Modifications to the tide equations to account for mantle elasticity and oceanic loading and self-gravitation. The nearly ´exact´ formulation requires the [T-V/g] quantities in the equations of motion (see Figure 1) to be replaced as shown; k and h are Love numbers, k´ and h´ are load Love numbers, and $\bar{\rho}$ is mean Earth density. For comparison, the approximate formulation often employed in numerical tide studies is also shown.

When the oceans can load the solid earth, the wobble equation must be modified to include the products of inertia of the loaded mantle. It turns out that those products are proportional to the oceanic products of inertia c, with the load Love number k'_2 being the proportionality constant; the net effect is to replace c in the Liouville equation (see Figure 3) with $(1+k'_2)c$.

The effects of the loading and self-gravitating dynamic tide in non-global oceans, truncated and overconstrained as before, are summarized in Table 5--again adjusted so that the results would be a wobble period of 433.2 days, no damping, and an ellipticity of 0.0078 if the tide were exactly static. Now the tide is even a bit closer to equilibrium, with the dynamic contribution to the Chandler period a little more than 1 day and the dissipation very weak.

TABLE 5. Pole Tide Effects for Non-Global Loading and Self-gravitating Oceans[1]

P	σ		ε
	Period	Damping Time	
1.5×10^{-5} sec^{-1}	434.4 days	898 years	0.00885
1.5×10^{-6}	434.4	841	0.00860
1.5×10^{-7}	434.5	785	0.00855
1.5×10^{-8}	434.5	781	0.00855
0	434.5	781	0.00855

[1] Adjusted using $2\pi/\sigma_e = 402.6$ days so that the effects would be a Chandler period of 433.2 days, infinite damping time, and an ellipticity of 0.0078 if the tide were static. These solutions are overconstrained and truncated, as described in the text.

A Variety of Results

Table 6 illustrates for the sake of curiosity how various factors modify the dynamic tide's effects on wobble. In all these cases the oceans are non-global, and the solutions are overconstrained and truncated. Making the mantle elastic (scaling the tide potential by $1+k_2-h_2$) increases the dynamic oceanic response, in that the tide contributes more to the Chandler period, although it dissipates less energy. Including loading and self-gravitation reduces the dynamic effects on both period and dissipation.

Table 7 illustrates (for non-loading oceans) the role of boundary conditions in determining the tide solution. For both bottom friction strengths considered, we see that the unconstrained solutions (36 tide equations, 0 boundary condition equations) imply a bigger dynamic effect on the Chandler period than the overconstrained solutions (36 boundary condition equations). However, the unconstrained solutions also predict significant enhancement (negative damping) of wobble. [In part, the rather dramatic effects of the unconstrained tide probably result from its large T_0^0, which feeds into all truncated tide height coefficients.]

TABLE 6. Pole Tide Effects
on Various Earth Models[1]

Model[2]	σ		ε
	Period	Damping Time	
1a	434.8 days,	599 years	0.0150
1b	435.7 ,	866	0.0104
1c	434.4 ,	898	0.0089
2a	434.9 days,	463 years	0.0145
2b	435.7 ,	669	0.0101
2c	434.5 ,	781	0.0086

[1] Adjusted so that effects would be a Chandler period of 433.2 days and infinite damping time if the tide were static. Overconstrained and truncated solutions.
[2] Model case 1 has $P=1.5 \times 10^{-5}$; case 2 has $P=1.5 \times 10^{-8}$. Model case a has rigid mantle, fluid core; case b has elastic mantle, fluid core; in all cases the oceans are non-global, in case c the oceans are loading and self-gravitating.

TABLE 7. Pole Tide Effects:
Role of Boundary Constraints

Model[1]	σ		ε
	Period	Damping Time	
	----- $P = 1.5 \times 10^{-5}$ -----		
uncon. (36, 0)	438.8 days,	-11 years	0.0275
con. (25, 11)	436.1 ,	50	0.0103
overcon. (36, 36)	434.8 ,	599	0.0150
	----- $P = 1.5 \times 10^{-8}$ -----		
uncon. (36, 0)	436.1 days,	-188 years	0.0141
con. (25, 11)	433.8 ,	30	0.0106
overcon. (36, 36)	434.9 ,	463	0.0145

[1] Unconstrained, constrained, and overconstrained solutions, all truncated, based on the number of tide equations and boundary constraints given in parentheses. For non-loading, non-gravitating oceans, with σ_e chosen so that the period would be 433.2 days and the damping time would be infinite if the tide were exactly static.

Till now I have been presenting results based on an ocean-continent distribution described by a fairly limited spherical harmonic expansion, through degree and order 10. For a bottom friction strength of $P = 1.5 \times 10^{-7}$, characteristic of the deep ocean, the dynamic tide effects on wobble have also been computed using an ocean function expansion through degree and order MAXHT=24 (those calculations take about 5 times longer, on the IBM 3090 vector facility). In both situations the effects are found to be very similar:

MAXHT	σ		ε
	Period	Damping Time	
10	434.45 days	785 yr	0.00855
24	434.47	821	0.00851

This suggests that we would be inferring essentially the same tidal effects --especially on wobble period--if we employed an even higher degree and order expansion.

Implications

The effects for MAXHT=24 are also my "final" results. It is evident that the dynamic pole tide in the deep ocean contributes only negligibly to Chandler wobble damping, and lengthens the Chandler period by about 1 day more than the static lengthening. As shown in Table 8, pole tide effects on wobble are traditionally expressed in terms of the lengthening needed to bring the wobble period up to the observed value. If the observed Chandler period is taken to be 433.2 days, the dynamic pole tide contributes 31.7 days. By subtraction, the oceanless earth wobble period is 401.5 days, almost 5 days greater than that predicted by Smith & Dahlen [1981]; since pole tide dynamics have now been accounted for, such a discrepancy would most likely be explained by mantle anelasticity. If

TABLE 8. Traditional Formulation
of Pole Tide Effects

Chandler Period	Oceanic Effect	Inferred σ_e	Discrepancy[1]	
			S & D	Y & I
433.2 days	31.7	401.5	4.6	-0.7
435.2	32.0	403.2	6.3	1.0

[1] Calculated oceanless elastic earth wobble frequency from S & D = Smith & Dahlen [1981], or from Y & I = Yoder & Ivins [1987].

the observed Chandler period is taken to be 435.2 days, all estimates are increased as shown and the discrepancy exceeds 6 days. On the other hand, recent calculations by Yoder & Ivins [1987] suggest an oceanless earth wobble period of 402.2 days, implying a discrepancy of only $\pm$ 1 day; in this case mantle anelasticity at the Chandler frequency would be almost non-existent.

Our dynamic effect of 1 day is certainly not a large percentage of the total oceanic contribution to the wobble period; nevertheless, from the oceans' point of view it may indicate an unusually vigorous response at low frequency. Such indications follow from a comparison of our final result with the work of Carton & Wahr [1986]. Carton and Wahr used a numerical approach to determine the characteristics of tides in realistic oceans forced tesserally at 28- and 50-day periods; by linear extrapolation to the pole tide's $\sim$ 435-day period they concluded that the tide's dynamical contribution to the Chandler period was only fractions, -0.043, of one day. However, such a linear extrapolation assumes that ocean dynamics depend smoothly and simply on frequency, and it neglects the possibility of any anomalous responses or resonances. I can argue that the differences between their ocean model and mine are not substantial (in terms of their effects on the tide solution), so that the difference between their results and mine is principally a consequence of their extrapolation procedure. The fact that I obtained an effect on wobble period roughly 30 times theirs (and with an opposite sign) suggests that the oceans are indeed responding resonantly at sub-annual frequencies.

Turbulence

The final oceanic model I will discuss is one incorporating turbulence. Turbulent drag forces are customarily treated as diffusive, of the form $A \nabla^2 \vec{u}$ with an eddy viscosity coefficient A typically much larger than that for molecular viscosity. Representation of the force, whose meridional and azimuthal components are

$$A_H [\nabla_H^2 u_\theta - \frac{u_\theta}{a^2 \sin^2\theta} - \frac{2\cos\theta}{a^2 \sin^2\theta} \frac{\partial u_\lambda}{\partial \lambda}]$$

$$A_H [\nabla_H^2 u_\lambda - \frac{u_\lambda}{a^2 \sin^2\theta} + \frac{2\cos\theta}{a^2 \sin^2\theta} \frac{\partial u_\theta}{\partial \lambda}]$$

is complicated enough that the previous spherical harmonic approach cannot be followed. I found it necessary in this case to expand the time-independent tide velocities (as well as the tide height) in spherical harmonics. Matrix equations involving the collections of velocity coefficients could be expressed in terms of the tide height coefficients, viz.

$$\boldsymbol{w} \cdot \vec{U} + \boldsymbol{\flat} \cdot \vec{V} = \vec{\Delta}(T, \underline{M}_P, \underline{M}_R^*)$$

$$\boldsymbol{w} \cdot \vec{V} - \boldsymbol{\flat} \cdot \vec{U} = \vec{\Gamma}(T, \underline{M}_P, \underline{M}_R^*)$$

where $\hat{u}_\theta = \Sigma u_\ell^n Y_\ell^n$, $\hat{u}_\lambda = \Sigma v_\ell^n Y_\ell^n$ and $\vec{U} = \{u_\ell^n\}$, $\vec{V} = \{v_\ell^n\}$. A single non-differential matrix equation--involving the collection of tide height coefficients, and of the same form as before but with more complicated elements--was ultimately obtained. Once again, the fact that the tide height depends linearly on wobble amplitude implies that a dynamic pole tide cannot produce a variable Chandler period.

My final table, Table 9, presents the solutions I have obtained for turbulent, loading and self-gravitating non-global oceans. As before, these are adjusted so that the results would be a Chandler period of 433.2 days, infinite damping time, and an ellipticity of 0.0078 if the solution represented a static tide. These solutions are for oceans with weak bottom friction, and a range of eddy viscosities spanning the extremes, 10^5 to 10^9 cm^2/sec, thought to actually characterize the oceans [Pedlosky, 1979]. These results are preliminary because I have some concerns about their accuracy. Nevertheless, it appears that turbulence has a negligible effect on the dynamic

TABLE 9. Pole Tide Effects (Preliminary) for Turbulent Non-Global Loading and Self-gravitating Oceans[1]

A	σ		ε
	Period	Damping Time	
0 cm^2/sec	434.5 days,	(> 10^5 years)	0.0088
*1.5x10^{10}	434.5 ,	(> 10^4)	0.0088
1.5x10^{14}	434.5 ,	40	0.0075
1.5x10^{15}	421.9 ,	7	0.0017
*1.5x10^{15}	421.5 ,	7	0.0018

[1]Adjusted using $2\pi/\sigma_e$ = 402.6 days so that the effects would be a Chandler period of 433.2 days, infinite damping time, and an ellipticity of 0.0078 if the tide were static. These solutions are overconstrained and truncated, as described in the text, and based on P = 1.5x10^8 sec^{-1}. Solutions marked with * are self-consistent, otherwise are direct solutions -- forced by a nominal Chandler frequency (2π/433.2 rad/day + i/100 yr) and ellipticity (0.0078) rather than a joint solution with the Liouville equation.

pole tide--not unreasonable, given the weak currents of that nearly static tide, and the low frequencies involved.

Out of curiosity I also computed the solutions in the case of ridiculously high viscosities (A $\sim$ 10^{14} to 10^{15}). It is interesting to note that, under such unrealistic conditions, the super-strong turbulence reduces the oceanic contribution to the Chandler period drastically, and succeeds in damping the wobble completely....

Thank you.

Acknowledgments. This research was supported by NASA grant NAG 5-145. Conversations with J. Wahr and M. Rochester were helpful. I would like to thank D. McCarthy for the invitation to speak at the IUGG.

References

Carton, J.A., and J.M. Wahr, Modelling the pole tide and its effect on the Earth's rotation, Geophys. J. Roy. Astr. Soc., 84, 121-137, 1986.

Dickman, S.R., and D.J. Steinberg, New aspects of the equilibrium pole tide, Geophys. J. Roy. Astr. Soc., 86, 515-529, 1986.

Lennon, G.W., Monthly and Annual Mean Heights of Sea Level, 3 vol., Permanent Service for Mean Sea Level, Institute of Oceanographic Sciences, Birkenhead, England, 1978.

Pedlosky, J., Geophysical Fluid Dynamics, Springer-Verlag, 624 pp., 1979.

Smith, M.L., and F.A. Dahlen, The period and Q of the Chandler wobble, Geophys. J. Roy. Astr. Soc., 64, 223-281, 1981.

Yoder, C.F., and E.R. Ivins, Improved analytic nutation model, Proc. IAU Symp. 129, 1987.

TIDAL PARAMETERS AND NUTATION: INFLUENCE FROM THE EARTH INTERIOR.

Véronique Dehant[1]

Institut d'Astronomie et de Géophysique G. Lemaître, Université Catholique de Louvain, 2, Chemin du Cyclotron, B 1348 Louvain-la-Neuve, Belgique

<u>Abstract.</u> A 1.5 percent difference is found between the observed tidal gravimetric factor (δ) and the computed factor obtained from Wahr's theory. This difference is reduced to 0.7 percent when considering the same definition for δ in the theory as in the observations analyzed by the International Center for Earth Tides. The introduction of mantle inelasticity in the model further reduces the discrepancy by 0.1 percent.

The effect of the mantle inelasticity is also considered on the Love numbers h and k. For O_1 and M_2 waves, this effect amounts to 1.4 percent for h and 1.6 percent for k respectively, and for the fortnightly wave M_f to 2.6 percent and 3 percent.

On the contrary with respect to the tidal parameters, the effect of the mantle inelasticity on the nutations increases the difference between the observations and the computed IAU nutation series. This difference can be reduced either by slightly increasing the core flattening in the integration process, or by considering an additional pressure at the core-mantle boundary.

Introduction

The possibilities offered by observational geophysics, allow us presently to compare thoroughly theory and observations, to point out discrepancies between them, to check the validity of both and to find out a closer and closer agreement. For example, Herring and al. [1986] deduced the forced nutations from VLBI observations and pointed out a difference with the adopted IAU nutations series based on an elliptical uniformly rotating Earth model [Wahr, 1979]. In 1983, Melchior and De Becker [1983] also pointed out a difference of about 1.5 percent between the computed tidal gravimetric factor and the observed one. In 1987, Dehant and Ducarme [1987] explained a part of this difference by reconsidering the definition of the tidal gravimetric factor (δ). Based on the International Center for Earth Tides (ICET) definition, they define δ as the Earth's transfer function (i.e., a coefficient in the frequency domain) between what is measured by the gravimeter and the external tidal force along the local vertical. New theoretical values were then recomputed using this definition. The remaining gap is about 0.7 percent. Let us mention that new Love numbers have also been redefined in the case of an elliptical, uniformly rotating Earth by Dehant [1987 b]. The second section will summarize the theory for the computation of tides and nutations for an elliptical, uniformly rotating Earth. The third section will give results that account for a new rheological profile in the mantle including dispersion effects. The effect of inelasticity on the new Love numbers and the tidal gravimetric factor will be discussed. Concerning the nutations, we shall point out an increase of the discrepancy between the theory and the observations. Gwinn et al. [1986] propose to solve this by increasing the Earth's equatorial radius at the Core Mantle boundary (CMb) by about 500 meters. Morelli and Dziewonski [1987] show in a recent paper that, within the error bars of the tomographic results, this is in good agreement with the topography obtained from the inversion of the combined PcP and PKP_{B_C} data set. For

---

[1]Senior Research Assistant - National Fund for Scientific Research (Belgium)

Copyright 1990 by
International Union of Geodesy and Geophysics
and American Geophysical Union.

this reason, we decided to compute the effects on the nutations of a change in the ellipticity at the CMb with respect to the hydrostatic equilibrium value (section 4). In section 5, the effects, on the nutations, of a pressure change at this boundary are also investigated.

Smith and Wahr's Theory

The last adopted nutation theory is the theory of Smith [1974] and of Wahr [1979, 1981a, 1981b and 1982]. In this theory, the reference frame is uniformly rotating, so that the nutations appear as toroidal motions. Indeed one can show that the toroidal displacement τ_1^1, component of the total displacement field describing the response of the Earth to a diurnal external potential of the order $(l = 2, m = 1)$, is exactly a rigid rotation around an equatorial axis (see Wahr [1979 or 1982], or Dehant [1986]), i.e., τ_1^1 is assimilated to the nutations.

In order to evaluate the tides, the effects of the nutations and the variations of the length of day must be removed from the total displacement field. The tidal parameters, as the gravimetric factor (δ) and the Love numbers k and h, can be computed from the tidal surface displacement field (for more details, see Dehant and Ducarme [1987] and Dehant [1987b]). Because this computation is done in the diurnal band, it must be mentioned that Wahr accounts for resonance effects due to the presence of normal modes such as, for example, the Nearly Diurnal Free Wobble (NDFW) [due to an angle between the rotation axis of the core and the rotation axis of the mantle, the Chandler Wobble (CW) [due to an angle between the figure axis and the rotation axis of the mantle and the Tilt-Over Mode (TOM) [due the possibility of having a tilt of the Earth.

Effect of Mantle Inelasticity

The developments on the computations of the tidal parameters are given in full details in my previous papers [Dehant 1987a and 1987b]. For the Love numbers h and k, the results are summarized in Figure 1. The effects of mantle inelasticity on the Love number h is about 1.4 percent in the semi-diurnal and diurnal band. On the Love number k, it is 1.6 percent in the semi-diurnal band and 1.7 percent in the diurnal band. The effects on the tidal gravimetric factor is about 0.1 percent as shown in Figures 2 and 3, for O_1 and M_2 respectively. The full curve shows the result using model PREM [Dziewonski and Anderson, 1981] and the corresponding dashed curve shows the effect of mantle inelasticity.

We compute a new rheological profile based on the model of Zschau described in Zschau and Wang [1985], Dehant [1986] and Dehant [1987a]. Because the equations are expressed in the frequency domain, the principle of correspondence of Biot [1954] can be used. This allows the use of the same stress-strain relationship as in

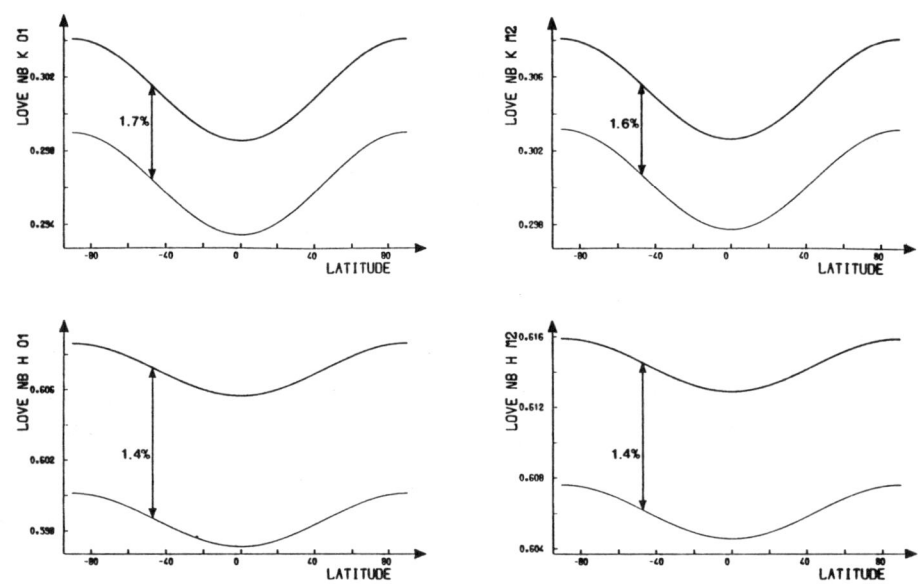

Fig. 1. Love number h and k for O_1 and M_2 [Dehant, 1987b]. Fine curve = elastic value using PREM; thick curve = corresponding inelastic effect using Zschau's μ profile for the Earth's mantle.

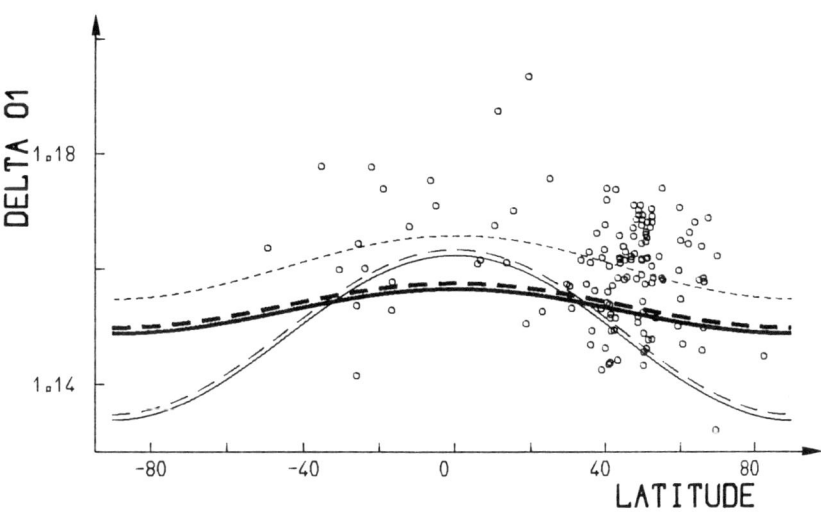

Fig. 2. Gravimetric factor for O_1 [Dehant, 1987b]. Points = individual observations (ICET data bank); thick curve = δ from the present paper definition using the elastic PREM model; fine curve = δ from Wahr's definition using the elastic PREM model; the corresponding dashed curves = results when adding mantle inelasticity; dashed curve = δ deduced by regression from the observations. (----)

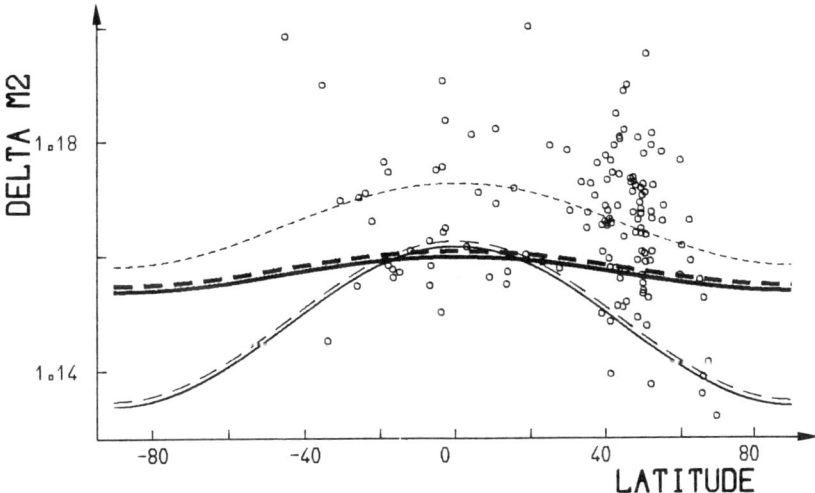

Fig. 3. Gravimetric factor for M_2 [Dehant, 1987b]. The points and the curves have the same meaning as in Figure 2.

the elastic case but with complex and frequency dependent coefficients, i.e., the Lamé parameters or the shear (μ) and the bulk (k) moduli are frequency dependent and complex. Zschau computes a complex shear modulus profile based on the hypothesis of diffusion controlled processes and on a gaussian distribution for the stress relaxation times. He could then account for a depth and a frequency dependence of the shear modulus (for more details, see [Zschau and Wang, 1985]). He kindly provided us with these profiles for different frequencies.

We first recompute the nutations using PREM model [Dziewonski and Anderson, 1981] and then use Zschau's profiles. The effects of mantle inelasticity are displayed in Table 1 for the four principal nutations, for the year 2000. One can see that the difference between the theory and the observations is even worse than previously, as already

TABLE 1. Nutation amplitudes for different Earth models.

Nutation periods	Elastic model PREM		Inelastic model using Zschau's μ	
	in obliquity	in longitude	in obliquity	in longitude
18.6 years				
without KC[1]	9.2021	-17.1966	9.2026	-17.1992
with KC	9.2024	-17.1975	9.2028	-17.1991
1 year				
without KC	0.0056	0.1433	0.0053	0.1425
with KC	0.0055	0.1431	0.0053	0.1423
$\frac{1}{2}$ year				
without KC	0.5737	-1.3186	0.5731	-1.3173
with KC	0.5735	-1.3182	0.5729	-1.3169
13.66 days				
without KC	0.0975	-0.2265	0.0971	-0.2255
with KC	0.0975	-0.2265	0.0971	-0.2254

[1] In this table, "without KC" means without using the precessional constant of Kinoshita (adopted value) but using $\frac{C-A}{C}$ computed from the seismic model; and "with KC" means using Kinoshita adopted constant.

shown by Wahr and Bergen [1987] who use a perturbation theory. Following the hypothesis of Gwinn et al. [1986], this might be due to resonance effects on the value of the NDFW frequency. Indeed, in space, there is a difference of 27 days between the computed period (460 days) and the observed one (433 days from VLBI results [Herring et al., 1986] and from data from two superconducting gravimeters [Richter and Zürn, 1987 and Neuberg et al., 1987]. When adding mantle inelasticity in the model, this difference increases (33 days) because the new theoretical value for the NDFW period is 466 days in space. This can be understood physically because, as already mentioned, the NDFW is due to an angle between the rotation axis of the core and the mantle. There exists a torque which tends to bring the two axes closer to each other. This induces a motion of one axis around the other. If the Earth is deformable, the torque will decrease and thus the period increases. Due to inelasticity, the Earth deforms more, the torque decreases more, and thus the period increases. The problem then is to reduce the theoretical value to the observed one, i.e., one must find something to increase the torque.

Variation of the Core Ellipticity

We agree with Gwinn et al. [1986] for a change in the core ellipticity which is equivalent to an increase of the equatorial radius at the CMb with respect to the hydrostatic equilibrium figure of the Earth. This is reinforced by the investigation of Yoder and Irvins [1987] who account for the existence of a particular density layer at the CMb (i.e., a boundary layer, Melchior, 1986], which could either nutate with the core or which is glued to the mantle. They use constraints from gravity to show that the deviation from hydrostatic equilibrium deduced by Gwinn et al. [1986] "seems secure." This is the reason why we do change arbitrarily the ellipticity at the CMb, adding some percentage up to about 15 percent and using each new ellipticity profile to compute new periods of the normal modes and the new nutations. The results of this experiment for the NDFW are given in Figure 4 and Table 2. On this figure, we also point out that there is good agreement between the observed results deduced from VLBI [Gwinn et al., 1986], LLR [Eubanks et al., 1985] and superconducting gravimeters data [Neuberg et

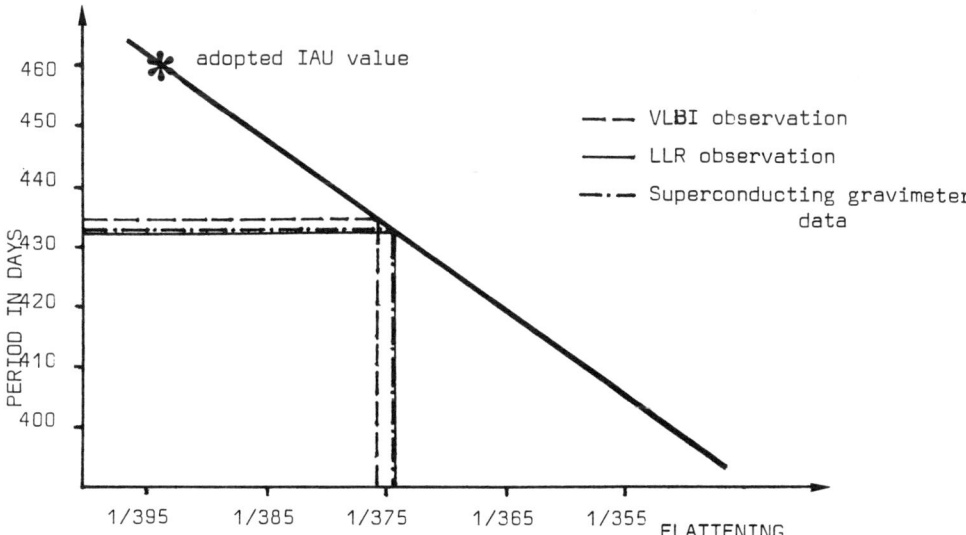

Fig. 4. Inertial space period of the NDFW in function of the core flattening.

TABLE 2. NDFW and FCN periods for different Earth models.

percentage of flattening change	flattening at CMb[1]		NDFW period (in days)		CW period (in days)	
	1066A	PREM	1066A	PREM	1066A	PREM
0%	$\frac{1}{393.0}$	$\frac{1}{392.5}$	459.9	457.8	403.3	402.6
4%	$\frac{1}{377.9}$	$\frac{1}{377.4}$	438.1	435.9	403.9	403.0
7%	$\frac{1}{367.3}$	$\frac{1}{366.8}$	423.1	420.8	404.2	403.0
10%	$\frac{1}{357.3}$	$\frac{1}{356.8}$	409.0	406.8	404.5	402.3
13%	$\frac{1}{347.0}$	$\frac{1}{347.3}$	395.6	393.6	404.6	402.7

[1] As an example, $\frac{1}{393}$ corresponds to a radius of 8868m at CMb,
$\frac{1}{377.9}$ corresponds to a radius of 9222 m,
i.e., 354m between both situations.

al., 1987 and Richter and Zürn, 1987] and the change in the ellipticity suggested by Gwinn et al. [1986]. The new periods of the CW are also reported in Table 2.

It must be mentioned that this variation of the ellipticity cannot be the consequence of a second order effect in the ellipticity, because the second order internal theory of Nakiboglu [1979] for a self-gravitating and slowly rotating planet in equilibrium, gives variations of the order of 0.1 percent. This is far away from what must be explained. Arguments based on hydrostatic equilibrium in a broad or strict sense, like the work of Tonn [1985], are also to be rejected because the effects are smaller than 0.3 percent. The changes in the nutations in function of the Earth's flattening are displayed in Figure 5 for the nutations in obliquity and in longitude. An agreement can be found between our results and the observations for the annual and the semi-annual nutations, while the differences between the observations and the results for the 13.66 days nutation increase. In this last case, the induced variations are very small. The observed results for the 18.6 years nutation are presently not sufficiently accurate to draw a definite conclusion.

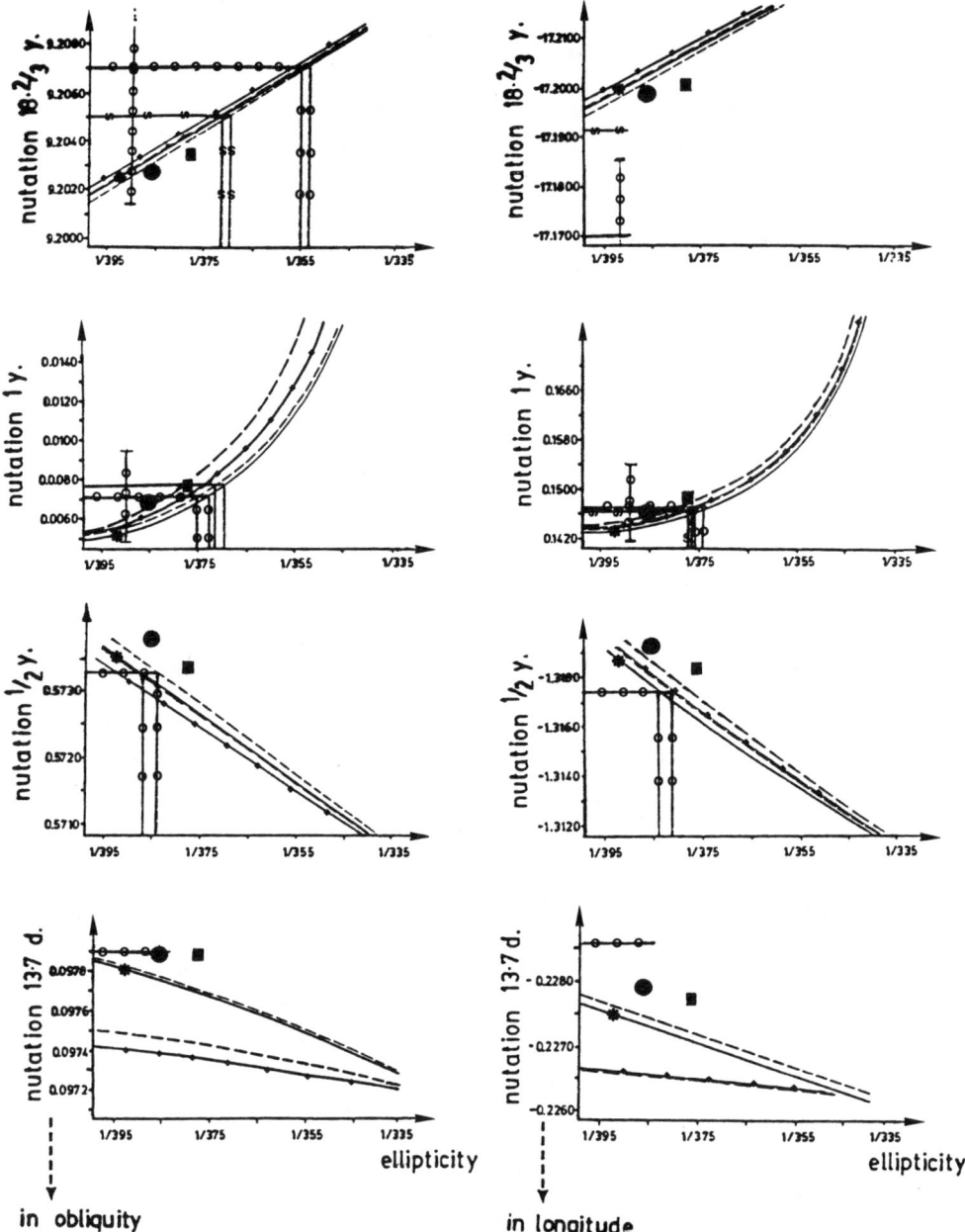

Fig. 5. Variations of the nutation amplitudes in obliquity and in longitude in function of the variation of the core flattening, for the four principal nutations. The thin curve (———) = results using 1066A model and the adopted precession constant; the thick curve (———) = results using PREM; the corresponding dashed lines are computed using the precession constant deduced from the model;

✱ = adopted IAU nutation; ⊶⊶ = VLBI observed nutation; ⌶ = corresponding confidence interval; ⊷⊷ = LLR observed nutation;

● = nutation corresponding to 5 percent pressure variation and 2 percent ellipticity variation;

■ = nutation corresponding to 5 percent pressure variation and 4 percent ellipticity variation.

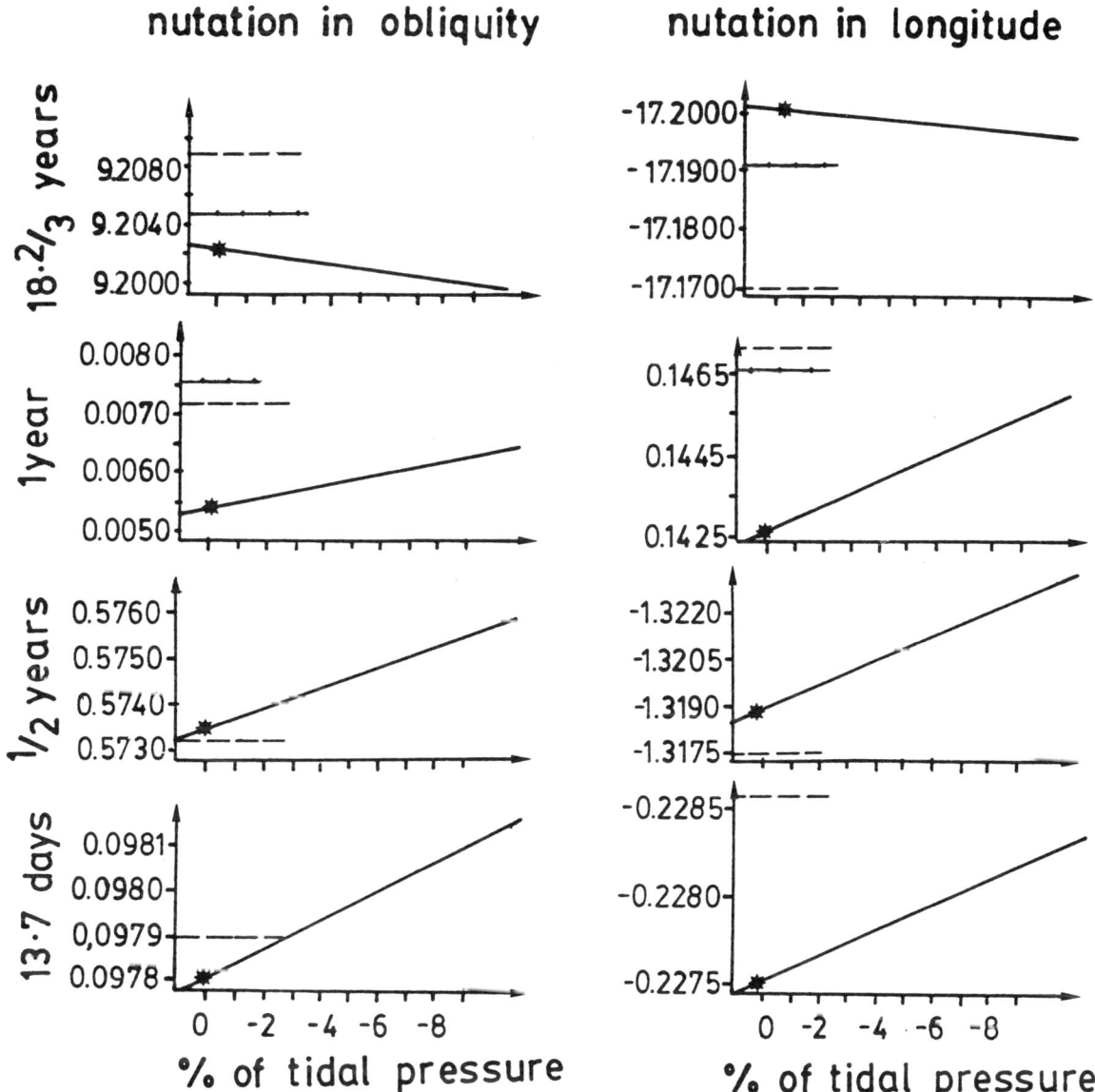

Fig. 6. Variations of the nutation amplitudes in obliquity and in longitude in function of the variation of the tidal induced pressure at the CMb, for the four principal nutations.

✱ = adopted IAU nutation; - - - = VLBI observed nutation; ••• = LLR observed nutation.

Variation of the Pressure at the CMb

As we are working in the tidal frequency band, the Earth is not hydrostatically prestressed. We believe that before the tidal deformations, some long term motions associated with convection or magnetic field drift, and so some pressure at the CMb, may exist. When the tidal deformations are computed, the pre-equilibrium state of the Earth is subtracted from the new instantaneous equilibrium associated with the instantaneous luni-solar attraction. In Smith and Wahr's theory, the Earth was considered in hydrostatic equilibrium and no additional pressure is subtracted. The induced 'tidal' pressure as computed by the program of Smith and Wahr must then be reduced. We have reduced the induced tidal pressure up to 10 percent inside the core and in particular at the CMb. We could then compute new nutations. Figure 6 displays the amplitudes of the nutations in obliquity and in longitude, in function of the variation of the induced tidal pressure, for the four principal nutations ,i.e., the 18.6 years nutation, the annual nutation, the semi-annual nutation and the 13.66 days nutation. In this last case, the amplitude varies in the right sense with respect to the observations. The same conclusion remains for the annual nutations, although it is not in the right sense for the semi-annual nutation!

Discussion and Conclusion

It is obvious that if one could combine both effects, the variation of the core flattening and the variation in the pressure at the CMb, we would reach a good compromise that should lead to a better fitting theory. We took as an example two particular cases corresponding to a variation of the ellipticity of respectively 2 percent and 4 percent and to a variation of the pressure of about 5 percent. This choice of pressure corresponds to the amplitude of the pressure induced by radial motion at the CMb as deduced from the non dipole magnetic drift [Voorhies, personal communication, 1987]. The results of these two examples are very close to the observations for all the nutations except for the long period 18.6 years nutation, but the corresponding observations are still imprecise as already mentioned.

To summarize our results, sensitivity tests on the amplitudes of the nutations have been performed introducing a change of the core flattening and of the pressure at the CMb. The high sensitivity on the amplitudes is obvious, although it is not always in the expected sense with respect to the observations. Nevertheless, the combination of both variations gives theoretical results consistent with the observations. A compromise must then be found. Our future work will concentrate on a new "pre-equilibrium theory."

Acknowledgments. We are thankful to G. van Marcke de Lummen for drawing the figures. We also wish to thank Prof. J.M. Wahr for providing us with his original programs.

References

Biot, M. A., Theory of stress-strain relations in anisotropic viscoelasticity and relaxation phenomena., J. Appl. Phys., 25, 11, 1385-1391, 1954.

Dehant, V., Intégration des équations différentielles aux déformations d'une Terre ellipsoïdale, inélastique, en rotation uniforme, avec un noyau liquide, Ph. D. thesis, Université Catholique de Louvain, Belgium, 298 pp., 1986.

Dehant, V., Integration of the gravitational motion equations for an elliptical uniformly rotating Earth with an inelastic mantle., Phys. Earth Planet. Int., 49, 242-258, 1987a.

Dehant, V., Tidal parameters for an inelastic Earth., Phys. Earth Planet. Int., 49, 97-116, 1987b.

Dehant, V., Nutations and Inelasticity of the Earth., Proceedings of the 128th Symposium of IAU/IAG on Earth Rotation and Reference Frames, Washington, USA, Proceedings published in 1988, ed. A. Babcock and G. A. Wilkins, 323-329, 1987c.

Dehant, V. and B. Ducarme, Comparison between the theoretical and observed tidal gravimetric factors., Phys. Earth Planet. Int., 49, 192-212, 1987.

Dziewonski, A. M. and D. L. Anderson, Preliminary reference Earth model., Phys. Earth Planet. Int., 25, 297-356, 1981.

Eubanks, T. M., J. A. Steppe and O. J. Sovers, An analysis and intercomparison of VLBI nutation estimates., JPL Geodesy and Geophysics preprint n° 135, 1985.

Gwinn, C. R., T. A. Herring and I. I. Shapiro, Geodesy by radio interferometry : studies of the forced nutations of the Earth, 2. interpretation., J. Geophys. Res., 91, B5, 4755-4765, 1986.

Herring, T. A., C. R. Gwinn and I. I. Shapiro, Geodesy by radio interferometry : studies of the forced nutations of the Earth, 1. data analysis., J. Geophys. Res., 91, B5, 4745-4754, 1986.

Melchior, P., The Physics of the Earth Core. An Introduction., Pergamon Press, Oxford, 256 pp., 1986.

Melchior, P. and M. De Becker, A discussion of world-wide measurements of tidal gravity with respect to oceanic interactions, lithosphere heterogeneities, Earth's flattening and inertial forces., Phys. Earth Planet. Int., 31, 27-53, 1983.

Morelli, A. and A. M. Dziewonski, Topography of the Core-Mantle boundary and lateral homogeneity of the liquid core., Nature, 325, 678-683, 1987.

Nakiboglu, S.M., Hydrostatic figure and related properties of the Earth., Geophys. J. R. astr. Soc., 57, 639-648, 1979.

Neuberg, J., J. Hinderer and W. Zürn, Stacking gravity tide observations in Central Europe for the retrieval of the complex eigenfrequency of the Nearly Diurnal Free Wobble., Geophys. J. R. astr. Soc., 91, 853-868, 1987.

Richter, B. and W. Zürn, Chandler effect and Nearly Diurnal Free Wobble as determined from observations with a superconducting gravimeter., Proceedings of the 128th Symposium of IAU/IAG on Earth Rotation Earth Rotationand Reference Frames, Washington, USA, Proceedings published in 1988, ed. A. Babcock and G. A. Wilkins, 309-315, 1987.

Seidelmann, P. K., 1980 IAU theory of nutation : the final report of the IAU working group on Nutation., Celestial Mechanics, 27, 79-106, 1982.

Smith, M. L., Scalar equations of infinitesimal elastic-gravitational motion for a rotating, slightly elliptical Earth., Geophys. J. R. astr. Soc., 37, 491-526, 1974.

Tonn, R., On the figures of the Earth., Proceedings of the 10th International Symposium on Earth Tides, Madrid, Spain, Proceedings published in 1987, 415-422, 1985.

Wahr, J. M., The tidal motions of a rotating, elliptical, elastic and oceanless Earth., Ph. D. thesis, University of Colorado, 216 pp., 1979.

Wahr, J. M., A normal mode expansion for the forced response of a rotating Earth., Geophys. J. R. astr. Soc., 64, 651-675, 1981a.

Wahr, J. M., Body tides on an elliptical, rotating, elastic and oceanless Earth., Geophys. J. R. astr. Soc., 64, 677-703, 1981b.

Wahr, J. M., The forced nutations of an elliptical, rotating, elastic and oceanless Earth., Geophys. J. R. astr. Soc., 64, 705-727, 1981c.

Wahr, J. M. and Z. Bergen, The effects of mantle anelasticity on nutations, Earth tides and tidal variations on rotation rate, Geophys. J. Roy. astr. Soc., 87, 633-668, 1987.

Wahr, J. M., Computing tides, nutations and tidally-induced variations in the Earth's rotation rate for a rotating elliptical Earth., Proceedings of the 3th International Summer School in the Mountains, Geodesy and Global Geodynamics, Admont, Austria, ed. Moritz H. and Sünkel H., 327-379, 1982.

Yoder, C. F. and E. R. Irvins, On the ellipticity of the Core-Mantle boundary from Earth nutations and gravity., Proceedings of the 128th IAU/IAG International Symposium on Earth Rotation and Reference Frames, Washington, USA, Proceedings published in 1988, ed. A. Babcock and G. A. Wilkins, 317-322, 1987.

Zschau, J. and R. Wang, Imperfect elasticity in the Earth's mantle, implications for Earth tides and long period deformations., Proceedings of the 10th International Symposium on Earth Tides, Madrid, Spain, Proceedings published in 1987, 379-384, 1985.

THE EARTH'S FORCED NUTATIONS: GEOPHYSICAL IMPLICATIONS

J.M. Wahr and D. de Vries

Department of Physics and
Cooperative Institute for Research in Environmental Sciences,
University of Colorado, Boulder, Colorado

Introduction

The Earth's nutational motion consists of periodic tipping of the Earth in space and is caused by the gravitational attraction of the Sun and Moon. The motion occurs at discrete frequencies of, as seen from the Earth, one cycle per day modulated by the orbital frequencies of the Sun and Moon. The periods are close to diurnal because the Sun and Moon rise and set once a day.

For many applications nutations are a nuisance since if they are not adequately removed they can degrade solutions for other parameters. However, recent observational results from Very-Long-Baseline-Interferometry (VLBI) have demonstrated that nutations can be important in their own right in providing a probe of the Earth's interior. In this paper, we review nutation theories and, in an appendix, we extend those theories to include an Earth with a non-hydrostatic equilibrium state. We discuss the geophysical implications of the observational results.

A Review of the Theory

The goal of a nutation theory is to estimate the nutation amplitude as a function of frequency. The amplitude depends on frequency because the forcing from the Sun and Moon depends on frequency, and because the Earth's response to forcing is different at different frequencies. Finding the forcing as a function of frequency (in fact, finding the frequencies themselves) is a celestial mechanics problem, and requires accurate solutions for the orbital motion of the Moon and Earth [Kinoshita, 1977]. Modelling the Earth's response as a function of frequency is a geophysical problem, and it is that problem that will be addressed in this paper.

The gravitational potential energy from the Sun and Moon can be expanded in Earth-based coordinates as a sum of complex spherical harmonics (Y_l^m). Each coefficient in that sum can be further expanded as a sum of terms which vary harmonically with time. Only Y_2^1 terms contribute to the nutational motion. We write an individual Y_2^1 term as

$$V = f(\omega)e^{i\omega t}r^2 Y_2^1(\theta,\lambda) \quad (1)$$

where ω is the frequency, θ and λ are the co-latitude and eastward longitude, $f(\omega)$ is a scalar amplitude, and Y_2^1 is normalized so that the integral of $|Y_2^1|^2$ over the unit sphere is 1. The values of ω and $f(\omega)$ are determined by the orbital motion.

Under very general conditions, the response of the Earth to the potential (1) can be expanded as a sum of the Earth's normal mode eigenfunctions [Wahr, 1981a], so that the nutation amplitude, ζ is

$$\zeta = e^{i\omega t}\sum_i \frac{B_i}{\omega-\omega_i} \quad (2)$$

Here, the sum over i is over all normal modes, ω_i is the eigenfrequency of the i'th mode, and B_i depends on the i'th eigenfunction and (linearly) on f. One implication of (2) is that the amplitude is large if the forcing frequency, ω, is close to an eigenfrequency.

Although the sum in (2) is, in principle, over every one of the Earth's normal modes, almost all of the important contributions come from just two modes, both with frequencies in the diurnal band. One of these modes, called the tilt-over-mode (TOM), is simply a tipping of the Earth in space with no associated deformation. The motion is exactly equivalent to tipping the coordinate system in the opposite direction. The period of the TOM is infinite as seen from non-rotating inertial space, and so is exactly one sidereal day as seen from a sidereally rotating system.

The more interesting of the two diurnal modes contributing to (2) is the free core nutation (FCN). This mode involves tipping the mantle and fluid core in opposite directions. Because the core-mantle boundary is not exactly spherical, the mantle and core push against each other when they tip, and the resulting pressure acts to restore the core and mantle to their untipped state. There are also gravitational restoring forces due to the interaction between the aspherical density distributions of the core and mantle. The result is a free periodic motion, the FCN, with a frequency equal to one cycle per day plus a small term dependent on the strength of the core-mantle coupling. That coupling depends on the aspherical shape and density distribution of the core, as described below. For an Earth where the shape and density are assumed to be consistent with a state of hydrostatic pre-stress, the frequency is $\omega_{FCN} = 1+1/460$ c/d [Wahr, 1981b].

The TOM contributions to the nutation amplitudes in (2) are much larger than the FCN contributions. The reason is that for the TOM, the factor B_i in (2) depends on the total torque on the Earth from the Sun and Moon at the frequency ω. For the FCN, B_i is sensitive, instead, to the difference between the torques per unit moment of inertia, acting on the core and mantle (and to deformation terms of about the same order). The Sun and Moon do not provide nearly as large a differential torque as they do a total torque.

The TOM contributions are not particularly interesting. The TOM resonance merely reflects the fact that the longer (as seen from inertial space) you torque an object, the more it tips. The FCN contributions,

Copyright 1990 by
International Union of Geodesy and Geophysics
and American Geophysical Union.

though, are of geophysical importance since they are sensitive to the poorly known shape and aspherical density distribution of the core. To consider those contributions in detail, it is usual to remove the TOM contribution from the sum in (2), and then to divide the remainder by the TOM contribution. This last division removes the scalar amplitude, $f(\omega)$, from the remainder, so that the results reflect only the geophysically interesting part of the signal. This normalized "admittance" is shown in Figures 1 and 2 as a function of frequency. The theoretical results (the solid line) use eigenfrequencies and eigenfunctions computed for a hydrostatically pre-stressed Earth. Although the results shown are computed using several modes in the sum (2) [Wahr, 1981b], almost all of the contributions come from the FCN. The resonance at the FCN eigenfrequency of 1+1/460 c/d is clearly evident.

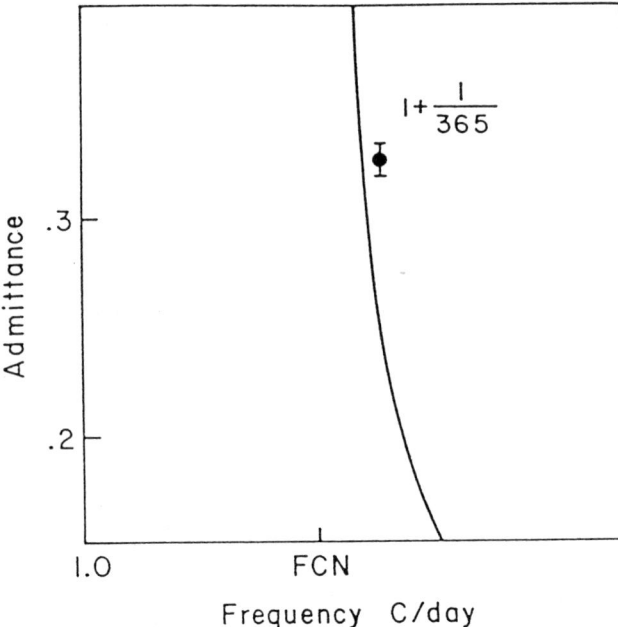

Fig. 2. An expanded view of the 1+1/365 term in Figure 1. There is significant disagreement between theory and observation. The discrepancy suggests the FCN eigenfrequency should be larger than the theoretical value.

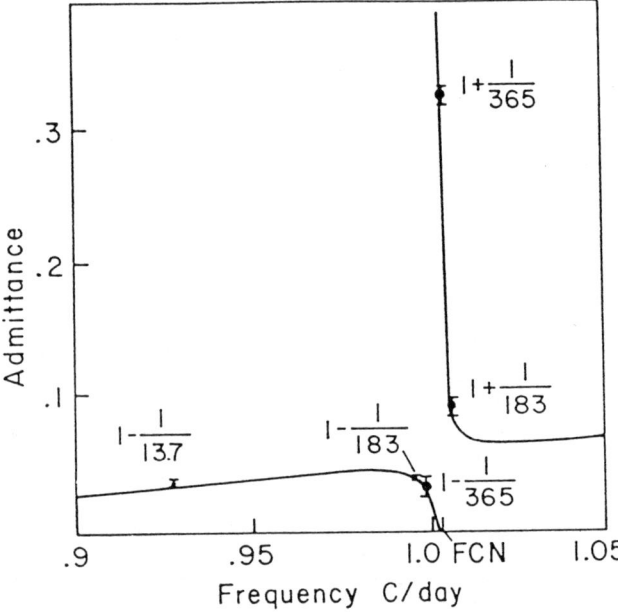

Fig. 1. A comparison between theoretical nutation results from Wahr (1981b) (solid line) and VLBI observational results from Herring, et al. (1986) (vertical bars). The results are normalized, as described in the text, to form admittances. The lengths of the vertical bars represent the observational errors.

Also shown in Figures 1 and 2 are recent VLBI observational results [Herring et al., 1986] for the nutation amplitudes at a few important frequencies. The TOM resonance has been removed from the observational results and then divided into the remainder, in order to compare with the theoretical results (the observations have also been corrected for the effects of oceans as described below). The vertical bars on the observational results reflect the published errors.

The agreement between the theory and the observations is, in general, good. The disagreement at the prograde fortnightly frequency is observationally significant but, in absolute terms, is only a few tenths of a milli-arcsecond. The agreement at the prograde semi-annual frequency looks from Figure 1 to be reasonably good, but, as we shall see below, there could be a disagreement of up to 1 milli-arcsecond after correcting for mantle anelasticity and non-hydrostatic core structure.

First, however, there is an even larger discrepancy at the retrograde annual frequency (see the enlarged comparison in Figure 2). The observed annual admittance lies well above the theoretical result, and the difference in absolute terms is about 2 milli-arcseconds. The results in Figures 1 and 2 suggest that the FCN eigenfrequency should be larger than the value of 1+1/460 c/d predicted using the hydrostatic assumption. In fact, Gwinn, et al. [1986] used the observational results to conclude that $\omega_{FCN} = 1+1/433$ c/d. (Although the annual discrepancy could also be resolved by adjusting the FCN value of B_i in (2), that adjustment would cause substantial discrepancies at other frequencies.)

What does this increase in eigenfrequency imply about the Earth? The theoretical results shown in Figures 1 and 2 do not include the effects of oceans (although the observational results have been corrected for the oceans), of mantle anelasticity, or of non-hydrostatic pre-stress and structure. These effects are discussed below.

The Oceans

Oceans affect nutation amplitudes through surface loading. The Sun and Moon cause diurnal tides in the oceans at exactly the nutation periods. Those ocean tides load the Earth and cause further nutational motion. Thus, the effects of oceans can be perceived as modifying the driving force for nutations. In that case, the force can no longer be written in terms of a Y_2^1 potential, as in (1). Oceanic corrections require some understanding of the loading force, and that requires, at the very least, reliable ocean tide models. Wahr and Sasao [1981] used diurnal tide models to estimate the oceanic corrections, and their results have been removed from the VLBI observations to give the results shown in Figures 1 and 2.

Mantle Anelasticity

Mantle anelasticity and non-hydrostatic structure affect the nutations by modifying the Earth's response to external forcing, rather than by contributing to the forcing itself. To understand their contributions, it is necessary to describe nutation models in more detail.

Existing nutation models fall into two categories, referred to here as numerical and semi-analytical. Both types of models involve the solution of the same infinite set of coupled ordinary differential equations. And, both types of models derive approximate solutions by truncating the equations. In the numerical method [Wahr, 1981b; see also Smith, 1977] the truncation is less severe and more terms are kept than in the semi-analytical method (see, for example, Jeffreys and Vicente [1957]; Molodensky [1961]; Sasao et al. [1980]). Thus, the advantage of the numerical method is that it is apt to be more accurate. The advantages of the semi-analytical methods are (1) they are easier to implement; and (2) the results are more readily understood in terms of the Earth's physical parameters. Furthermore, the results from the semi-analytical models appear to agree well with those from the numerical method.

The semi-analytical results of Sasao et al. [1980] for B_{FCN} and ω_{FCN} are:

$$B_{FCN} = -\sqrt{15/2\pi}\, \frac{A_f}{A_m}\, (e-\gamma) \frac{f(\omega)}{\Omega} \qquad (3)$$

$$\omega_{FCN} = \left[1 + \frac{A_f}{A_m}(e_f - \beta)\right] \Omega \qquad (4)$$

where A, A_f, A_m are the equatorial moments of inertia of the Earth, core, and mantle; $e = (C-A)/A$ and $e_f = (C_f - A_f)/A_f$ are the dynamical ellipticities of the Earth and core (C and C_f are the polar moments of inertia of the Earth and core); and γ and β represent the effects of deformation within the mantle and core ($\gamma \ll e$ and $\beta \ll e_f$). The results shown in (3) and (4) were derived by Sasao, et al. [1980] assuming the inner core is fluid, and that the Earth's equilibrium state is one of hydrostatic pre-stress.

Mantle anelasticity affects these results by modifying γ and β. Those parameters become complex, leading to phase lags in the nutations. Even larger are the effects on the real part of those parameters. The main source of uncertainty when modelling the anelastic corrections is the uncertainty in mantle Q at diurnal periods. Seismic information is pertinent to much shorter periods. On the other hand, this suggests that perhaps the nutation results could be used to learn about mantle Q in this frequency regime.

Wahr and Bergen [1986] and Dehant [1988] used a variety of anelastic models and assumptions to estimate the effects on nutation amplitudes. They found that, in general, the contributions to B_{FCN} in (2) were more important than the contributions to ω_{FCN}. The model results shown in Figures 1 and 2 are corrected for Wahr and Bergen's anelastic estimates and the results are shown in Figures 3 and 4. Also shown, in Figure 5, is a comparison between the out-of-phase components inferred from the VLBI results (corrected for the effects of oceans) and the predicted results due to anelasticity. For a dissipationless Earth, the out-of-phase components would be zero. The vertical bars associated with the anelastic results in these figures reflect the uncertainty in the value of mantle Q at diurnal periods. The figures suggest that the present VLBI results are, in principle, nearly accurate enough to discriminate between various mantle Q results. However, the effects of anelasticity do not resolve the disagreements between the observations and the theory. In fact, they tend to make the agreement for the in-phase amplitudes (Figs. 3 and 4) worse. Thus, before the nutation results can be used to learn about anelasticity, it will first be necessary to understand the reasons for the large discrepancy.

Non-hydrostatic Structure

The most likely explanation for the difference between the observed and theoretical ω_{FCN} is that the Earth is not hydrostatically

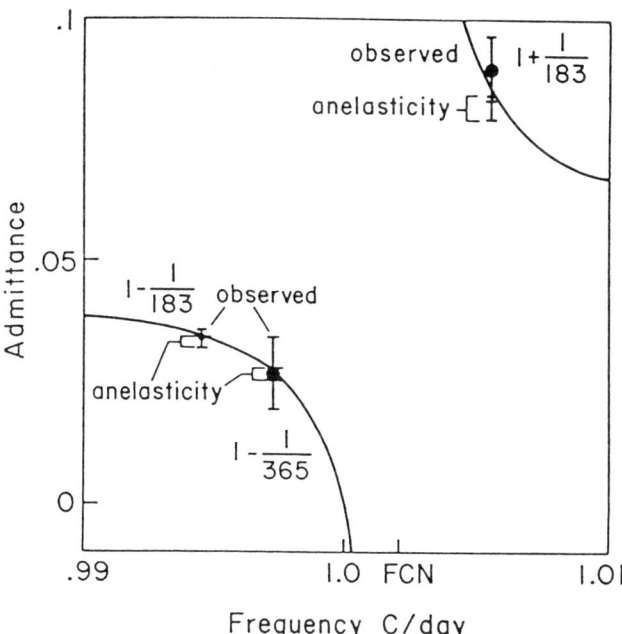

Fig. 3. The theoretical nutation admittances for an anelastic Earth from Wahr and Bergen (1986). The vertical bars for the anelastic results reflect the uncertainty in mantle Q at diurnal periods. Also shown are the theoretical results for an elastic Earth (solid line) and the VLBI results.

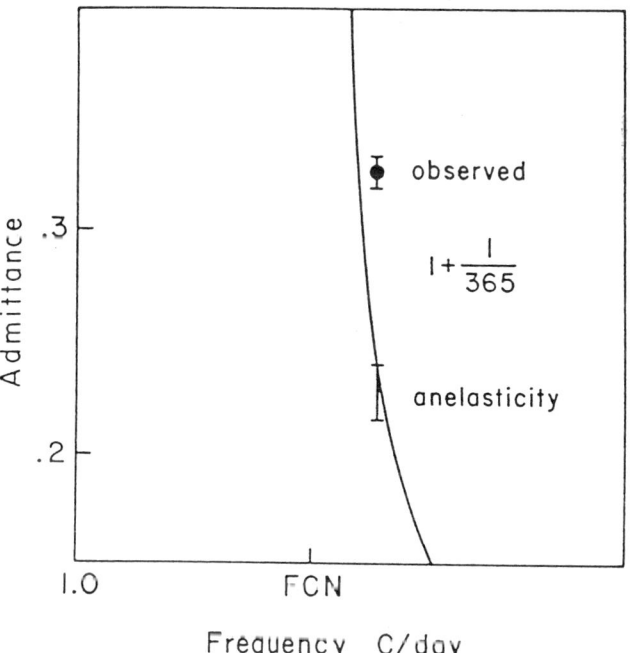

Fig. 4. The anelastic admittance and observational result for the 1+1/365 nutation. Anelasticity worsens the agreement between observation and theory.

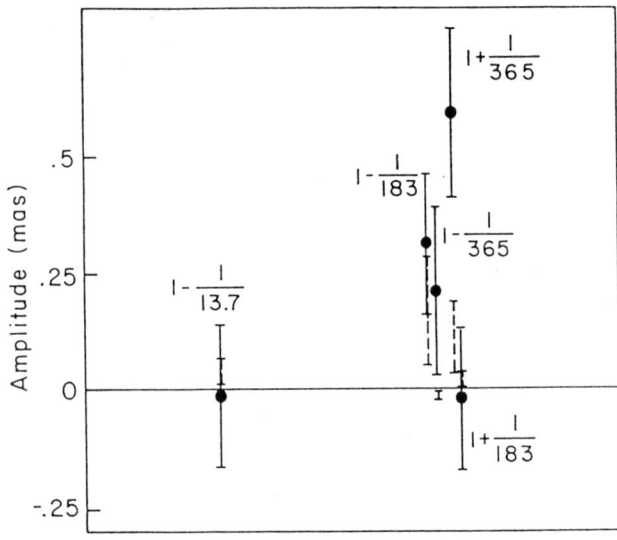

Fig. 5. The out-of-phase components for the nutations. Both the VLBI results (solid lines) corrected for the oceans and the anelastic contributions (dashed lines) are shown. The vertical bars for the anelastic results reflect uncertainty in mantle Q at diurnal periods. The out-of-phase components should be zero for a dissipationless Earth.

pre-stressed. In the Appendix, we extend the semi-analytical model of Sasao et al. to include an Earth which is not hydrostatically pre-stressed (although we, too, assume the inner core is fluid). In particular, the core-mantle boundary and density distribution within the Earth can have an arbitrary spherical harmonic expansion. We find, though, that B_{FCN} and ω_{FCN} are still given by equations (3) and (4). Thus, the effect of the non-hydrostatic pre-stress on ω_{FCN} can be computed by estimating the non-hydrostatic contributions to e_f in (4). Since e_f depends only on the Y_2^0 spherical harmonic components of the core-mantle boundary shape and of the core's internal density, the observational results for ω_{FCN} constrain only those particular spherical harmonic coefficients.

The VLBI results for ω_{FCN} suggest that e_f is about 5% larger than the hydrostatic value [Gwinn et al., 1986]. Suppose the radius of the core-mantle boundary as a function of θ and λ is

$$r(\theta,\lambda) = a \left[1 + \sum_{l=1}^{} \sum_{m=-l}^{m=l} a_l^m Y_l^m(\theta,\lambda) \right] \quad (5)$$

where a is the mean core radius, and the Y_l^m are normalized so that the integral of $|Y_l^m|^2$ over the unit sphere is 1. We have found numerically [J. Wahr and D. de Vries, unpublished manuscript], that unless there is a thin, low density boundary layer at the top of the core, e_f is approximately equal to the ellipticity of the core mantle boundary, ε_c,

where $a_2^0 = -(2/3)\varepsilon_c$. (If there is a low density boundary layer, then e_f could also depend critically on the shapes of the equi-density surfaces inside the core.) Thus, in the absence of such a boundary layer, the nutation results constrain the Y_2^0 component of the boundary. They imply, in fact, that ε_c is about 5% larger than the hydrostatic value, and suggest a non-hydrostatic topography on the boundary of about 1/2 km.

When the value $\omega_{FCN} = 1 + 1/433$ c/d is used in (2) in place of the hydrostatic result $1 + 1/460$ c/d, large differences occur at other frequencies, as well. For example [Gwinn, et al., 1986], the theoretical retrograde 18.6-year amplitude is decreased relative to the IAU adopted value by about 2 milli-arcsecond. This amplitude is also decreased by about 1 milli-arcsecond due to the oceans and about .5 milli-arcsecond by anelasticity.

The theoretical prograde six-month term is increased by about .4 milli-arcseconds due to the change in ω_{FCN}. In fact, the VLBI result for this term disagrees with the theoretical amplitude by about 1 milli-arcsecond, after correcting for the effects of the oceans and anelasticity, and using $\omega_{FCN} = 1 + 1/433$ c/d. Similarly, there is a discrepancy for the prograde fortnightly term of about .3 milli-arcseconds between the VLBI results and the adjusted theory. It is not clear, at present, what is responsible for the fortnightly and semi-annual discrepancies.

Acknowledgements. This work was supported in part by NASA (grant NAG5-485), the Air Force Geophysical Laboratory (contract F19628-86-k-0011), and NSF (grant EAR-8407110).

References

Dahlen, F.A., *Geophys. J. Roy. Astr. Soc.*, **32**, 203, 1973.
Dehant, V., in *Earth Rotation and Reference Frames*, eds. Babcock, A., and G.A. Wilkins, 1988.
Gwinn, C.R., T.A. Herring, and I.I. Shapiro, *J. Geophys. Res.*, **91**, 4755, 1986.
Herring, T.A., C.R. Gwinn, and I.I. Shapiro, *J. Geophys. Res.*, **91**, 4745, 1986.
Jeffreys, H., and R.O. Vicente, *Mon. Not. R. Astr. Soc.* **117**, 162, 1957.
Kinoshita, H., *Cel. Mech.*, **15**, 277, 1977.
Lieske, J., T. Lederle, W. Fricke, and B. Morando, *Astron. Astrophys.*, **58**, 1, 1977.
Molodensky, M.S., *Commun. Obs. R. Belg.*, **288**, 25, 1961.
Saito, M., *J. Phys. Earth*, **22**, 123, 1974.
Sasao, T., S. Okubo, and M. Saito, in *Proceedings of IAU Symposium no. 78 "Nutation and the Earth's Rotation"*, eds. Fedorov, E., M. Smith, and P. Bender, D. Reidel Publishing Co., Dordrecht, 1980.
Smith, M.L., *Geophys. J. Roy. Astr. Soc.*, **50**, 103, 1977.
Smith, M.L., and F.A. Dahlen, *Geophys. J. Roy. Astr. Soc.*, **64**, 223, 1981.
Wahr, J.M., *Geophys. J. Roy. Astr. Soc.*, **64**, 651, 1981a.
Wahr, J.M., *Geophys. J. Roy. Astr. Soc.*, **64**, 705, 1981b.
Wahr, J.M. and T. Sasao, *Geophys. J. Roy. Astr. Soc.*, **64**, 747, 1981.
Wahr, J., and Z. Bergen, *Geophys. J. Roy. Astr. Soc.*, **87**, 633, 1986.
Wahr, J., and D. de Vries, *Geophys. J. Roy. Astr. Soc.*, 1989 (submitted).
Woodhouse, J.H., and F.A. Dahlen, *Geophys. J. Roy. Astr. Soc.*, **53**, 335, 1978.

APPENDIX: Nutation Amplitudes for a Non-hydrostatically Pre-stressed Earth

Previous nutation models have assumed the Earth is hydrostatically pre-stressed. Among the implications of this assumption is that the core-mantle boundary and all surfaces of constant density within the core and mantle are elliptical. Thus, the departure of those surfaces from spherical symmetry is proportional to the single spherical harmonic $Y_2^0(\theta,\lambda)$.

It is quite likely, though, that surfaces of constant density in the Earth and surfaces of discontinuity, such as the core-mantle boundary, have non-hydrostatic shapes. The Y_2^0 components of these surfaces are apt to differ from the hydrostatic values, and there are apt to be components with other Y_l^m angular dependence. Since the FCN is critically dependent on pressure coupling acting across the core-mantle boundary (and, to a lesser extent, gravitational coupling between the core and mantle), the FCN contribution to the eigenfunction expansion (2) could be affected.

In this appendix, the FCN contribution is modelled without assuming hydrostatic equilibrium in the mantle. Instead, the mantle is allowed to possess a small, arbitrary deviatoric pre-stress, and so surfaces of constant material properties and surfaces of discontinuity (such as the core/mantle boundary) within the mantle can have arbitrary shape. The core, however, is not permitted to have a deviatoric pre-stress (the core, after all, is a fluid even at seismic periods). As a result, the density distribution in the core is uniquely determined by the shape of the core-mantle boundary and by the gravitational potential from the mantle acting on the core. Thus, the constant density surfaces in the core need not coincide with the surfaces predicted for an everywhere hydrostatically pre-stressed Earth. The derivation, below, extends the semi-analytical model of Sasao, et al. [1980] to include these non-hydrostatic modifications, and to more fully justify a number of the necessary approximations. As in the model of Sasao, et al. [1980], we assume the inner core is fluid.

Consider a reference frame fixed to the mantle and rotating with it. For nutational motion at frequency ω, the rotation vector of the frame is defined as $\Omega\hat{z} + \Omega m_o(\hat{x}+i\hat{y})e^{i\omega t}$, where the nutational motion of the mantle is described by m_o, and where $m_o \ll 1$. For the forced nutations, the ratio $(\omega-\Omega)/\Omega$ is small. We will consider this ratio to be first order in ε, where ε is some measure of the departure of the Earth from spherical symmetry. This assumption is reasonable for the annual nutations (the terms with the largest discrepancy between theory and observation), where $|(\omega-\Omega)/\Omega| \cong 1/365$. It is less accurate for the fortnightly nutations, where $|(\omega-\Omega)/\Omega| \cong 1/13.7$.

Denote the luni-solar tidal force by $-\rho\nabla V$, where V is the luni-solar tidal potential given by equation (1) in the text. The equations of motion in the hydrostatically pre-stressed core are [Sasao, et al., 1980]:

$$\rho[-\omega^2 s + 2i\omega\Omega\times s] = -\rho\nabla(\phi_1+V) - \nabla P_E - \rho_1\nabla\Phi - \Omega^2\rho m_o[(\hat{x}+i\hat{y})z(\Omega-\omega)$$
$$+\hat{z}(x+iy)(\Omega+\omega)] \quad (A1)$$
$$\nabla^2\phi_1 = 4\pi G\rho_1$$

where s is the displacement field in the core; ϕ_1, P_E, and ρ_1 are the Eulerian perturbations in gravitational potential energy, pressure, and density, respectively; $\boldsymbol{\Omega}=\Omega\hat{z}$ is the mean rotation vector of the Earth; and ρ and Φ are the initial density and total potential energy (gravitational plus centrifugal) and can have arbitrary angular dependence, except that surfaces of constant ρ must be surfaces of constant Φ in the core. There are, also, equations relating ρ_1 and P_E to s, which are not shown here. And, there are similar, although more complicated, differential equations describing displacements in the mantle, which are also omitted here. The mantle equations must explicitly include any assumed non-hydrostatic pre-stress (see Woodhouse and Dahlen [1978] for the complete mantle and core equations and for the boundary conditions).

For nutational motion, the displacement field in the core has the form

$$\mathbf{s}(\mathbf{r}) = \Omega\theta_o(\hat{x}+i\hat{y})\times\mathbf{r} + \mathbf{s}_d(\mathbf{r}) \quad (A2)$$

where θ_o represents the mean nutational motion relative to the mantle, and $\mathbf{s}_d$, which represents the deformation of the core, is first order in ε compared with the θ_o term (see, for example, Sasao, et al. [1980]). Taking $(\hat{x}-i\hat{y})\cdot[\mathbf{r}\times(\text{equation}(A1))]$ and integrating through the core, and using (A2) for $\mathbf{s}$, gives an angular momentum equation for the core:

$$2i\Omega^2 A_f m_o + 2\omega\Omega^2 A_f \theta_o\frac{\Omega-\omega}{\Omega} + i\omega\Omega c_-^f$$
$$= (\hat{x}-i\hat{y})\cdot\int_{core}\mathbf{r}\times\left[-\rho\nabla(\phi_1+V)-\nabla P_E-\rho_1\nabla\Phi\right]. \quad (A3)$$

Here, A_f is the core's principal moment of inertia in the equatorial plane; and $c_-^f = c_{13}^f - ic_{23}^f$, where the c_{ij}^f are perturbations in the core's inertia tensor due both to the deformational displacements, $\mathbf{s}_d$, and to the rotational motion, θ_o (the components of the inertia tensor change when the aspherical core is rotated with respect to the coordinate system). The terms on the right hand side of (A3) include the pressure and gravitational torques on the core from the mantle, and the gravitational torque on the core from the luni-solar tidal force.

All terms in (A3) are second order in ε relative to θ_o. As an example, ϕ_1 has contributions from the deformation, $\mathbf{s}_d$, and from the rotational motion, described by θ_o. The contributions to ϕ_1 from the deformation are first order, because $\mathbf{s}_d$ is first order. The contributions from θ_o are caused by rotating the Earth's unperturbed gravity field through the angle θ_o. These rotational contributions are also first order, because there is no change in gravity if the unperturbed field is spherically symmetric. Thus, ϕ_1 is a first order quantity. Furthermore, the integral of $\rho\mathbf{r}\times\nabla\phi_1$ is 0 no matter what ϕ_1 is, if both the core shape and the internal density in the core are spherically symmetric. Thus, $\int_{core}\mathbf{r}\times\rho\nabla\phi_1$ is second order.

As a second example, $\Omega\omega\cong\Omega^2$ is a first order quantity (the unperturbed centrifugal force is order ε times the unperturbed gravitational force), as is $(\Omega-\omega)/\Omega$. Thus, the θ_o term in (A3) is second order. And, as a third example, m_o is first order compared with $\Omega\theta_o$ (m_o represents motion of the mantle rotation axis as seen from the mantle and is of order $(\Omega-\omega)\theta_o$ so that $\Omega^2 m_o$ is second order. Finally, for completeness, we note here that V, P_E, ρ_1, Φ, and c_-^f are all first order relative to θ_o.

A comparable angular momentum equation for the entire Earth, after dropping all terms third order or smaller, is

$$2i\Omega^2 A m_o = 2(\omega-\Omega)A_f\Omega^2\theta_o + \sqrt{15/2\pi}\,i\,(C-A)\,f(\omega) \quad (A4)$$

where C and A are the principal moments of inertia for the entire Earth, $f(\omega)$ is the scalar amplitude of the tidal potential (see equation (1) in the text), and the last term on the right hand side of (A4) represents the luni-solar torque on the Earth.

Next, c_-^f and the right hand side of (A3) can be related to θ_o. Use (A2) in (A1) and separate the resulting vector equations into spheroidal and toroidal scalar equations. Similarly, separate the mantle differential equations and the boundary conditions into toroidal and spheroidal scalar equations. Then, consider only the spheroidal equations, ignore all terms in these equations that are second order or smaller in ε, and solve the entire system on a computer. In the core, for example, the first order spheroidal equations derived from (A1) are the scalar components of

$$\rho_0\nabla(\phi_1+V) + \nabla P_E + \rho_1\nabla\Phi_0 = \left[i\theta_o\Omega^3\sqrt{8\pi/15}\right]\rho_0\nabla(r^2 Y_2^1) \quad (A5)$$

where ρ_0 and Φ_0 are the spherically symmetric parts of ρ and Φ. (One of the consequences of the truncation to first order is that m_o does not appear in (A5).)

The first order differential equations in the mantle corresponding to (A5) in the core, are the usual set of spheroidal equations describing tidal deformation of a spherical, non-rotating, static, and hydrostatically pre-stressed mantle (see, for example, Saito [1974]). Although, in principle, there is an apparent spheroidal force in the mantle which depends on m_o, that force is second order and so can be ignored in the first order deformation equations. All first order boundary conditions within the core and mantle and at the outer surface are also equivalent to the boundary conditions for a spherical, non-rotating, hydrostatically pre-stressed Earth.

In effect, then, all deformation terms can be computed by solving the static equations of motion for a spherical, non-rotating, hydrostatically pre-stressed Earth, subject to an apparent force proportional to $\rho_0 \nabla(r^2 Y_2^1)$, but with different proportionality constants in the core and mantle. In the core, the apparent force is proportional to θ_o and $f(\omega)$ (see (1) for a definition of $f(\omega)$). In the mantle, the apparent force is proportional only to $f(\omega)$. Neither the non-hydrostatic pre-stress in the mantle nor any of the Earth's aspherical structure enters explicitly into any of the first order deformation equations. Their effects are included only through the integrals on the right hand side of (A3).

Because the apparent force in the core is proportional to $\rho_0 \nabla(r^2 Y_2^1)$ and because the first order deformation equations are spherically symmetric, ρ_1, P_E, and ϕ_1 have $Y_2^1(\theta,\lambda)$ angular dependence. Using this angular dependence, (A5) yields directly

$$P_E(r,\theta,\lambda) = \rho_0(r)\left[\left[i\Omega^3\theta_o\sqrt{8\pi/15}-f(\omega)\right]r^2 Y_2^1(\theta,\lambda)-\phi_1(r,\theta,\lambda)\right] \quad (A6)$$

$$\rho_1(r,\theta,\lambda) = \frac{\partial_r \rho_0(r)}{\partial_r \Phi_0(r)}\left[\phi_1(r,\theta,\lambda)+\left[f(\omega)-i\Omega^3\theta_o\sqrt{8\pi/15}\right]r^2 Y_2^1(\theta,\lambda)\right].$$

Using (A6) in (A3) and doing the integrals gives a result accurate to second order of

$$2i\Omega^2 m_o + 2\Omega^2\theta_o(\Omega-\omega) + \frac{ic_-^f \Omega^2}{A_f} = -2\Omega^3\theta_o e_f \quad (A7)$$

where $e_f = (C_f - A_f)/A_f$ is the dynamical ellipticity of the core (C_f is the polar moment of inertia of the core).

Define the dimensionless, real parameters β and γ so that

$$c_-^f = \beta 2 A_f i \Omega \theta_o - \gamma A_f \sqrt{15/2\pi}\, f(\omega)/\Omega^2 . \quad (A8)$$

β and γ can be determined by solving the deformation equations on a computer. Using (A4) and (A8) in (A7) and solving for the core rotation angle θ_o, gives

$$\theta_o = \frac{i\sqrt{15/8\pi}\dfrac{A}{A_m}(e-\gamma)f(\omega)/\Omega^2}{\omega-\Omega\left[1+\dfrac{A}{A_m}(e_f-\beta)\right]} \quad (A9)$$

where $e = (C-A)/A$ is, to lowest order, the dynamical ellipticity of the Earth, and A_m is the principal moment of inertia of the mantle.

The nutation amplitude observed at the Earth's outer surface is (see Sasao and Wahr [1981, eq. 3.20])

$$\zeta = \frac{\Omega}{\Omega-\omega} m_o . \quad (A10)$$

Using (A4) to relate m_o to θ_o, and (A9) to relate θ_o to $f(\omega)$, (A10) reduces to

$$\zeta = \left[\frac{e}{\Omega-\omega} - \frac{(A_f/A_m)(e-\gamma)}{\omega-\left[1+\dfrac{A}{A_m}(e_f-\beta)\right]}\right]\sqrt{15/8\pi}\frac{f(\omega)}{\Omega} . \quad (A11)$$

The $e\Omega/(\Omega-\omega)$ term in (A11) represents the TOM resonance. The other term in (A11) is the FCN resonance, and can be written as $B_{FCN}/(\omega-\omega_{FCN})$, where B_{FCN} and ω_{FCN} are given by equations (3) and (4) in the text.

Although this result was derived here without assuming a hydrostatically pre-stressed mantle, it is identical in form to the hydrostatic result. The dynamical ellipticity, e_f, depends on the Y_2^0 component of the core-mantle boundary shape, and on the Y_2^0 terms in the density stratification inside the core. The dependence of ω_{FCN} on the internal density stratification is due to the effects of gravitational torques between the core and mantle, represented by the $\rho\nabla\phi_1$ and $\rho_1\nabla\phi$ terms on the right hand side of (A3). The dependence on the boundary structure is due to pressure torques at the core/mantle boundary, represented by the ∇P_E term in (A5). There is no dependence, to this order of approximation, on any other Y_l^m terms in the aspherical structure. Similarly, e depends on the Y_2^0 density structure throughout the Earth, and is well determined from independent observations of the Earth's precession (see, for example, Lieske et al. [1977]). The factors β and γ represent the effects of deformation and are insensitive, to this order, to aspherical structure. Sasao et al. [1980] found that β is about 25% of the hydrostatic value of e_f. For a hydrostatically pre-stressed Earth, $\omega_{FCN} \cong (1+\frac{1}{460})$ cycles per day.

STUDY OF FLUID-SOLID EARTH COUPLING PROCESS USING
SATELLITE ALTIMETER DATA[1]

Wooil M. Moon, Roger Tang[2] and B.H. Choi[3]

Department of Geological Sciences
The University of Manitoba, Winnipeg, Canada, R3T 2N2

Abstract. In the ocean-solid Earth system, there are geodynamic parameters which can characterize the coupling and energy transfer processes across the Earth's fluid-solid discontinuities. One such parameter is the ocean bottom coupling coefficient. In this study, the hydrodynamic modelling method is used to analyze and to study the ocean bottom friction coefficient with respect to the sea surface elevations measured from the SEASAT altimetry over the East China Sea and Yellow Sea ($25°-38°N$; $120°-130°E$). Periods of significant atmospheric disturbances during the SEASAT mission were selected for this study. These include the periods of July 28-August 2 and August 18-21. Meteorological forcing functions, which are needed for the sea model, are derived by a 2-dimensional grid that is governed by a set of theoretical and empirical meteorological relations over the study area. Ocean tides in this area are known to be significant and introduce a large spatial and time variability in the sea surface elevation. Consequently major tidal constituents of M_2, S_2, K_1 and O_1 are included in the computation. With some knowledge of other known sea surface phenomena (body tide, loading tide, and steric variation of the ocean), the time-dependent sea surface variation is predicted to compare statistically with the satellite altimetric measurements and to achieve the objective of ocean bottom friction study. From a total of 10 SEASAT orbit tracks, the quadratic ocean bottom friction coefficient was found ranging from 0.0023 to 0.0027.

---

[1]The University of Manitoba, Center for Precambrian Studies Publications No. xxxx.

[2]On leave from Sung Kyun Kwan University, Su-Won, Korea.

[3]Now with Texaco Resources Canada Ltd., Calgary, Canada, T2P 2P8.

Copyright 1990 by
International Union of Geodesy and Geophysics
and American Geophysical Union.

Introduction

For many years, efforts in geodynamics have been spent on the development of the global ocean tide model [Schwiderski, 1980; Pekeris and Accad, 1969], the study of the lunar orbit, and the investigation of the variable rotation of the Earth [Munk and MacDonald, 1960; Lambeck, 1975; Denis, 1986]. Most of these problems require some degree of knowledge about the ocean bottom friction and, even more importantly, the understanding of the coupling mechanism between the fluid ocean and the solid earth. In the development of the global ocean tide model, for example, one arrives at the bottom friction term through the integration of the Navier-Stokes equations with the Boussinesq replacement of the Reynolds stress tensor. A certain form of the ocean bottom friction term has to be assumed to complete the problem. Another well known example is the evaluation of tidal dissipation by finding the rate of work per unit surface done by the current at the seafloor. The success of this approach relies on the knowledge of the frictional coefficient (or constant) which links the frictional force at the ocean floor, either linearly or quadratically, with the current velocity. Not only is the frictional coefficient important in these contexts just described, but a variety of storm surge and local ocean tide modelling problems requires a similar dissipation mechanism to integrate the results more accurately [e.g., Heaps, 1969; Tang and Moon, 1984; Moon and Tang, 1985; Stock, 1976; Grace, 1930].

In the earlier study by Moon and Tang [1987], in the Hudson Bay area, a limited range of values for the quadratic friction coefficient (0.0019-0.00465) was estimated through the least squares optimization process between the numerically computed sea surface elevation and the SEASAT-ALT data, at wind speeds ranging from 1 m/s to 10 m/s. The sea model algorithm employed was that of the linear shelf model [Heaps, 1969; Tang and Moon, 1984; Moon and Tang, 1985]. In the present investigation in the East China Sea and Yellow Sea area, a slightly different version which includes the modelling of advection phenomena, ocean tide and meteorological-induced motion is used. This model is employed in conjunction, with the observed ocean tidal phenomena in the area and is capable of simulating surge and its interaction with tide. One of the distinguishing features of this paper that is markedly different from the previous study is that the investigation was carried out during a severe weather condition with wind speeds up to 40 m/s. This implies that the propagation of storm surge must be carefully accounted for in the study area in order to derive meaningful results. An atmospheric model is required to compute the meteorological forcings prior to the computation of the sea surface response. The 2-dimensional atmospheric model is based on the finite difference form of the geostrophic wind-pressure gradient balance equation. This model operates at some height above the sea model and provides the necessary meteorological driving force as a function of time and space in the sea model.

The satellite orbit tracks for the study periods are plotted in Figure 1. These data, including some correction and processing algorithms, were obtained from the U.S. Department of Commerce, National Oceanic and Atmospheric Administration (NOAA). By tuning the ocean bottom friction coefficient in the equation of motion, adjustments of the computer sea surface elevation with respect to the altimeter observations from these passes can be made. This procedure is performed through the use of the time varying sea surface equation [Cartwright and Alcock, 1981; Le Provost, 1983; Tang, 1985]. The results of these adjustments allow a set of frictional coefficients to be obtained. This set of values is then subjected to an mean-square-error type analysis to find the optimum value of ocean bottom friction coefficient, which gives the best agreement between modelled and observed sea surface elevation profiles.

Time Dependent Sea Surface Equation

The time varying part of the sea surface can be broadly classified into two types based on their periodicity of occurrences. The periodical components include the solid Earth and ocean tides whereas the transient components are due to the varying sea surface wind and pressure gradient fields. The sea surface height $h'(\phi,x,t)$ above the standard ellipsoid (semimajor axis = 6378136m and flattening = 1/298.257) in satellite recording geometry is basically of type [Cartwright and Alcock, 1981]:

$$h'(\phi,x,t) = h_{or}(\phi,x,t) - h_a(\phi,x,t) - h_{re} \quad (1)$$

where (ϕ,x,t) are the spatial and time coordinates, $h_{or}(\phi,x,t)$ is the estimated orbital radius of satellite by ground tracking system, $h_a(\phi,x,t)$ is the altimeter height after removing instrumen-

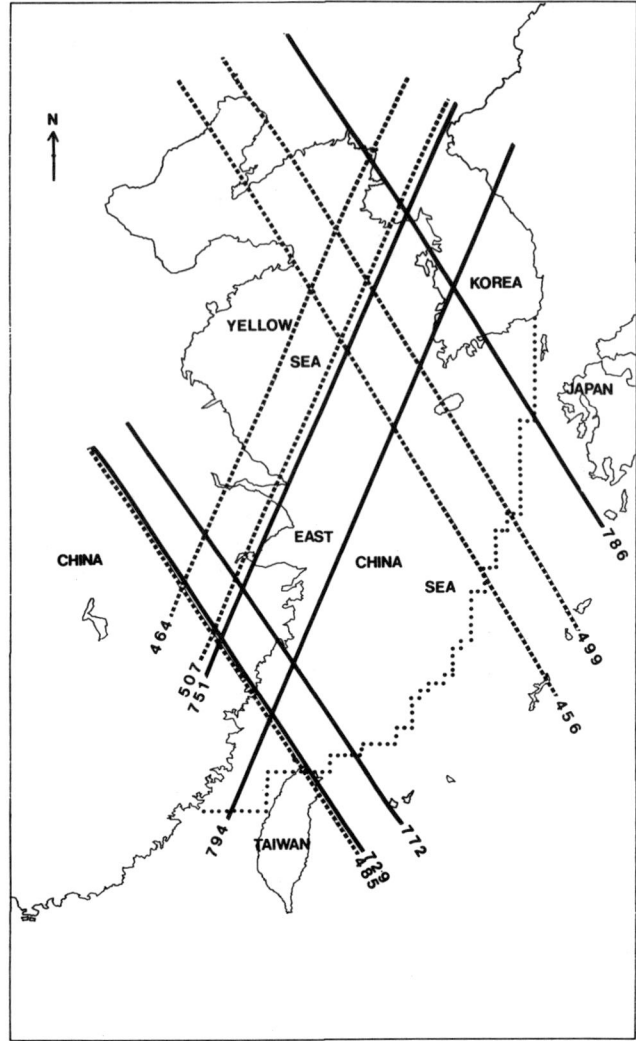

Fig. 1. SEASAT orbit tracks over the area of East China Sea and Yellow Sea during two selected time periods (July 28 - Aug 2 and Aug 15 - Aug 21, 1978).

tal error and atmospheric delay, and h_{re} is the adopted reference (or standard) ellipsoid. This equation is closely approximated by the geoid, plus other known time varying sea surface phenomena [Le Provost, 1983; Cartwright and Alcock, 1981]:

$$G(\phi,x,t)+h_b+h_{ot}+h_l+h_s+h_w+h_{pq}+\epsilon = h'(\phi,x,t) \quad (2)$$

where

$G(\phi,x,t)$ - geoid above the standard ellipsoid
h_b - body tide of the solid Earth
h_{ot} - ocean tide
h_l - ocean induced loading tide of the Earth crust

h_s — long wavelength (low frequency) steric variations of the ocean
h_w — sea surface fluctuation caused by wind
h_{pq} — sea surface fluctuation caused by pressure gradient
ε — contains unmodelled error of both satellite measurements and the time invariant part of the steric ocean surface set up by quasi-steady currents.

The time varying part of the sea surface $T(t)$ can then be expressed by:

$$T(t) = h_b + h_{ot} + h_l + h_s + h_w + h_{pq} \qquad (3)$$

Most of the items in the above equation are given in the SEASAT-ALT GDR except for the effects due to the meteorological forcings which require accurate information of local weather conditions. In continental shelf areas or coastal sea basins, some of these given items are considered to be insufficiently precise for specific application. For example, the two global ocean tide models in the GDR prepared by Schwiderski [1978] and Parke and Hendershott [1980] are primarily for the deeper part of the ocean, and their values in shelf areas may need to be recomputed. In the study of the ocean bottom friction using the SEASAT-ALT data, equation (2) is slightly rearranged as follows:

$$h'(\phi,x,t) - G(\phi,x,t) + \varepsilon = T(t) + \varepsilon_c \qquad (4)$$

where ε_c is introduced to represent error in the computation of different components. Equation (4) implies that the altimetric-derived sea surface height with respect to a mean equilibrium signature, normally the geoid plus errors in altimeter data processing algorithms, is balanced by the total time varying sea surface height with respect to an initial state of rest, with some computational uncertainties.

The variable $G(\phi,x,t)$ can be substituted by an existing geoid model, although a locally constructed geoid will be the best to use. In SEASAT-ALT GDR, a fairly accurate geoid surface proposed by the Goddard Space Flight Center is included. The GEM 10B gravimetric geoid was constructed using the global set of GEOS 3 altimeter data with GEM 9 data plus a global set of $1° \times 1°$ surface gravity data. Although this model was completed to a degree and order of 36, it was found to have up to a few meters of error in some areas [Marsh and Martin, 1982]. These uncertainties are significant since they may be included in the measurements if they were true anomalies. Thus an altimetric-derived reference surface, SS3 surface, is chosen as the equilibrium sea surface in equations (2) and (4). This mean ocean surface is given above the standard ellipsoid and was constructed by Marsh and Martin [1982] using the SEASAT ephemeris calculated by the Preliminary Gravity Solution - SEASAT 3 (PGS-S3) and SEASAT altimeter data from the period 28 July to 14 August. PGS-S3 contains data from GEM 9, SEASAT laser, S band, global gravimetry and GEOS 3 [Lerch et al., 1981] altimetry. In their analysis, both data sets are combined by accurate gridding techniques to yield global contour maps of the mean sea surface topography (SS3). The data of SS3 surface north of 60 degree and south of 63 degree are set to zero due to the possibility of icebergs. This SS3 surface is believed to give a more accurate representation of the equilibrium sea surface than those computed previously [Marsh and Martin, 1982; Fu, 1983]. Hence, the SEASAT altimeter GDR tape also accompanies the SS3 data set as an alternative reference geoid to the Goddard Earth Model 10B (GEM10B) Earth gravity model already adopted for the GDR production. Figure 2 is a schematic diagram showing the SEASAT data collecting geometry.

Equations (2) and (4) are inevitably contaminated by a number of errors which include: (1) the SEASAT ephemeris error which was up to 1.5 m rms, (2) the numerical errors of gridding and interpolation for the production of the SS3 surface, and (3) some unknown systematic errors which can not be minimized by the gridding procedure. Equations (2) and (4) represent all time-dependent sea surface phenomena, as far as they are known, but obviously, the importance of each individual component varies significantly with locality. In the study of sea surface elevations and slopes of the northeast Atlantic Ocean [Cartwright and Alcock, 1981], for example, loading tide was completely ignored over the North Sea region, but its effect on open ocean was taken as a constant fraction of the ocean tide. Also, the meteorologically induced sea surface variation was calculated differently for the shelf area and open ocean. In the North Sea and adjacent shelf areas, it was computed by a surge-forecasting model while the same component was estimated by simple hydrostatic law in open sea. Similar simplifications were introduced by Le Provost [1983] in the English Channel where the body tide, loading tide, and steric variations of the ocean were ignored simply because of their small magnitudes compared to other oceanic events.

Employing the gridding technique developed by Schwiderski [1978], the loading tide and body tide were calculated during the two selected mission times to examine their contributions to the sea surface elevation change. As a first attempt, a $0.5° \times 0.5°$ grid was used with all available tidal constituents (11 of them). This grid required excess computational time. Thus it was cut down to two $1° \times 1°$ grids for the East China Sea and the Yellow Sea, separately, with the M_2, S_2, K_1 and O_1 tidal constituents. The loading tide in this part of the continental shelf was found to be less than 4.5 cm, a value which is small compared to the ocean tide itself. Ocean tide is a well known ocean phenomenon in that area and it dominates all other components in the time varying sea surface equation. The

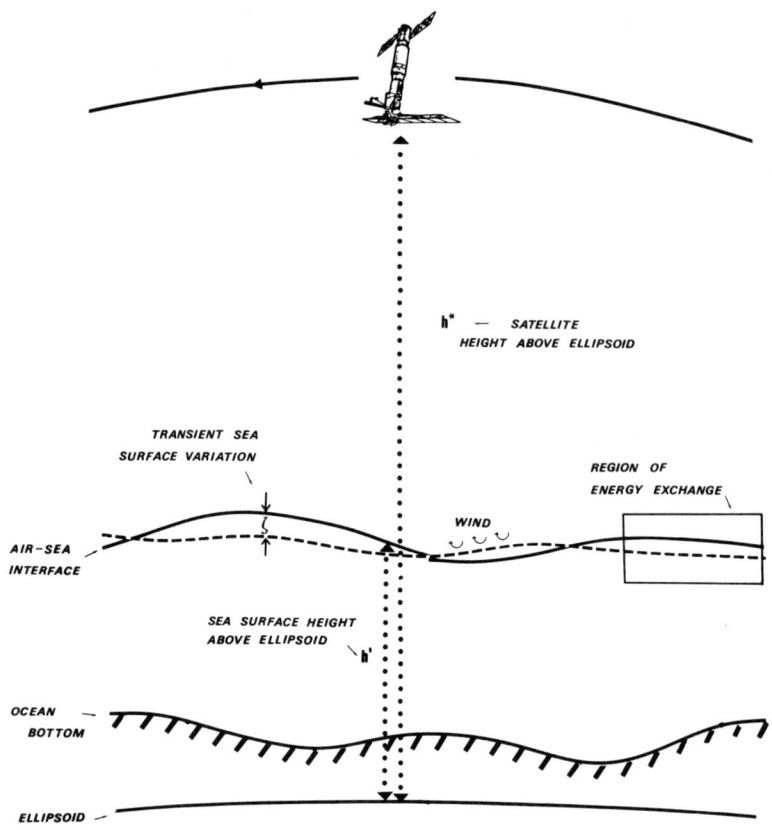

Fig. 2. Geometry of the SEASAT-ALT measurement. h' is the sea surface height above the stan- dard ellipsoid (defined by semimajor axis 6378.137 km and flattening of 1/298.257) after all corrections have been applied except for the transient effects and the ocean tides.

body tide values along the satellite tracks will be presented later, but their effects, similar to the steric variations of the ocean, are also negligible due to their extremely low frequency characteristic along each orbit track. The largest and smallest magnitude of the body tide are about -12 m and 3 cm according to the computations. Hence the time varying sea surface equation can be expressed as:

$$h'(\phi,x,t) - G(\phi,x,t) = h_{ot} + h_w + h_{pq} + \epsilon_c \quad (5)$$

Before the application of equation (5) which incorporates the corrected SEASAT altimeter data with the model results, short wavelength geoid anomalies have to be investigated over the area. The sea surface height given in GDR is an averaged measurement of 1 sec interval; this is a sampling of a surface distance of 7 km. Therefore, the smallest resolvable wavelength in the measurements along the tracks is approximately 14 km (a distance of 3 consecutive points), according to the Nyquist criteria. Small scale geoid anomalies of this order could be very important because they can be attributed to observations which can not be accounted for in the sea model. The geoid profiles of six up-track and four down-track orbits during the two selected time periods are shown in Figure 3. They are mostly characterized by linear functions (extremely low frequency). The possible high frequency components in the geoid were also extracted along the satellite tracks by removing the geoid profiles from the SS3 reference surface profiles. The residual also indicates a low frequency content along the tracks. Thus the geoid, at least along these tracks, is basically long wavelength, and the reduction of the satellite data by a higher frequency reference surface (SS3 surface) should further suppress somewhat the small scale geoid anomalies if they exist.

Ocean Tide and Surge Model

In the past decade, considerable development has been undertaken on the linear shelf model [Heaps, 1969] and the result is a much improved non-linear version capable of simulating the main tides, surge, and tidal interaction [Heaps, 1983]. The major difference between the vertically integrated model and the non-linear version is the inclusion of the advective terms, which have been shown to be important in shallow water bodies [Charnock and Crease, 1957]. Including both the atmospheric driving force terms and the non-linear advective terms in the equation of motion, the non-linear version in vector form is given by:

$$\frac{\partial \bar{V}}{\partial t} + \bar{V} \cdot \nabla \bar{V} + 2\omega \wedge \bar{V} = -g \nabla \zeta + \bar{\tau}_s - \bar{\tau}_B - \nabla \bar{P}_a \quad (6)$$

The equation of continuity is given by:

$$\frac{\partial \zeta}{\partial t} + \nabla \cdot \zeta \bar{V} = 0 \quad (7)$$

where $\bar{V}$ is the vertically-averaged horizontal velocity, ζ is the sea surface elevation, ω is the angular speed, g is the gravity acceleration, τ_B is bottom friction term, τ_s is the surface friction due to wind stress, and P_a is the atmospheric pressure. The equivalent scalar forms of the equations of motion and continuity are given by Davies and Flather [1978] and Choi [1980].

These equations are solved iteratively through time for ζ and V starting at an initial state of rest. Because of the capability of handling shallow water effects and modelling of different types of motion, this algorithm has been used for some time in the East China Sea and Yellow Sea continental shelves to study the nature of tides without meteorological input forcings, i.e. pressure gradient and wind stress at the sea surface [Choi, 1980]. Some agreeable comparisons between the model results with a number of measurements conducted across the continental shelf during the joint USA-China Marine Sedimentation Dynamics Study have been reported by Larson and Cannon [1983] and Choi [1983]. In this paper, the algorithm is modified to include the meteorlogical forcing in addition to the original tidal forcing. The tidal constituents considered are two semi-diurnal tides (M_2 and S_2) and two diurnal tides (K_1 and O_1). The model results will be utilized to compare with the corrected SEASAT observations to study the ocean bottom friction coefficient in that area.

The mechanism of bottom stress that takes place at the seafloor is not well understood and two alternative representations are acceptable in numerical modelling. If the linear stress law is assumed to be the bottom frictional dissipation mechanism for the transient surge, the bottom stress term is given by:

$$\bar{\tau}_B = \frac{\eta \rho}{H} \bar{V} \quad (8)$$

where η is the linear stress coefficient having a typical value of 0.24 cm/s [Weenink, 1958], ρ is the density of the sea water, and H is the water depth. For both surge and tide modelling, the linear law has been demonstrated to be a reasonable approximation [e.g. Grace, 1931; Heaps, 1969].

The second and more commonly used alternative of frictional dissipation is to express the bottom stress in terms of the square of the current velocity. This is expressed as:

$$\bar{\tau}_B = \frac{C \rho}{H^2} \bar{V} |\bar{V}| \quad (9)$$

where C is the quadratic stress coefficient which traditionally takes values close to 0.003. Although these values have been used in a variety of hydrodynamic modelling with success [Flather and Heaps, 1975; Flather and Davies, 1978; Choi, 1980], whether the particular numerical value used is an optimum one for global ocean or different continental shelf problems has been questionable.

The supply of atmospheric energy and the dissipation of frictional energy in the sea model occur continuously during the computation of the sea elevation, and after a period of time the influence of the initial conditions becomes negligible. To solve equations (6) and (7), the following initial conditions are used:

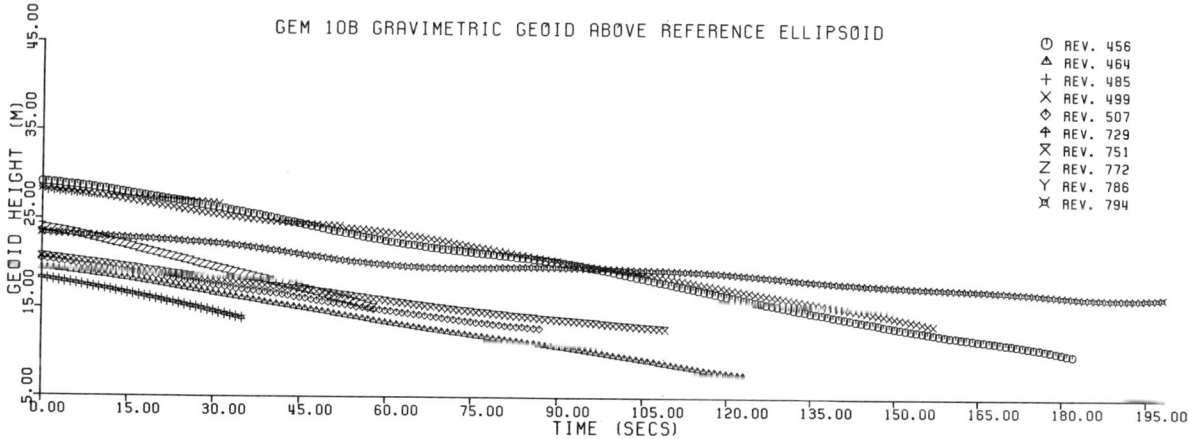

Fig. 3. Six up-track and four down-track geoid profiles from the GEM 10B gravimetric geoid over the East China Sea and Yellow Sea.

$$\zeta(\phi,x,t) = \bar{V}(\phi,x,t) = 0 \text{ at } t = 0 \qquad (10)$$

This condition implies that the sea surface elevation change is generated from an initial state of rest. The sea surface at this state is called the initial state of equilibrium. The coastal boundary condition is given by:

$$V_\phi \cos\psi + V_x \sin\psi = 0 \qquad (11)$$

where ψ denotes the inclination of the normal to the northerly direction and (V_ϕ, V_x) are the components of the depth-averaged current vector.

The finite difference technique used to approximate the nonlinear dynamical equations is the angled derivative method developed by Roberts and Weiss [1976]. This is an explicit method that relies on the sequential updating of the current values over the grid to evaluate derivatives in the advective terms at each middle time step. The method employed for implementing this experiment is described by Flather and Heaps [1975], Flather and Davies [1978], and Choi [1980]. The equations of motion and continuity [equations (6) and (7)] are discretized in time and space using a staggered spatial grid in which the current components are computed at different grid points (Hansen type).

The bottom topography of the model and the discrete water depth values were obtained earlier from various sources [Choi, 1980]; Korean Admiralty Charts #1262 and #2347; and Japanese Hydrographic Charts #182a, #182b, #187, #210 and #302. The open sea boundary of the model is chosen to be bound by a 200 m water depth contour and the entire continental shelf has a typical water depth value of about 80 m.

The grid resolution of the sea model is 0.2 degree in latitude and 0.25 degree in longitude. According to Flather [1972] and Choi [1983], the stability criterion of a linearized version of the scheme in rectangular coordinates is closely approximated by the CFL criterion as follows:

$$\Delta t < \frac{2}{gH}\left(\frac{\Delta x}{2}\right) \qquad (12)$$

where Δx is the grid spacing in Cartesian coordinates and Δt is the time step for the computation.

The above condition gives a minimum time step of 3.75 min. Even though it is obtained from a simplified assumption, it should be treated as a guide for numerical computation.

Input Forcing Functions for the Hydrodynamic Model

Tidal Forcing

There are several expressions for input tidal generating force depending on the type of modelling and the structure of the numerical scheme. For tide and surge simulation, Davies and Flather [1978] used the following open boundary condition for sea surface elevation in their North Sea model which was previously developed by Davies [1976]:

$$\zeta(\phi,x,t) = \zeta_m(\phi,x,t) + \zeta_\tau(\phi,x,t) \qquad (13)$$

where $\zeta_m(\phi,x,t)$ is the surge elevation due to the meteorological influence either observed or computed by simple hydrostatics law at the open boundaries and $\zeta_\tau(\phi,x,t)$ is the part due to tidal motion given by:

$$\zeta_\tau(\phi,x,t) = \zeta_0(\phi,x) + \Sigma_i f_i \tilde{H}_i(\phi,x)\cos(\tilde{V}_i + \tilde{\sigma}_i t + \tilde{U}_i - \tilde{g}_i(\phi,x)) \qquad (14)$$

where

$\zeta_0(\phi,x)$ — mean sea level taking to be zero
$f_i, \tilde{U}_i$ — nodal factors
$\tilde{H}_i(\phi,x)$ — the amplitude of constituent i
$\tilde{\sigma}_i$ — the speed of the constituent
$\tilde{V}_i$ — the phase corresponding equilibrium constituent at time = 0 at Greenwich
$\tilde{g}_i(\phi,x)$ — the phase lag of the tidal constituent behind the equilibrium constituent.

Alternatively, the tidal current and surge current can be specified at the open boundaries through the radiation condition used by Flather [1976]:

$$q = q_m + q_\tau + \sqrt{\frac{g}{H}}(\zeta - \zeta_m - \zeta_\tau) \qquad (15)$$

where

q_τ — ith constituent tidal current at the open boundary
ζ_τ — ith constituent tidal amplitude at the open boundary
q_m — surge current at the open boundary
ζ_m — surge amplitude at the open boundary.

The tidal part of the normal current q_τ is determined from the following:

$$q_\tau = \Sigma_i \tilde{f}_i \tilde{Q}_i \cos(\tilde{\sigma}_i t + \tilde{V}_i + \tilde{U}_i - \tilde{\gamma}_i) \qquad (16)$$

where

$\tilde{Q}_i$ — amplitude of the normal component of the depth-averaged current of constituent i
$\tilde{\gamma}_i$ — phase of the current of tidal constituent i.

Equations (13) to (16) supply the tidal forcings to the sea model through the open boundary. Slightly different expressions can be derived from the above conditions in a different type of simulation. For example, Flather [1979] used equations (15) and (16) with only M and S tidal constituents in storm surge modelling whereas Heaps and Jones [1979] and Davies [1976] used exactly the same constituents but employed equations (13) and (14) instead.

Meteorological Forcing

The major difficulty of predicting the surface wind speed of a typhoon using the approach of surface wind analysis [Moon and Tang, 1985; Tang and Moon, 1984; Hsueh and Romea, 1983; Heaps, 1983] has been the lack of an accurate method of converting geostrophic to surface wind speed. From the examination of weather charts obtained from the Korean and Japanese meteorological offices, it is found that not only the weather systems move with an unpredictable speed and direction, but that the pressure gradient near the center of the typhoon may also give unrealistically high geostrophic wind speed. To correct the predicted wind speed to typical values of typhoon, a number of numerical experiments were carried out in this study for the East China Sea and Yellow Sea area.

Several formulae for converting geostrophic approximation to surface wind are reviewed and plotted in Figure 4. With those proposed by Hasse and Wagner [1971], four more geostrophic to surface wind relations were tested; the first two of these were modified from Hasse and Wagner [1971], while the last two were derived by Hsueh and Romea [1983] during the wintertime experiment (December 1, 1980 through March 31, 1981) over the East China Sea. The proposals by Hsueh and Romea [1983] were obtained by comparing the geostrophically approximated wind speed from weather charts (supplied by the Japanese Meteorological Agency) with the observations made around the Korean and Japanese coastal stations. Based on a practical standpoint and the evaluations of testing results, the following geostrophic wind speed ($|W_g|$) to surface wind speed ($|W_s|$) conversion equation is adopted for the East China Sea and Yellow Sea area:

$$|\overline{W_s}| = 0.443 |\overline{W_g}| + 2.92 \text{ m/s} \quad (17)$$

The back angle is assumed to be 20 degrees. The adoption of equation (17) is aimed to reduce the magnitude of surface wind and, in addition, the inequality below is added to ensure a realistic range of deduced typhoon wind speed:

$$|\overline{W_s}| > 30 \text{m/s}, |\overline{W_s}| = (|\overline{W_s}| \pm 20 \text{m/s} * 0.1656 \pm 20 \text{m/s} \quad (18)$$

where $|\overline{W_s}|$ is the magnitude of sea surface windfield vector derived by the model. The inequality gives a reasonably acceptable pattern of the two dimensional windfield since it suppresses unrealistically high wind speeds near the storm center while it retains nearly normal wind speeds away from it.

To apply the method of extracting wind speed for the East China Sea and Yellow Sea model, an 18×23 atmospheric model grid is set up over the region with a resolution of 1° by 1°. Figure 5 shows some of the grid points in and near the study area. A smaller resolution of 0.5 degree (longitude) by 0.4 degree (latitude) was also examined; however, the computation time increases significantly with only a slight improvement in the results.

A total of 13 days of weather (2 periods) is considered over the East China Sea and Yellow Sea area during the SEASAT mission. Each period is characterized by a strong tropical weather disturbance (typhoon) with a duration of almost 6 days. Figure 6 depicts the tracks of the typhoons with Wendy (July 28-Aug 2) as a dashed line and Carmen (Aug 15-Aug 21) as a solid line. They enter the sea model at the south-east corner and become stationary over East China Sea region for about 4 days before they move away through the north-east corner of the model (Korea). Since the

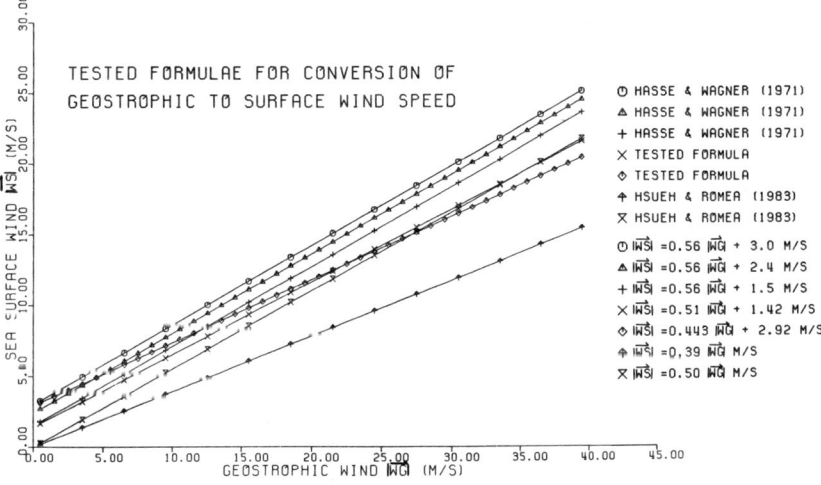

Fig. 4. Geostrophic wind speed ($|W_g|$) to surface wind speed ($|W_s|$) conversion formulae used in the test runs of the East China Sea and Yellow Sea atmospheric model.

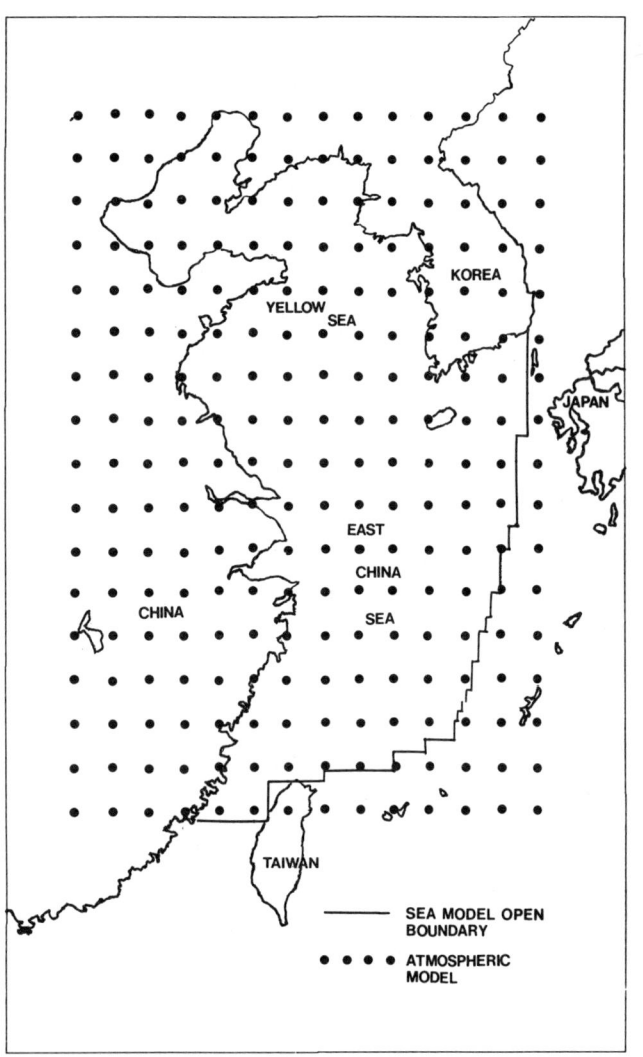

Fig. 5. Some geographical locations of the atmospheric model's grid points. The sea model boundary is shown as solid line segments.

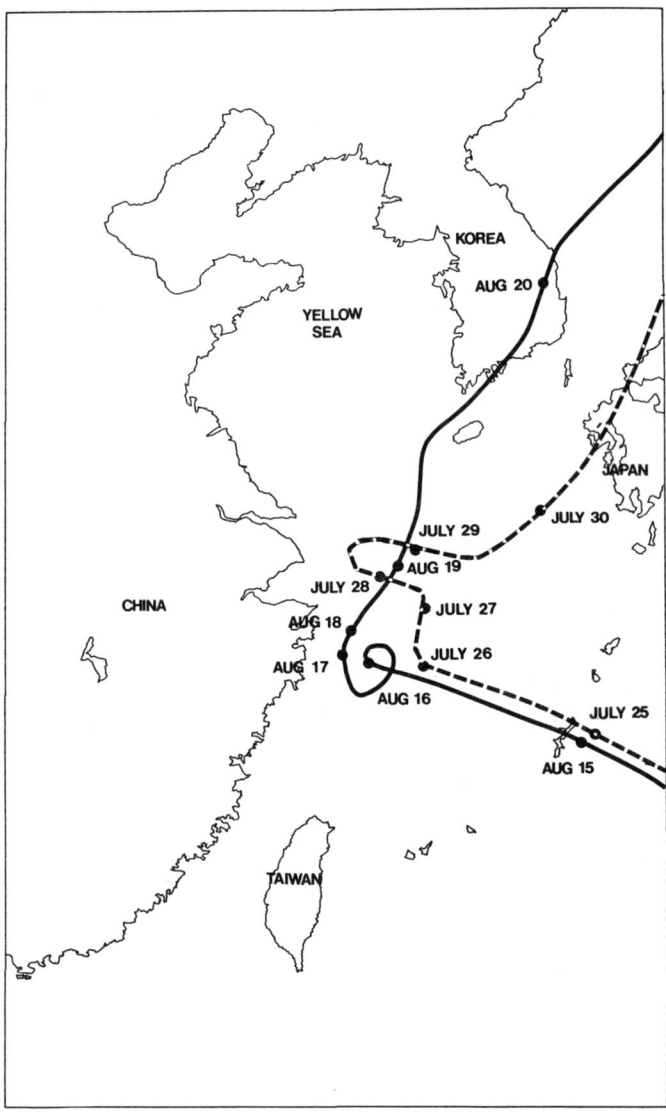

Fig. 6. Traces of the two tropical storms (typhoon): Wendy (dashed line) and Carmen (solid line), during the SEASAT mission time over this area.

weather information is collected every 6 hours, it may not be continuous enough to describe the changing weather pattern. Experiences from storm surge modelling in the North Sea and Irish Sea areas (from the research at the Institute of Oceanographic Sciences (IOS) at Bidston, UK) show that at least 3-hour intervals of weather input are preferred to produce realistic surge phenomena. Thus interpolation of weather charts in time and space is performed for both periods along the disturbance tracks. This procedure assumes that the typhoons move linearly between two points at which the weather information was made available. A windfield is then recreated evenly along the line defined by these positions at hourly intervals.

Wind stress that inputs into the sea model is computed by the standard quadratic law

$$\overline{\tau}_s = \rho_a C'(|\overline{W}_s|) \overline{W}_s |\overline{W}_s| \qquad (19)$$

where ρ_a is the density of air. Various representations of $C'(|W_s|)$ (Figure 7) were numerically examined before the adoption of Wu's proposal [Wu, 1980, 1982]. Previous investigation on the sea surface response caused by wind stress over the Hudson Bay area by Moon and Tang [1985] has also concluded the practical usefulness of Wu's equation during a strong wind environment. This equation is given by Wu [1980, 1982]:

$$C'(|\overline{W}_s|) = 0.8 + 0.065 |\overline{W}_s| \qquad (20)$$

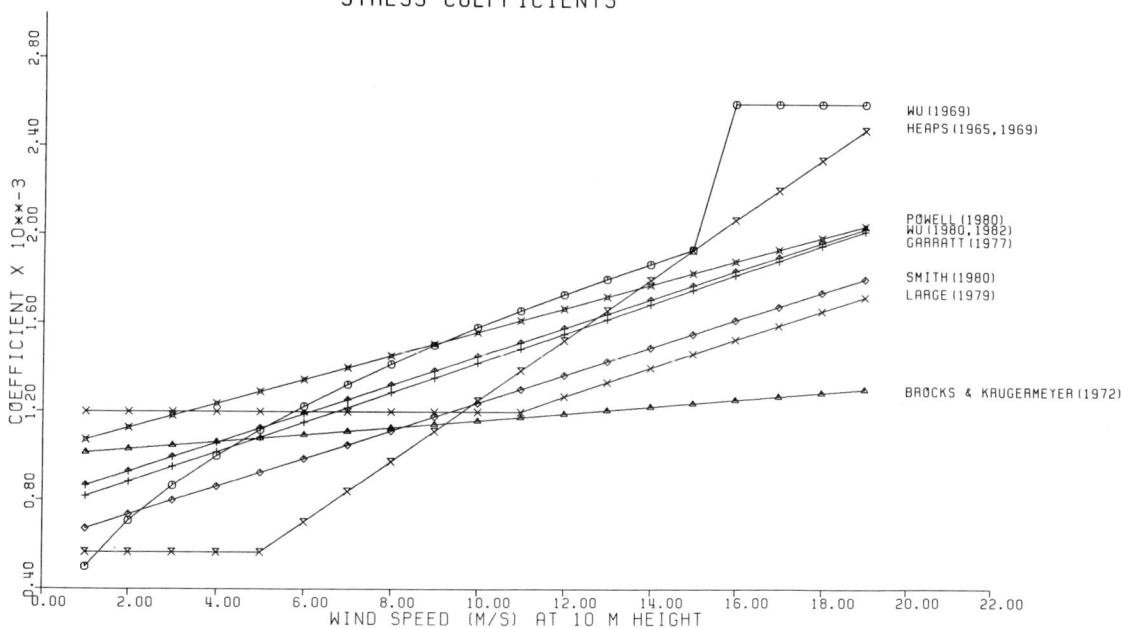

Fig. 7. Plot of surface wind speed to wind stress conversion formulae tested for the sea model.

It is interesting to note that the Powell's relation of wind stress coefficient [Powell, 1980] obtained from hurricane data is quite consistent with Wu's formula at high wind speeds (Figure 7). Figures 8 and 9 show some sample wind stress patterns calculated from the atmospheric model. Isobars are overlain in each diagram to show the direction of movement.

Since tidal phenomena have been known to be important in this area, ocean tide was calculated separately and also inclusively with surge analysis. Surges in this case have special importance because they were produced by typhoons. Figures 10a through 10j show the relative magnitudes of surge, ocean tide, and body tide along the satellite orbit tracks. Ocean tide was generated by 4 major tidal constituents (M_2, S_2, O_1 and K_1) and they have been shown to be sufficient for the reproduction of tide in this region [Choi, 1980]. In general, the variations of ocean tide along the tracks is about 1.75 m with the largest value occurring (in revolution 794) at about 3 m. Surge amplitude was smaller according to computation, with a typical value being about 0.5 m. The effect of computing surge and tide together may change the shape of these profiles due to the surge and tide interaction. Figure 11 illustrates examples of the surge plus tide simulation along four satellite orbit tracks using a value of 0.0025 as the quadratic friction coefficient. Also plotted in this diagram are the DC corrected residual sea surface elevation profiles (dashed lines) obtained by subtracting the SEASAT observations from the SS3 altimetric-derived reference sea surface. The amount of DC shifting for each orbit track is computed by the difference between the satellite observed value and the computed value at the mid-point of the profile. This point corresponds to the time when the simulated profile was produced. Comparing the surge and tide simulations with the results shown in Figure 10a through 10j, suggests that the sea surface topography (surge + tide) along the tracks is associated with the feature which has the largest relative magnitude, that is the ocean tide. The surge profiles modify the profiles of ocean tide slightly by adding to them with a small residual, probably caused by some interaction. It is also obvious from Figure 10a through 10j that low frequency, small magnitude body tides along the satellite tracks cause negligible vertical shifts of the profiles when they are subtracted for correction, and thus they can be safely ignored.

Quadratic Friction Coefficient of the Ocean Bottom

In general, the expression for the bottom friction law must satisfy the following criteria: (1) It should give a reasonable magnitude of bottom stress in shallow water (less than 1 m), since both the linear and quadratic laws approach infinity nearly exponentially as the depth approaches zero.

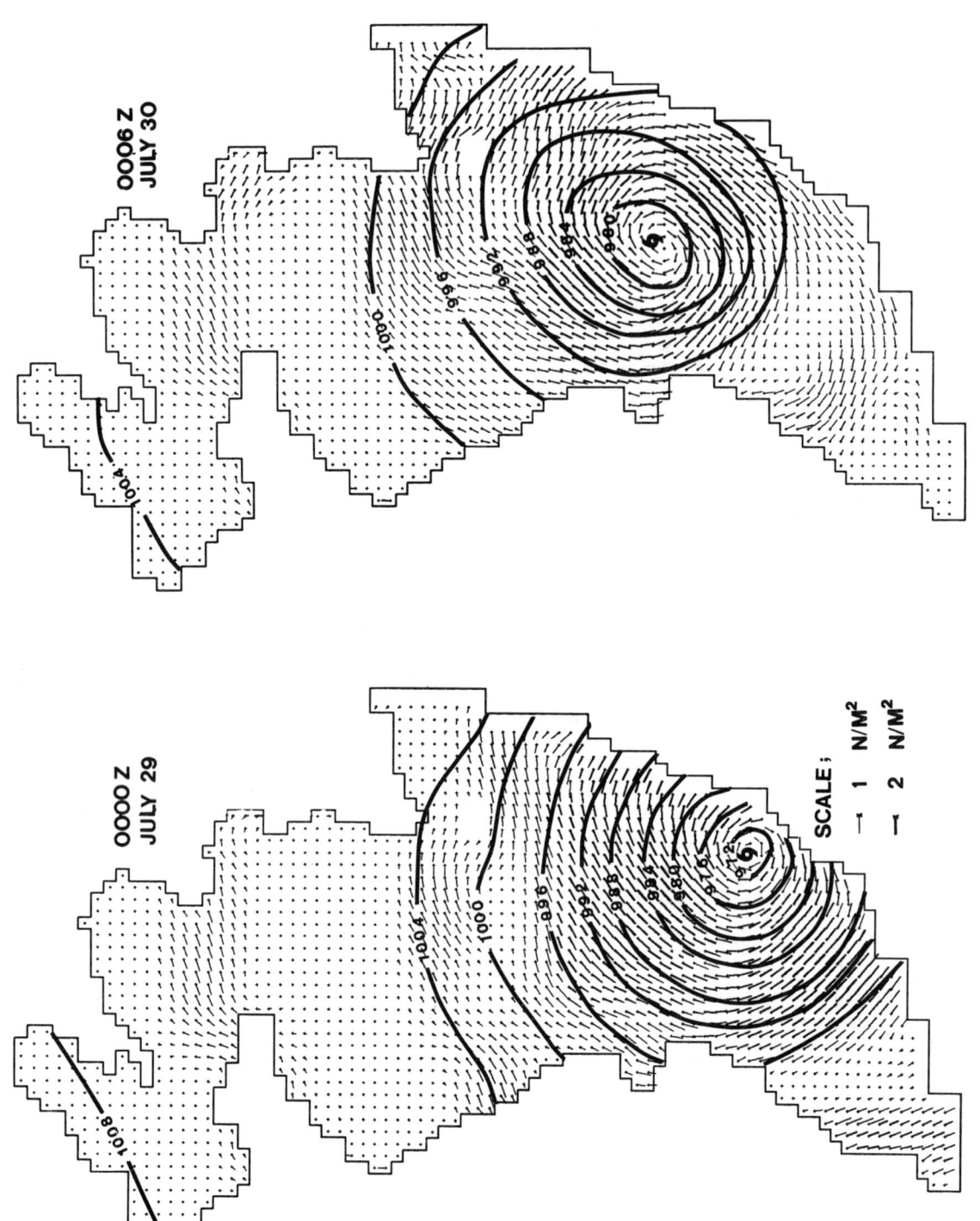

Fig. 8a. July 29, 0000Z.   Fig. 8b. July 30, 0006Z.

Fig. 8. Samples wind stress patterns as the results of typhoon Wendy over the East China Sea and Yellow Sea.

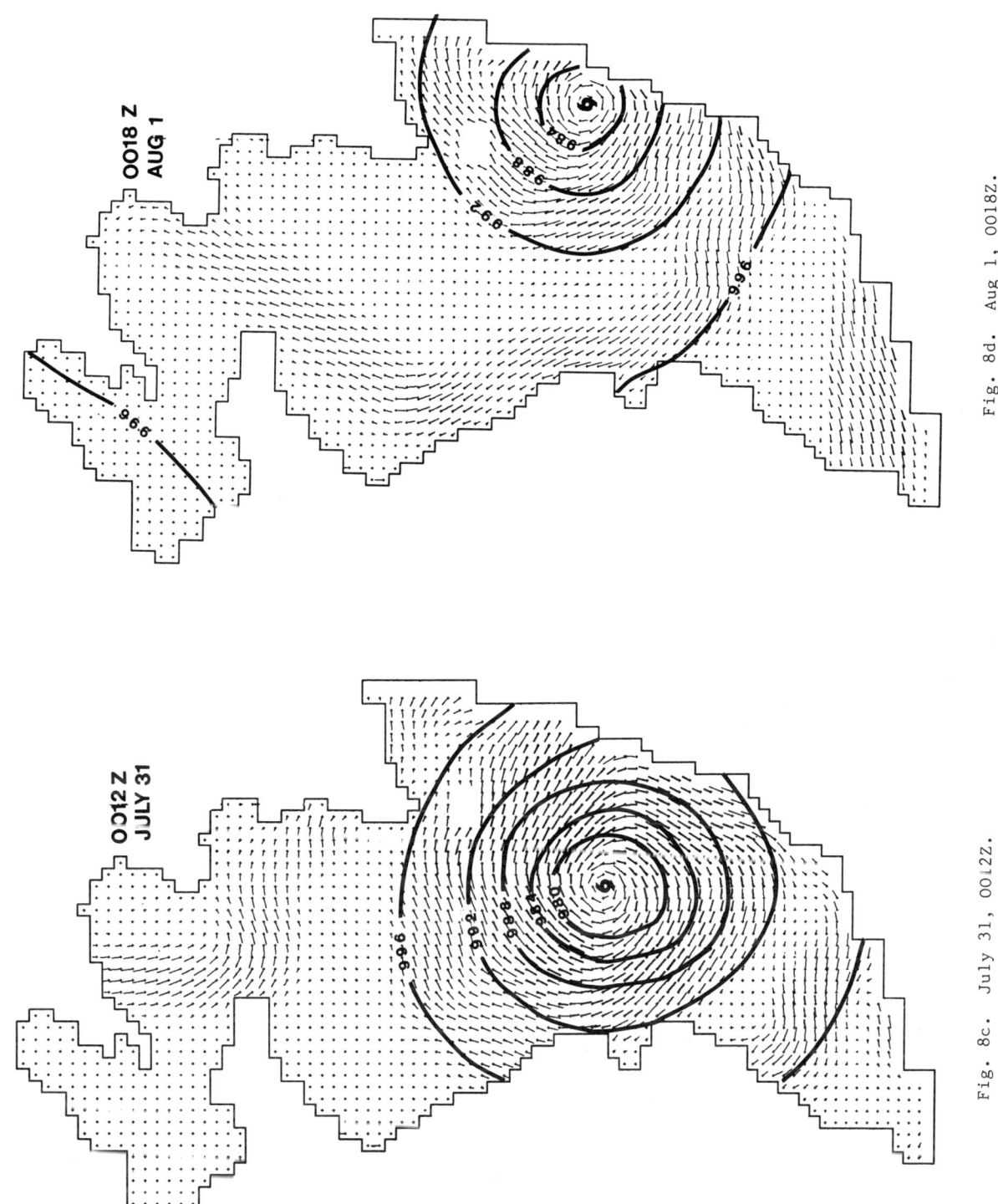

Fig. 8c. July 31, 0012Z.

Fig. 8d. Aug 1, 0018Z.

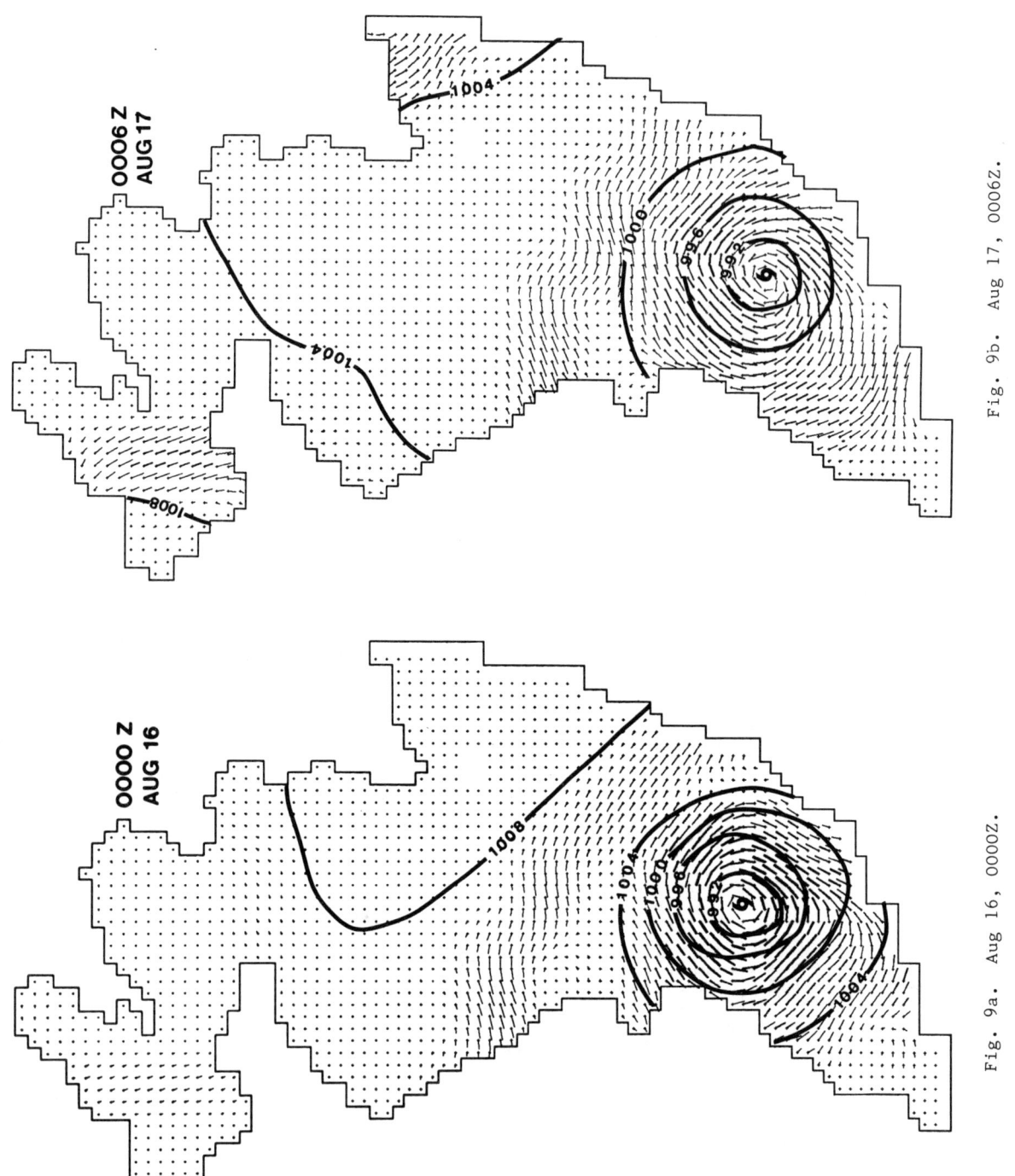

Fig. 9a. Aug 16, 0000Z.

Fig. 9b. Aug 17, 0006Z.

Fig. 9. Samples wind stress patterns as the results of typhoon Carmen over the East China Sea and Yellow Sea.

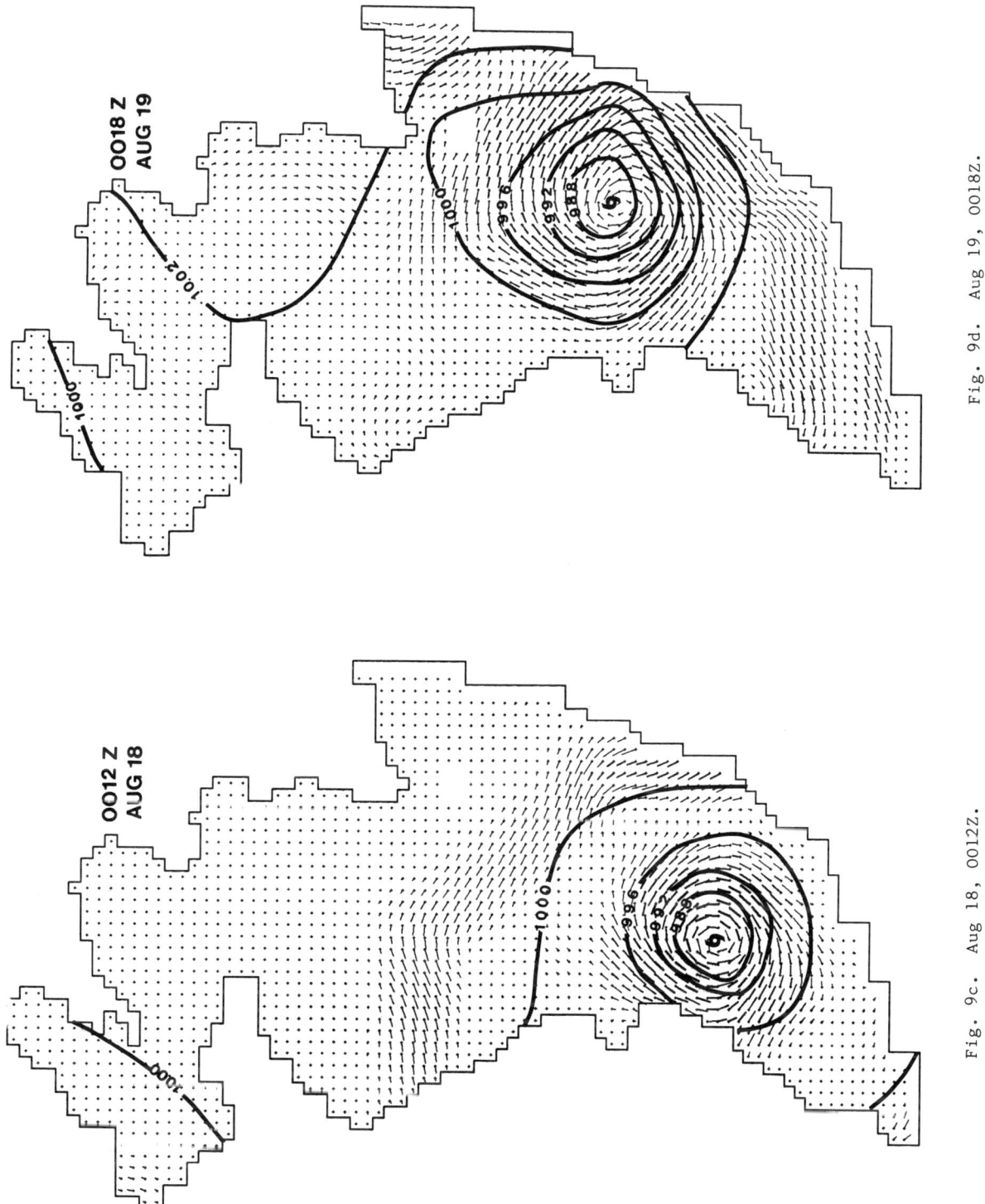

Fig. 9c. Aug 18, 0012Z.

Fig. 9d. Aug 19, 0018Z.

98 STUDY OF FLUID-SOLID EARTH COUPLING PROCESS

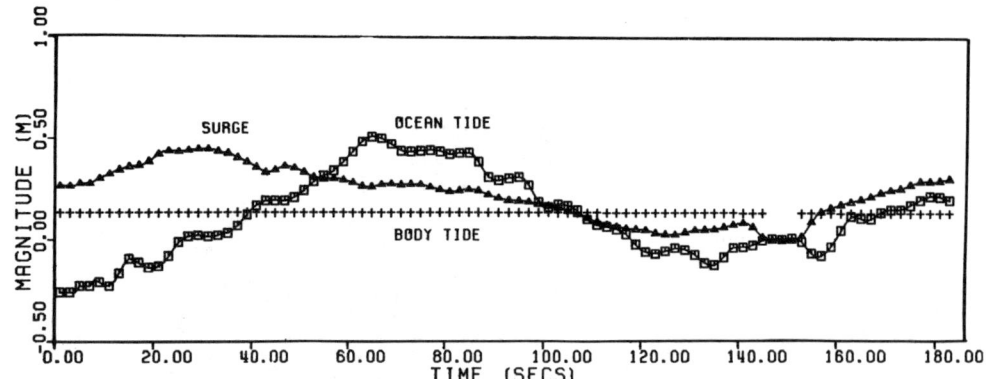

Fig. 10a. Revolution 456.

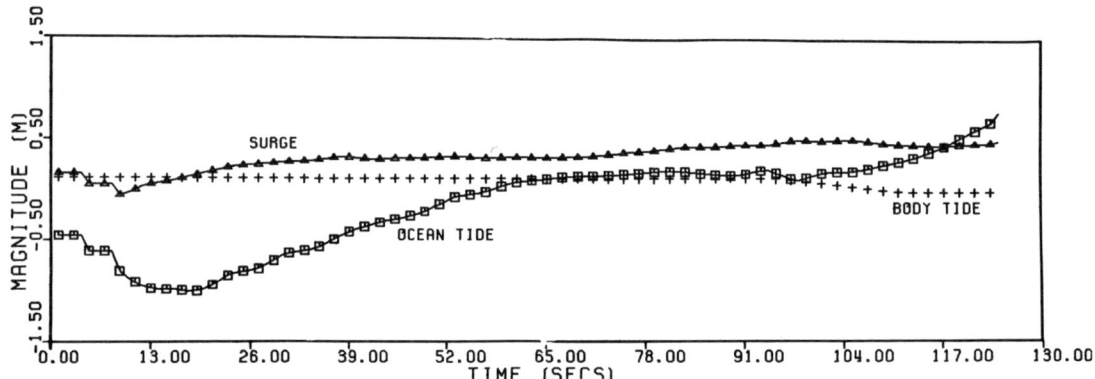

Fig. 10b. Revolution 464.

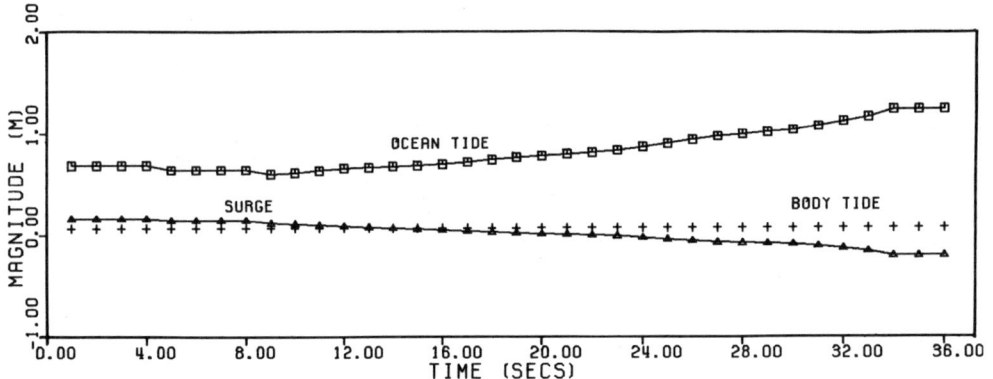

Fig. 10c. Revolution 485.

Fig. 10. Magnitude of ocean tide, storm surge and body tide along the satelite profiles shown in

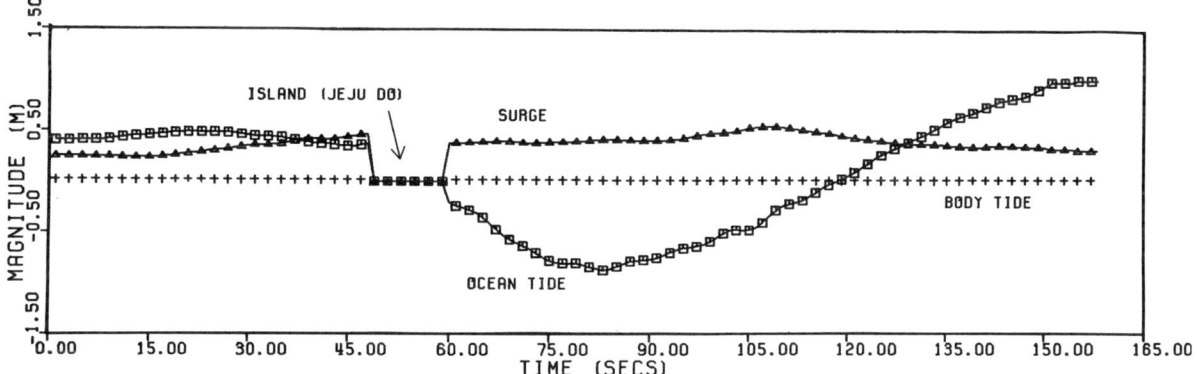

Fig. 10d. Revolution 499.

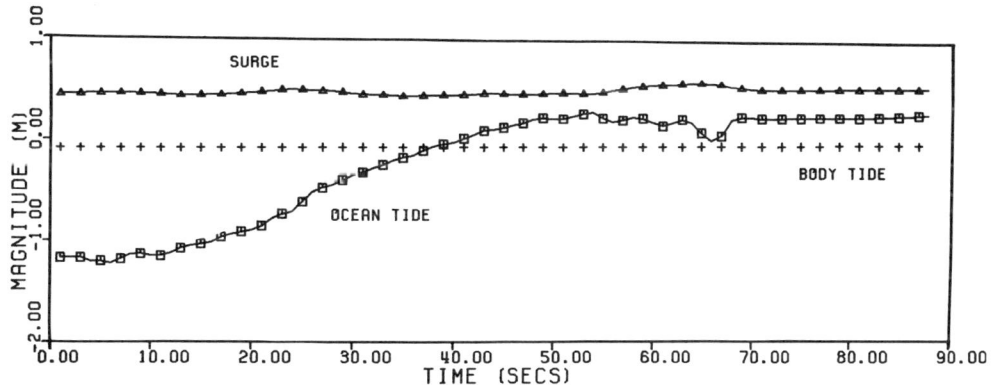

Fig. 10e. Revolution 507.

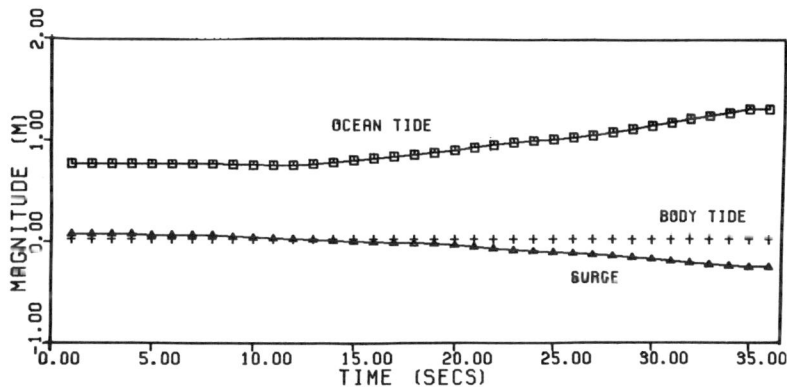

Fig. 10f. Revolution 729.

STUDY OF FLUID-SOLID EARTH COUPLING PROCESS

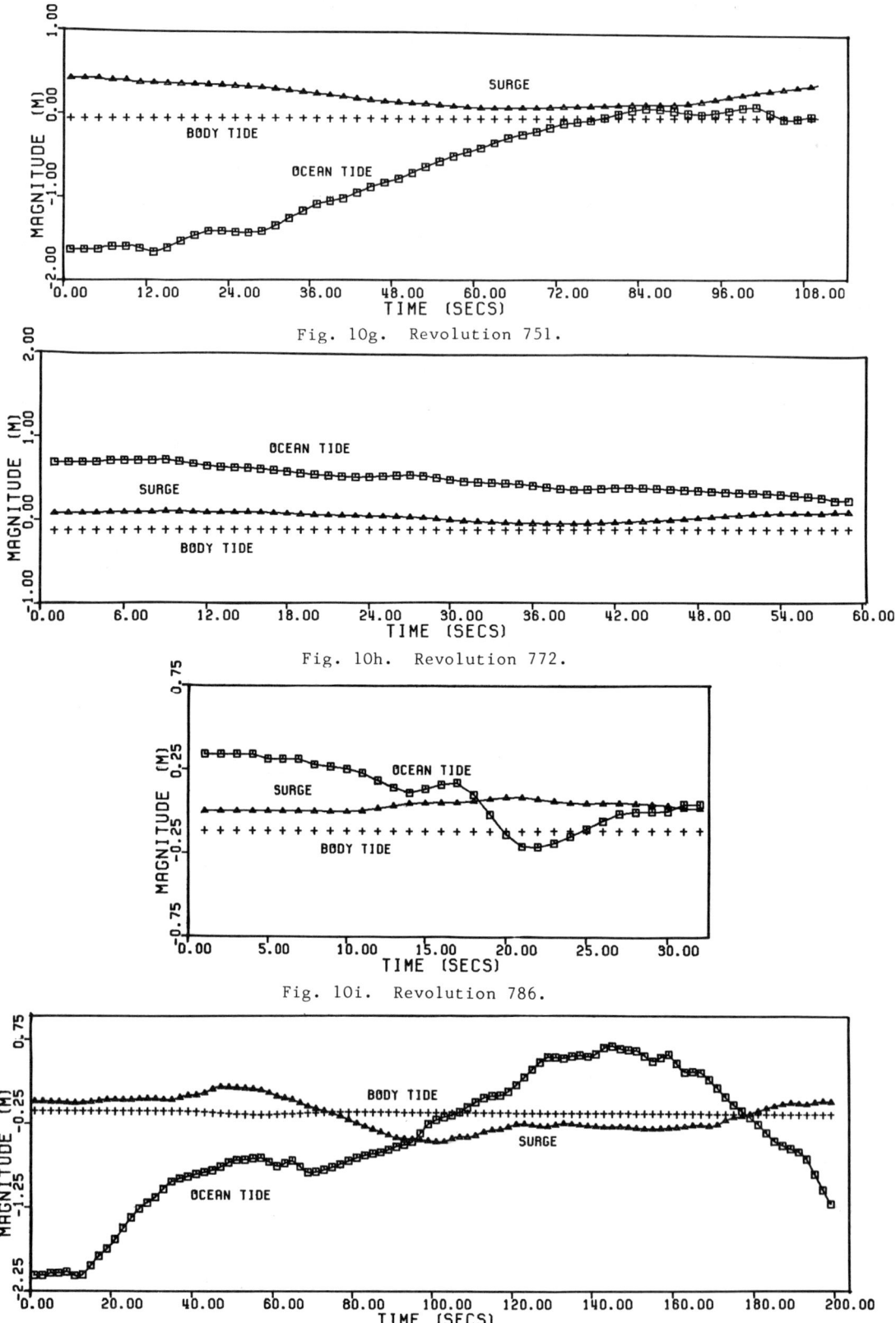

Fig. 10g. Revolution 751.

Fig. 10h. Revolution 772.

Fig. 10i. Revolution 786.

Fig. 10j. Revolution 794.

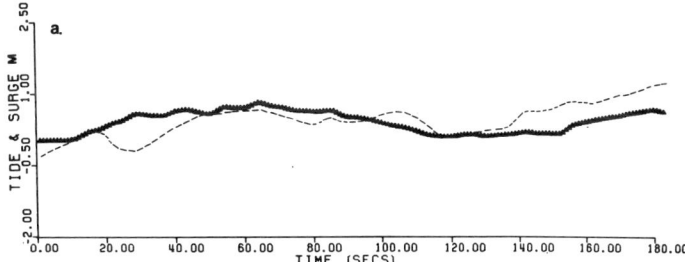

Fig. 11a. Revolution 456.

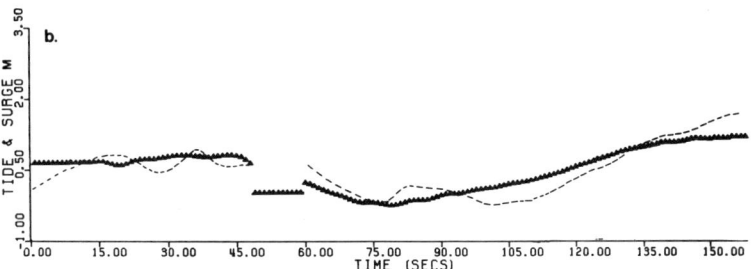

Fig. 11b. Revolution 499.

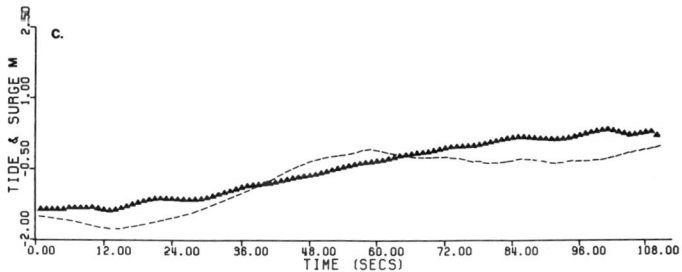

Fig. 11c. Revolution 751

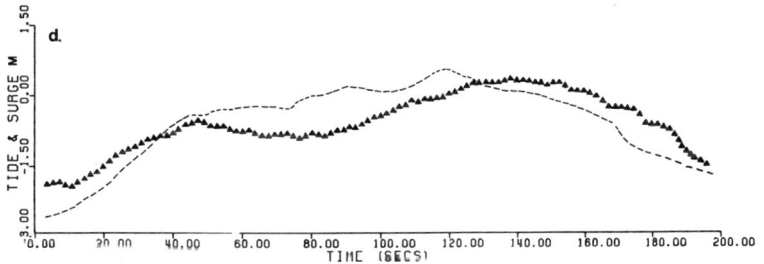

Fig. 11d. Revolution 794.

Fig. 11. Diagram showing the results of surge and tide simulations (solid triangles) taking into account their interactions. Dashed lines are the residual DC shifted sea surface elevation profiles obtained from the SEASAT-ALT data after removing from them the SS3 reference sea surface.

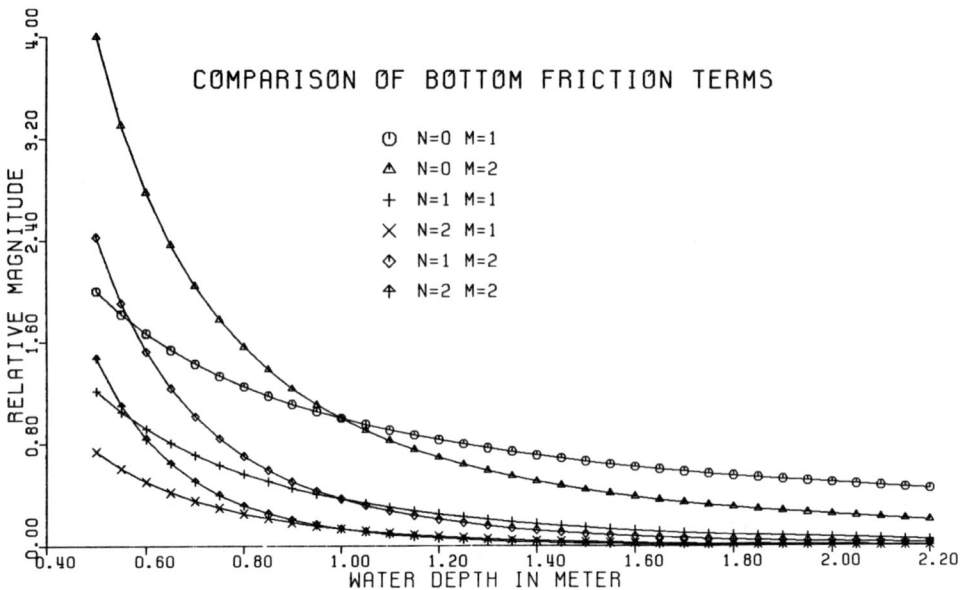

Fig. 12. Graphic representation of the relative magnitude of Equation (21) as a function of water depth.

(2) It must be computationally simple to use. A complex expression may be physically unrealistic in practice. Furthermore, a suitable representation should somehow vary with water depth, an important control parameter.
(3) It should be compatible with the conventional bottom friction laws (i.e., the linear or quadratic friction laws, as the water depth increases).

Given these conditions, the following generalization for the ocean bottom friction term can be used [Kinsman, 1965; Ramming, 1978]:

$$\tilde{\tau}_B = a_o \frac{e^{-NH} \rho}{H^M} \vec{V}^M \qquad (21)$$

where M = 1 or 2 and N = 0, 1, 2, ... The parameter M denotes the selection of frictional law with $\vec{V}$ corresponding to the magnitude of depth-averaged current. The parameter N denotes the selection of the damping effect in shallow water and a_o is a constant with or without dimension depending upon the selections. For N = 0, the generalized bottom friction law of equation (21) reduces to the familiar linear and quadratic friction laws for M = 1 and 2, respectively [equations (8) and (9)]. In these cases, a_o becomes the linear stress coefficient and quadratic stress coefficient, accordingly. To see the behavior of the generalized law in shallow water, equation (21) is plotted for a few combinations of N and M against the water depth H, assuming both the current and the constant to be unity (Figure 12). From this diagram, it is evident that the functions 1/H (linear friction) and 1/H**H (quadratic friction) which are associated with depth-averaged current are not suitable for shallow water (e.g. from 0.6 m, to 1.2 m). However, the magnitudes predicted by various values of N and M uniformly converge to a consistent value as the water depth increases. In general, the addition of the inverse exponential function exp(-NH) in equation (21) reduces the magnitude of the bottom stress when the water depth approaches zero. If the water depth field in the modelling area is restricted similarly to the number given in the figure, equation (21) with N= 1 or 2 would be adequate to minimize the instability of the hydrodynamic solution in shallow water. Further practical experiments have to be carried out to examine other treatments of the bottom friction defined by equation (21). This will require proper insitu measurements in coastal water areas. Further complications arise if the coefficient a_o in equation (21) is no longer simply a constant. From the experiences with the sea surface wind stress coefficient, one may assume that a_o may vary with current (for example, a linear proportionality of a_o to current speed). This is itself a very complicated matter to be studied with our present knowledge, although the possibility does exist for further experiment. The proposal of equation (21), however, overcomes the problem of shallow water and converges to usual friction law when water depth increases. Table 1 summarizes some of the earlier

TABLE 1. Values of Quadratic Friction Coefficients for Bottom Friction Process.

Author	Frictional Coefficient	Area
Grace [1930]	0.003	Gulf of Suez
Bowden & Fairbairn [1952]	0.0036	Red Wharf Bay, Anglesey
Bowden & Fairbairn [1956]	0.0024	Red Wharf Bay, Anglesey
Bowden et al. [1959]	0.0035	Red Wharf Bay, Anglesey
Charnock [1959]	0.0034	Red Wharf Bay, Anglesey
Sternberg [1968]	0.0031	Puget Sound, Washington
Hearthershaw [1976]	0.0015-0.0019	Irish Sea
Stock [1976]	0.0015	Gulf of California
Wolf [1980]	0.0014-0.0271	Northern Irish Sea
Weatherly & Wimbus [1980]	0.0056	Blake-Bahama Outer Ridge
Chriss & Caldwell [1982]	0.0032-0.008	Oregon Shelf
Bowden & Ferguson [1980]	0.0040-0.0047	Eastern Irish Sea
Moon and Tang [1987]	0.0019-0.0046	Hudson Bay, Canada
Moon et al. [1988]	0.0023-0.0027	East China and Yellow Seas (this study)

as well as recent works done on the estimation of the quadratic friction coefficients (N=0, M=2).

Most of these values are derived from the tidally-induced current observations through a combination of the quadratic stress law with one of the other methods, such as the velocity profile technique or the Reynold stress method. These values are largely confined by the number of measurements, so they can not be regarded as reliable representations. From Table 1 it is clear that the consistency of the quadratic friction coefficient is not satisfactory, although the values obtained from the Red Wharf Bay indicate a small deviation and fall in a range near the traditional value of 0.003. Our confidence with regard to the derived value can only be improved by increasing the number of observations and by extending the experiment to different areas with a consistent approach.

Only the quadratic friction term is tested in this area simply because the nonlinearity of the term is consistent with the sea model formulation (advection) and other phenomena such as the influence of nonlinear interactions among the tidal constituents. With the quadratic friction dissipation implemented in the equation of motion, the SEASAT-ALT data and the model results during the selected time periods are subjected for analysis through a function defined by Tang [1984]:

$$\text{Var}_i(C) = \frac{1}{N} \sum_{j=1}^{N} | \zeta_c^j(C) - \zeta_{oB}^j(\hat{C}) | \quad (22)$$

where
$\text{Var}_i(C)$ - variance function for the ith orbit track
C - experimental quadratic friction coefficient
$\hat{C}$ - optimum value of C when $\text{Var}_i(C)$ is minimized
N - number of recording points along the orbit track
ζ_c^j - computed sea surface elevation profile of ith orbit track
ζ_{oB}^j - observed sea surface elevation profile of ith orbit track.

The optimum frictional coefficient C in equation (22) is assumed to be the truth value that gives rise to the satellite recordings. Ideally, the optimization procedure can be carried out using equation (5) which links the satellite data collecting geometry with the sea model coordinates. It is, however, that the long wavelength unmodelled error in orbit computation due to insufficient accuracy in the gravity model and random noise are excluded. Since the unmodelled error is in general long wavelength of semi-cyclic per revolution [Marsh and Williamson, 1980], it would only have an effect on local 'DC' shifting in the SEASAT-ALT data segments such as those used in this research. In fact, 'DC' shifting can also be true for other uncorrected long wavelength small amplitude anomalies of tidal loading of the solid Earth, body tide, and steric ocean variations.

The variance curves ($\text{Var}_i(C)$) for the correlation between the model and the altimeter derived observations for each revolution are displayed in Figure 13a through 13j. The results of the optimization are very consistent with each other within a range of values from 0.0023 to 0.0027. An averaged value is obtained to be 0.002505 for the friction coefficient of the quadratic law.

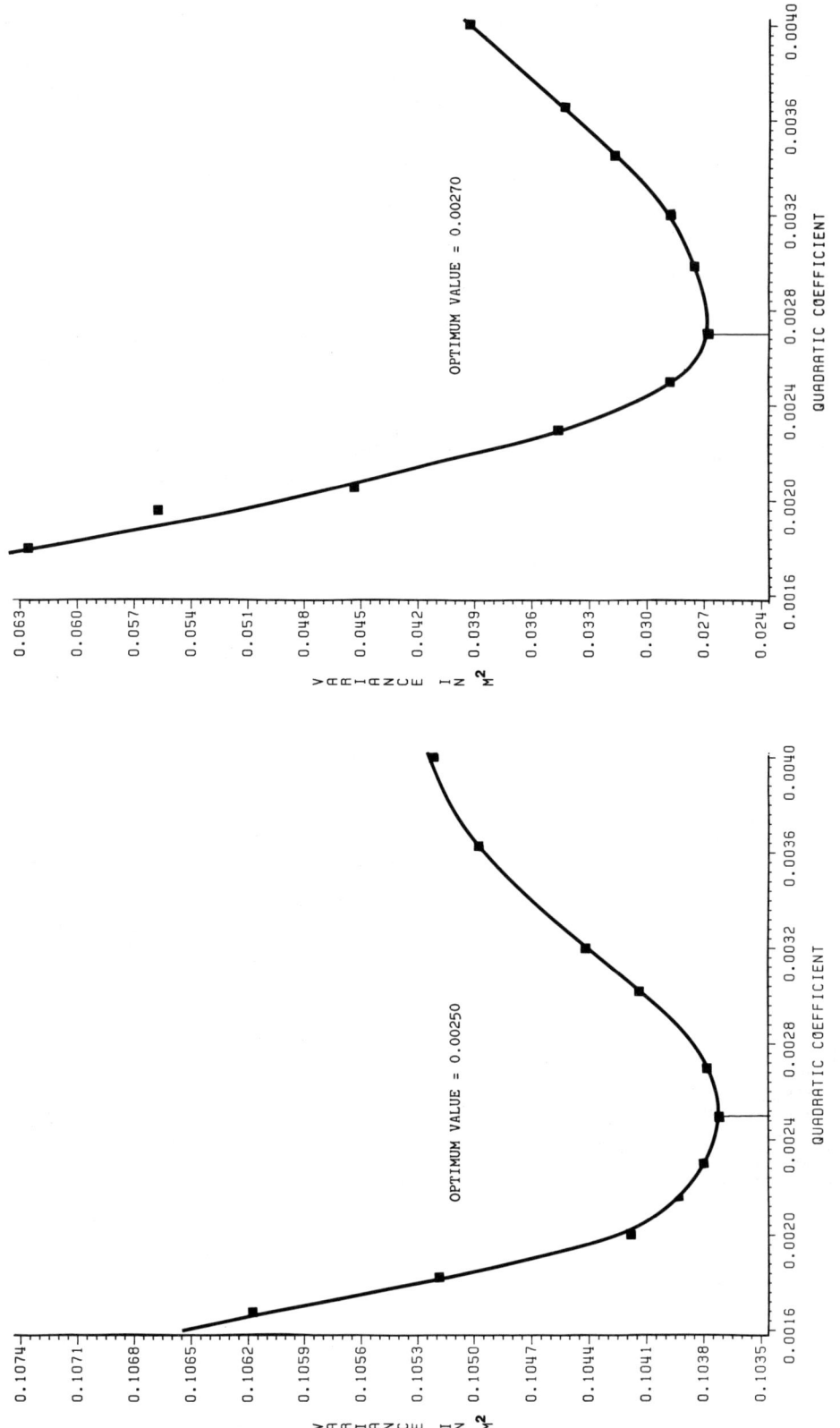

Fig. 13a. Revolution 456.

Fig. 13b. Revolution 464.

Fig. 13. East China Sea and Yellow Sea Experiment Variance Curves calculated from the observed and simulated sea surface profiles as a function of quadratic friction coefficient.

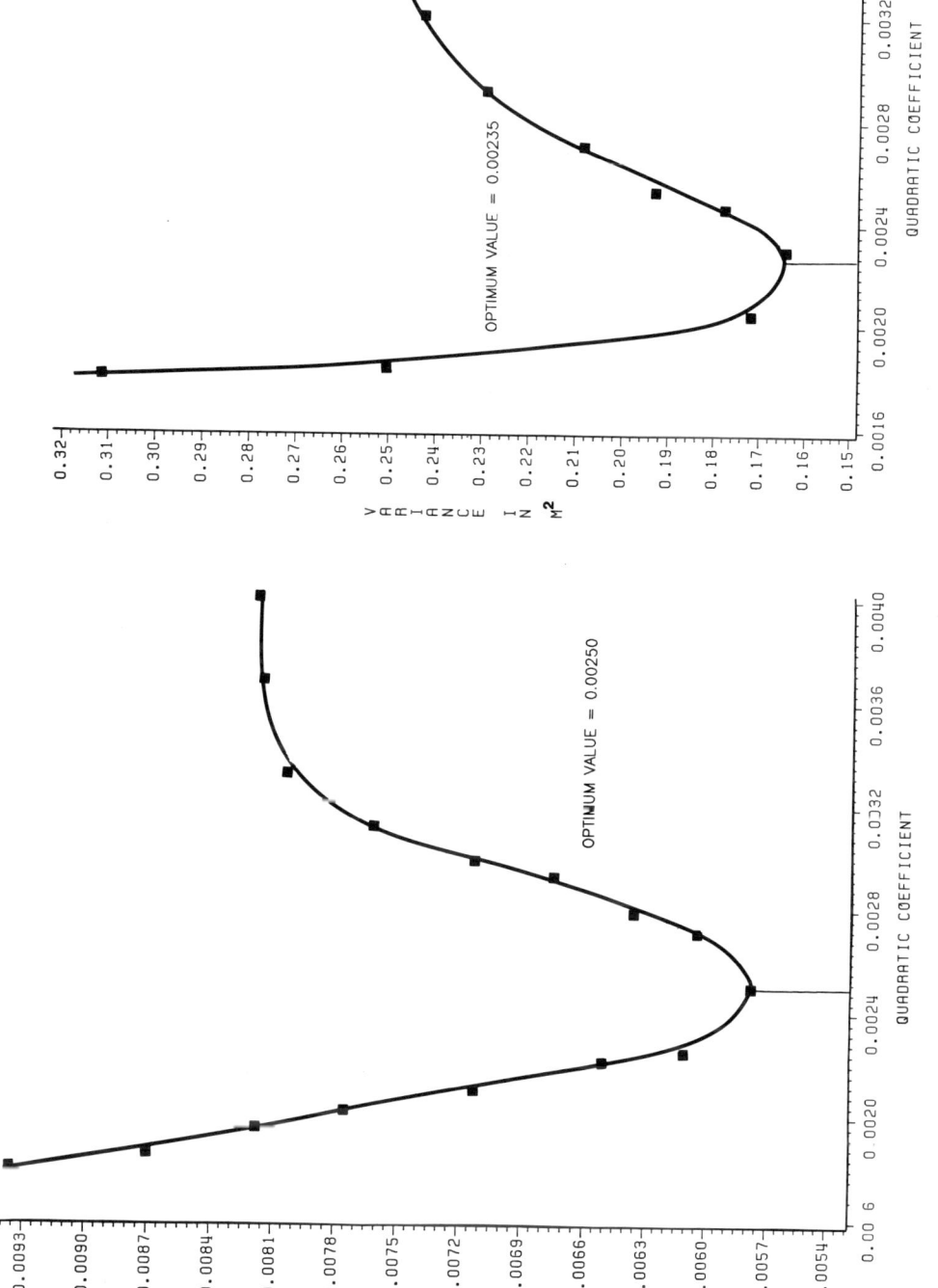

Fig. 13c. Revolution 485.

Fig. 13d. Revolution 499.

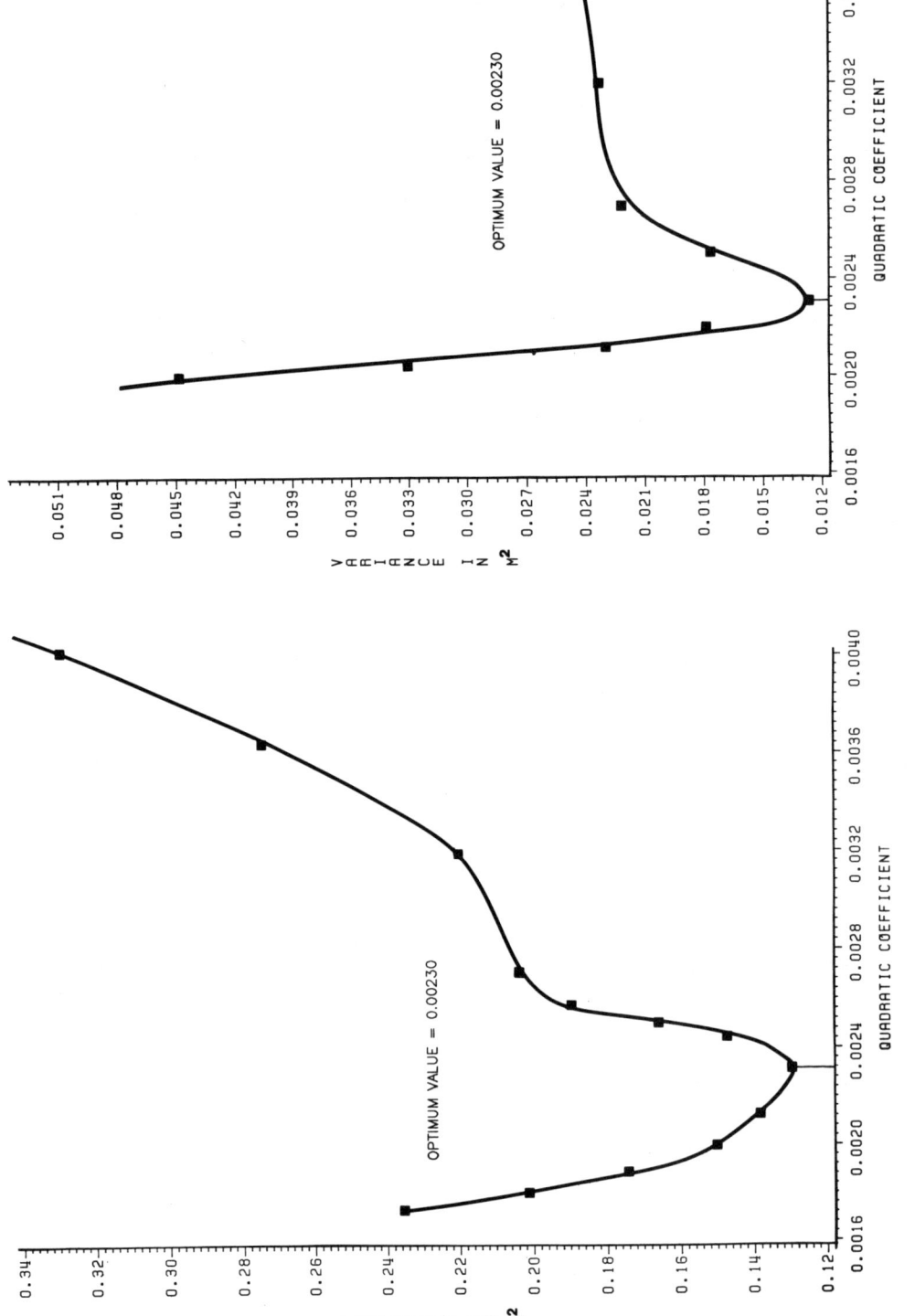

Fig. 13e. Revolution 507.

Fig. 13f. Revolution 729.

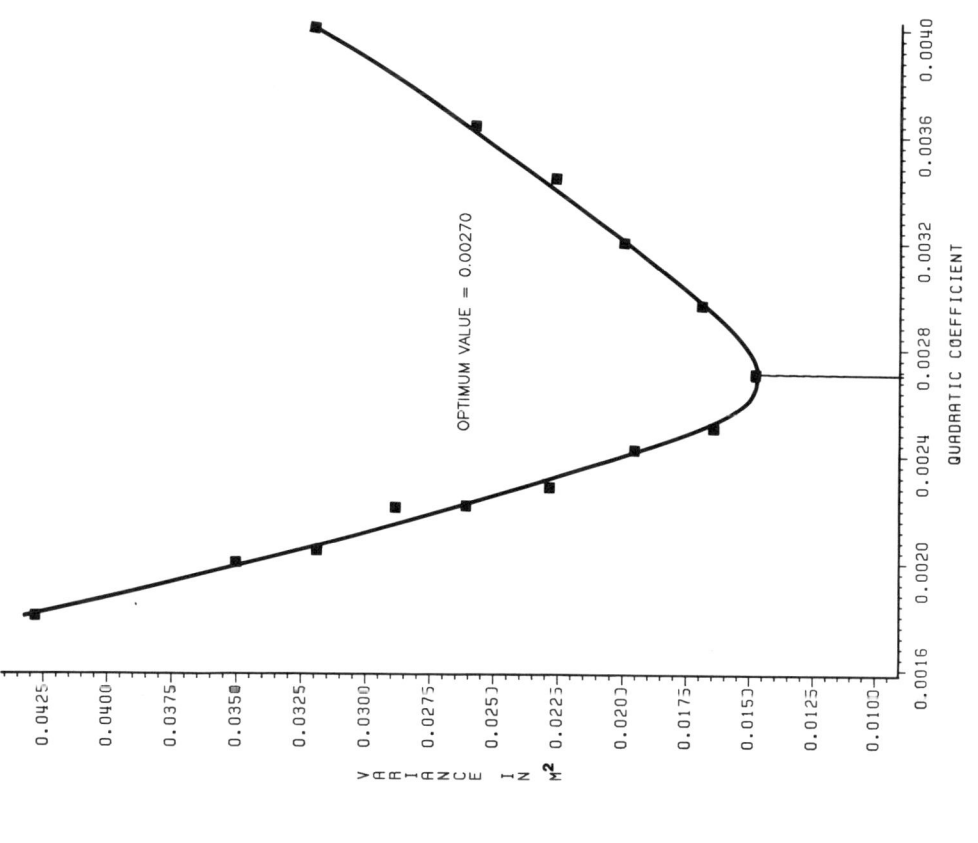

Fig. 13h. Revolution 772.

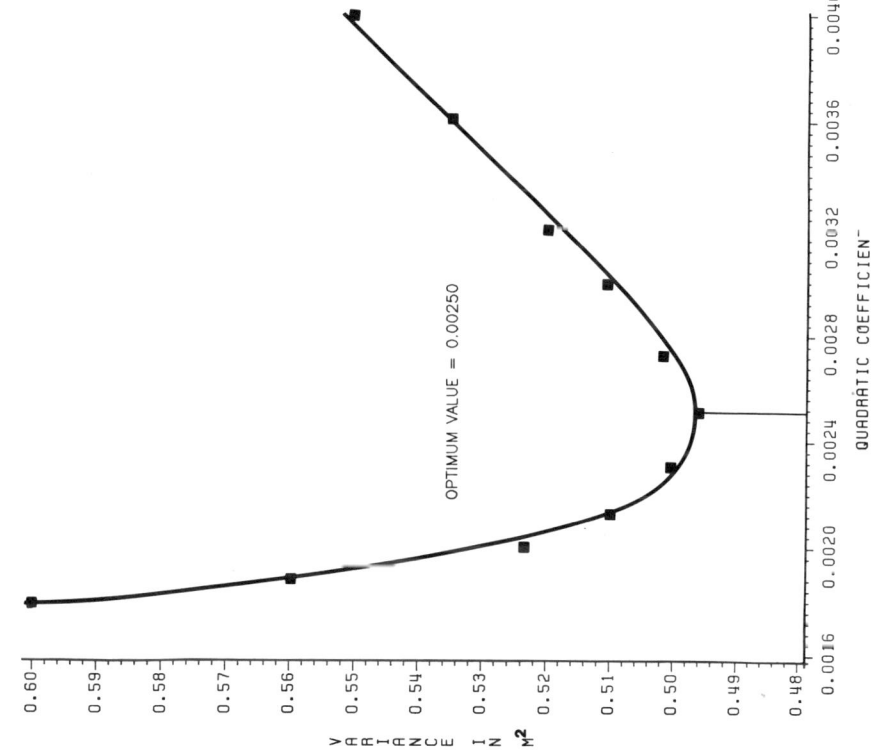

Fig. 13g. Revolution 751.

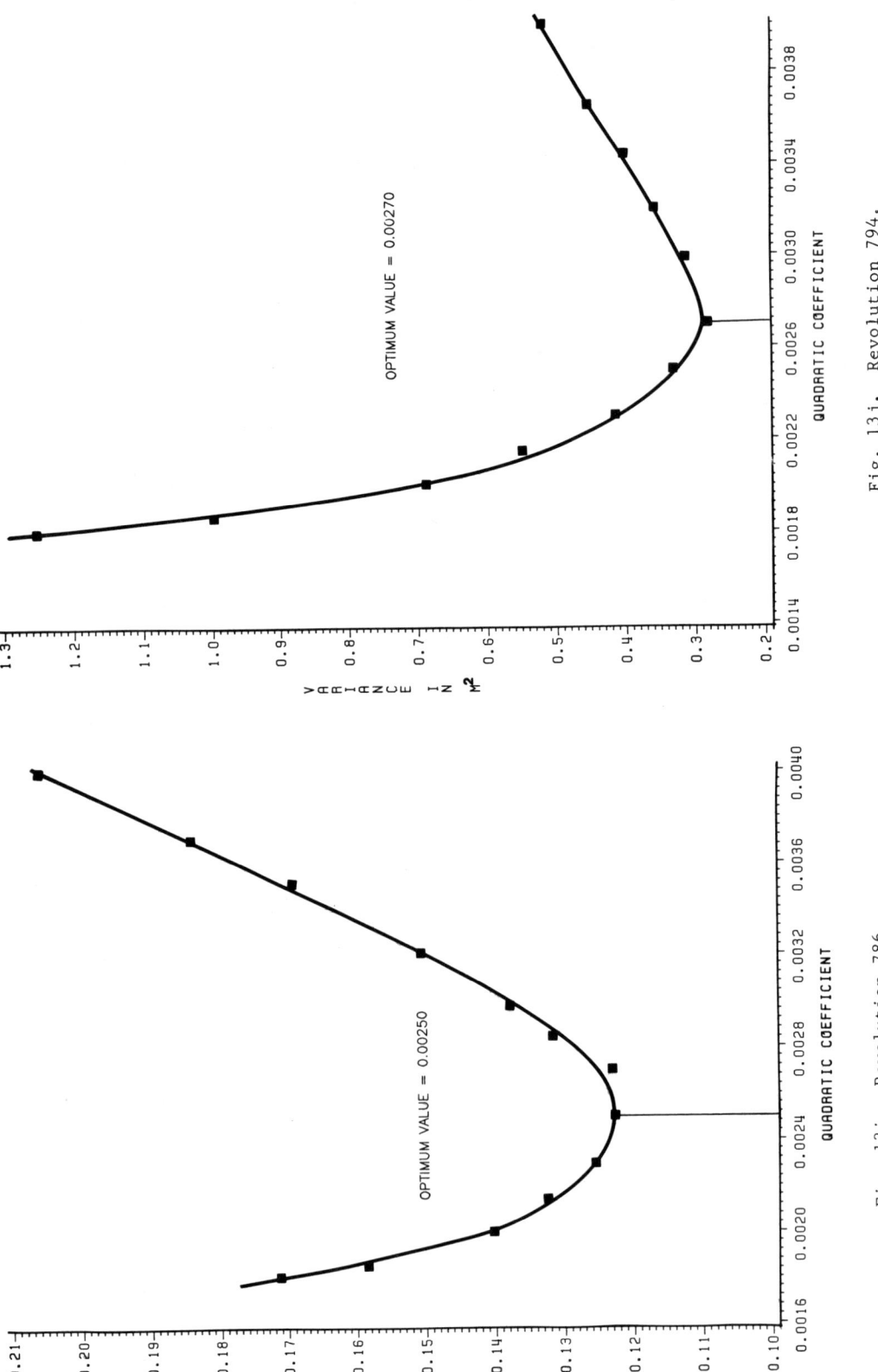

Fig. 13i. Revolution 786.

Fig. 13j. Revolution 794.

Conclusions

The quadratic ocean bottom friction law has been examined in the East China Sea and Yellow Sea area through numerical modelling and optimization procedures, in which various frictional coefficients were used in the equation of motion. The variance functions computed by measuring the error between the satellite observations and the theoretical values show surprisingly good agreement. From a total of 10 orbit tracks during two tropical storm periods, a range of quadratic coefficient between 0.0023 and 0.0027 was obtained. The correlation between the satellite measurement and the modelling conclusively show consistency of the estimated friction coefficient close to the conventionally adopted value of 0.0025.

The small range of coefficients obtained in this paper has several important meanings. First, it indicates the effectiveness of the modelling technique which is capable of handling ocean tide, storm surge, and (possibly) their interaction simultaneously. This thereby allows an accurate comparison with the satellite data to estimate a value for the quadratic friction coefficient. Secondly, it implies the validity of the empirical relations used to predict surface wind speed and wind stress. These relations appear to be able to produce reasonable and practical estimations of the magnitude of wind speed for the sea model. Finally, it shows some indications of qualitative improvements in the signal to numerical noise ratio (modelling accuracy) when the surface wind speed or stress increases. This is supported by the earlier investigations in Hudson Bay [Tang and Moon, 1984; Moon and Tang, 1987, 1985] during which the range of coefficients was found to be larger in a calm to moderate wind speed (1 m/s to 10 m/s) environment.

Acknowledgments. We would like to thank Dr. E. Schwiderski for making the Global Ocean Tide Tape available. This research was supported by the NSERC operating grant A-7400.

REFERENCES

Bowden, K.F. and L.A. Fairbairn, A determination of frictional forces in tidal current, Pro. Roy. Soc., A214, 371-392, 1952.

Bowden, K.F. and L.A. Fairbairn, Measurements of turbulent flucations and Reynolds stresses in a tidal current, Proc. Roy. Soc., A237, 442-438, 1956.

Bowden, K.F. and S.R. Ferguson, Variation with height of the turbulent in the tidally-induced bottom boundary layer, in Marine Turbulent, ed. Nihoul, J.C.J., Elsevier Oceanography Series 28, 1980.

Bowden, K.F., L.A. Fairbairn and P. Hudges, The distribution of shearing stress in a tidal current, Geophys. J.R. Astro. Soc., 2, 288-305, 1959.

Brocks, K. and L. Krugermeyer, The hydrodynamics roughness of the Sea surface, In Gordon (e.d.) Studies in Physical Oceanography, 1, 75-92, 1972.

Cartwright, D.E. and G.A. Alcock, On the precision of the sea surface elevations and slopes from SEASAT altimetry of the northeast Atlantic Ocean, Oceanography from Space, e.d., Gower, J.F.R., Plenum Publishing Corporation, 1981.

Charnock, H., Tidal friction from current near the North Sea bed, Geophys. J.R. Astro. Soc., 215-221, 1959.

Charnock, H. and J. Crease, North Sea surges, Sci. Prog. London. 45, 494-511, 1957.

Choi, B.H., A tidal model for East China Sea and Yellow Sea, Korea Ocean Research and Development Institute (KORDI), rep. 80-02, 1980.

Choi, B.H., Mathematical modelling of tides and surges in the East China Sea, Paper presented in the International Union of Geophysics and Geodesy, Hamburg, West Germany, 1983.

Chriss, T.M. and D.R. Caldwell, Evidence for influence of form drag on bottom boundary layer flow, J.G.R., 87, 4148-4154, 1982.

Davies, A.M., Application of a fine mesh numerical model of the North Sea to the calculation of storm surge elevations and currents, Rep. Inst. Oceanogr. Sci., No. 28, 30 pp., 1976.

Davies, A.M. and R.A. Flather, Application of numerical models of the north-west European continental shelf and the North Sea to the computation of the storm surges of November to December, Dt. Hydrogr. Z. Erganzungsheft, A., No. 14, 1978.

Denis, C., On the change of kinetical parameters of the Earth during geological times, Geophys. J. R. Astr. Soc., 87, 559-568, 1986.

Flather, R.A., Analytical and numerical studies in the theory of tides and storm surges, Ph.D. theses, University of Liverpool, 1972.

Flather, R.A., A tidal model of the north-west European continental shelf, Mem. Soc. R. Sci. Liege, 6 Ser., 10, 141-164, 1976.

Flather, R.A., Recent results from a storm surge prediction scheme for the North Sea, in Marine Forecasting, ed. Nihoul, J.C.J., Proc. 10th Int. Liege Coll. Ocean Hydrodynamics, 1978, 385-409, 1979.

Flather, R.A. and A.M. Davies, On the specification of meteorological in numerical models for North Sea storm surge prediction, with application to the surge of 2 to 4 January 1976, Dt. Hydrogr. Z. Erganzungsheft, A, No.15, 51, 1978.

Flather, R.A. and N.S. Heaps, Tidal computation for Morecambe Bay, Geophys. J.R. Astr. Soc., 42, 489-517, 1975.

Fu, L.L., Recent progress in the application of satellite altimetry to observing the mesoscale variability and general circulation of the oceans, Rev. Geophys. Space Phys., 21, 1657-1666, 1983.

Garratt, J.R., Review of drag coefficient over oceans and continents, Mon. Weather Rev., 105, 915-929, 1977.

Grace, S.F., The influence of friction on the tidal motion of the Gulf of Suez, Mon. Notic. Roy. Astron. Soc. Geophys., Suppl., 2, 316, 1930.

Grace, S.F., The influence of friction on the

tidal motion of the Gulf of Suez, Mon. Notic. Roy. Astron. Soc., Geophys., Suppl., 7, 309-318, 1931.

Groen, P. and Groves, G.W., Surges, in The Sea, 1, 611-646, e.d. Hill, M.N., Wiley, New York, 1962.

Hasse, L. and V. Wagner, On the relationship between geostrophic and surface wind on sea, Mon. Weather Rev., Wash., 99, 225-260, 1971.

Heaps, N.S., Storm surge on continental shelf, Phil. Trans. R. Soc., A, 256, 351-383, 1965.

Heaps, N.S., A two-dimensional numerical sea model, Phil. Trans. R. Soc., A, 265, 93-137, 1969.

Heaps, N.S., Storm surges, 1967-1982, Geophys. J.R. Astr. Soc., 74, 331-376, 1983.

Heaps, N.S. and J.E. Jones, Recent storm surge in Irish Sea, Marine Forcasting, Elsevier Scientific

Heathershaw, A.D., Measurement of turbulent in the Irish Sea benthic boundary layer, in The Benthic Boundary Layer, ed. McCave, I.N., Plenum Press, N.J., 11-31, 1976.

Hsueh, Y. and R.D. Romea, A comparison of observed and geostrophically calculated wintertime surface wind over the East China Sea, J. Geophys. Res., 88, C14, 9588-9594, 1983.

Kinsman, B., Wind Waves, Prentice Hall Inc., Englewood Cliffs, N.J., 1965.

Lambeck, K., Effects of tidal dissipation in the oceans on the Moon's orbit and the Earth's rotation, J. Geophys. Res., 80, 2917-2925, 1975.

Large, W.G., The turbulent fluxes of momentum and sensible heat over the open sea during moderate to strong wind, Ph.D. Thesis, Univ. of British Columbia, Vancouver, Canada, 1979.

Larson, L.H. and G.A. Cannon, Tides in the East China Sea, Paper presented to the Symposium on sedimentation on the continental shelf, Hangzhon, China, 1983.

Le Provost, C., An analysis of SEASAT altimeter measurements over a coastal area: The English Channel, J. Geophys. Res., 88, C3, 1647-1654, 1983.

Lerch, F.J., C.A. Wagner, S.M. Klosko, and B.H. Putney, Goddard earth models for oceanographic applications (GEM 10B and 10C), Marine Geodesy, 5(2), 2-43, 1981.

Marsh, J.G. and T.V. Martin, The SEASAT altimeter mean sea surface model, J. Geophys. Res., 87, C5, 3269-3280, 1982.

Marsh, J.G. and R.G. Williamson, Precision orbit analyses in support of the SEASAT altimeter experiment, J. Astronaut. Sci., XXVIII(4), 345-369, 1980.

Moon, W. and R. Tang, On the hydrodynamic correction of SEASAT altimeter data (Hudson Bay area of Canada), Marine Geodesy, 9, 291-333, 1985.

Moon, W. and R. Tang, Ocean bottom friction study using SEASAT-ALT data, Geophys. J. R. Astro. Soc., 88, 535-567, 1987.

Munk, W.H. and G.J.F. MacDonald, The Rotation of the Earth, a geophysical discussion, Cambridge University Press, 1960.

Parke, M.E. and M.C. Hendershott, M2, S2, K1 models of the global ocean tide on an elastic earth, Marine Geodesy, Vol 3, 379-408, 1980.

Powell, M.D., Evaluations of diagnostic marine boundary-layer model applied to hurricanes, Mon. Weather Rev., 108, 757-766, 1980.

Pekeris, C.L. and Y. Accad, Solution of Laplace's equation for the M2 tide in the world oceans, Phil. Trans. Roy. Soc. London, A265, 413, 1969.

Ramming, H.G., Numerical investigation of the influence of coastal structures upon the dynamic off-shore process by application of a nested tidal model, in Hydrodynamics of Estuaries and Fjords, ed. Nihoul J.C.J., Elsevier Oceanography, Series 23, 1978.

Roberts, K.V. and N.O. Weiss, Convective difference schemes, Math. comput., 20, 272-299, 1976.

Schwiderski, E.W., Global ocean tides, Part I: A detailed hydrodynamic interpolation model, NSWC/DL, Tr-3866, Naval Surface Weapons Center, Dahlgren, Va., 1978.

Schwiderski, E.W., On charting global ocean tides, Rev. of Geophys. and Space Phys., Vol. 18, 1, 1980.

Smith, S.D., Wind stress and heat flux over the ocean in gale force winds, J. Phys. Oceanogr., 10, 709-726, 1980.

Sternberg, R.W., Friction factors in tidal channels with differing bed roughness, Mar. Geol., 6, 243-260, 1968.

Stock, G.G., Modelling of tides and tidal dissipation in the Gulf of California, Ph.D. dissertation, University of California, San Diego, 1976.

Tang, R., Ocean bottom friction study using hydro-dynamic modelling and SEASAT-ALT data, M.Sc. thesis, University of Manitoba, Winnipeg, Manitoba, 1984.

Tang, R. and W. Moon, Finite difference transient sea surface modelling for the SEASAT altimeter data correction, Congressus Numerantium, Vol. 42, 299-312, 1984.

Weatherly, G.L. and M. Wimbus, Near-bottom speed and temperature observations on the Blake-Bahama Outer Ridge, J.G.R.,85, 3971-3981, 1980.

Weenink, M.P.H., A theory and method of calculation of wind effects on sea levels in a partly-enclosed sea, with special application to the southern coast of the North Sea, Koninklijk Nederlands Meteorologisch Instituut, Mededelingen en Verhandelingen, 73, 1958.

Wolf, J., Estimation of the shearing stress in a tidal current with application to the Irish Sea, in Marine Turbulent, ed. Nihoul, J.C.J., Elsevier Oceanography, Ser. 28, 1980.

Wu, J., Wind stress and surface roughness at air-sea interface, J. Geophys. Res., 74, 444-455, 1969.

Wu, J., Wind-stress coefficients over sea surface near neutral conditions: A revisit, J. Phys. Oceanogr., 10, 727-747, 1980.

Wu, J., Wind-stress coefficients over sea surface from breeze to hurricane, J. Geophys. Res., 87, C12, 9704-9706, 1982.

ORTHOGONAL STACK OF GLOBAL TIDE GAUGE SEA LEVEL DATA

A. Trupin and J. Wahr

(Department of Physics and Cooperative Institute
for Research in Environmental Sciences, University of Colorado,
Boulder, Colorado 80309 USA)

Abstract. Yearly and monthly tide gauge sea level data from around the globe are fitted to numerically generated equilibrium tidal data to search for the 18.6 year lunar tide and 14 month pole tide. Both tides are clearly evident in the results, and their amplitudes and phases are found to be consistent with a global equilibrium response. Global, monthly sea level data from outside the Baltic sea and Gulf of Bothnia are fitted to global atmospheric pressure data to study the response of the ocean to pressure fluctuations. The response is found to be inverted barometer at periods greater than two months. Global averages of tide gauge data, after correcting for the effects of post glacial rebound on individual station records, reveal an increase in sea level over the last 80 years of between 1.1 mm/yr and 1.9 mm/yr.

Introduction

Long period ocean tides affect estimates of certain geophysical parameters, in some cases through oceanic contributions to the Earth's inertia tensor and, in others, through crustal deformation caused by the weight of the ocean. Two examples of such estimates are the use of satellite solutions for J_2 to constrain anelasticity at the lunar tidal period of 18.6 years, and the use of the observed period and damping of the Chandler wobble to estimate mantle anelasticity at the 14 month wobble period. The 18.6 year solid-Earth and ocean tides, related to the precession of the lunar nodes, cause an 18.6 year variation in J_2. Lambeck and Nakiboglu [1983] assumed the ocean tide was equilibrium and used Rubincam's [1984] observed 18.6 year variability in J_2 to constrain the value of mantle anelasticity at this period.

The Chandler wobble results [see, for example, Smith and Dahlen, 1981] depend critically on the response of the ocean to the incremental centrifugal force caused by the wobble. That response, called the pole tide, is known to affect the period and, if it were non-equilibrium, could contribute to the damping. In fact, a global departure from equilibrium of only 10 percent could have observable consequences. Theoretical models for the pole tide in the deep ocean suggest the pole tide should be equilibrium [O'Connor and Starr, 1983; Carton and Wahr, 1986; Dickman, 1988].

The response of the ocean to variations in atmospheric pressure is another oceanographic disturbance with implications for solid earth geophysics. An inverted barometer response (ie. a 1 cm depression of sea level for every 1 mbar increase in pressure) is suggested by simple analytical models [see, eg., Munk and MacDonald, 1975]. The inverted barometer assumption has been invoked when studying the effects of atmospheric pressure on the Earth's rotation [see, eg., Munk and Hassan, 1961; Wilson and Haubrich, 1976; Merriam, 1982; Wahr, 1983] and when estimating the crustal deformation caused by pressure fluctuations [Rabbel and Zschau, 1985; Van Dam and Wahr, 1987]. The results in both cases are sensitive to the accuracy of this assumption.

Climate models that are used to study the effects of atmospheric greenhouse gasses predict an increase in the global temperature over the next century of from 1 to 4 degrees centigrade [Hansen, et al., 1981]. An increase of this magnitude could have numerous catastrophic effects, not the least of which would be a global rise in sea level due to a combination of melting polar ice caps and the thermal expansion of sea water. One of the important goals of global change studies is to improve our understanding of this variability in water storage.

There have been previous attempts to use tide gauge data to constrain the 18.6 year and 14 month tides [see, eg., Munk and Cartwright, 1966; Cartwright and Tayler, 1971; and Dickman, 1988] the response of the ocean to atmospheric pressure [Chelton and Enfield 1986]) and the global rise in sea level [Emery, 1980; Gornitz and Lebedeff, 1982; Barnett, 1983; Peltier, 1986]. These studies have primarily involved the analysis of data from individual tide gauges or, at most, from a small subset of all available stations.

In the present study, we combine tide gauge data from several hundred stations scattered around the globe, to test the hypotheses that the 18.6 year and 14 month tides are equilibrium, and that the response to pressure is inverted barometer. Our tide gauge data set consists of monthly sea level values obtained from the Permanent Service for Mean Sea-Level (PSMSL), at Bidston, England [see Pugh and Faull, 1983] In both cases, our observational results are consistent with these hypotheses. (Although the resolution of the monthly data limits the investigation of pressure-driven response at forcing periods of less than two months.)

We also use the PSMSL data to constrain the global rise in sea level. Although our estimates vary somewhat, depending on how we correct for post-glacial rebound [see eg., Peltier, 1986; Wagner and McAdoo, 1986], we infer a global sea level rise of approximately 1.75 mm/yr.

Methods of Analyses

In this section, we describe two analysis methods we have used to help identify small signals in the global tide gauge data. We used a

Copyright 1990 by
International Union of Geodesy and Geophysics
and American Geophysical Union.

simultaneous least squares fit technique to study the ocean's response to atmospheric pressure. And, we stacked the data to investigate the 18.6 year tide, the 14 month tide, and the global rise in sea level. We used the monthly PSMSL data to study the pole tide and the response to atmospheric pressure. We used yearly averages of the PSMSL data to study the 18.6 year tide and the global rise in sea level.

Least squares fit

Let $s_k(t)$ and $p_k(t)$ represent the time dependent fluctuations in sea level and in pressure at tide gauge k. Separate the observed sea level fluctuations into a component caused by the pressure, and a remainder, $\varepsilon_k(t)$, due to a combination of wind-driven fluctuations, tides, and observational noise. Suppose that the pressure-induced variability at an individual tide gauge depends on the past, present, and future values of pressure as measured at that gauge alone. (Causality precludes dependence on the future, but by solving for that dependence we can partially assess the accuracy of our results.) Then, the most general linear relationship between s_k and p_k is:

$$s_k(t) = \int_{-\infty}^{\infty} A(\tau)p_k(t-\tau)d\tau + \varepsilon_k(t) \qquad (1)$$

Here, $A(\tau)$ is identical to the cross-correlation function between s_k and p_k, and can be interpreted physically as a Green's function describing the response of the ocean to an impulsive pressure disturbance (the true Green's function is actually more complicated than $A(\tau)$, since eq. (1) assumes that sea level at any point is unaffected by pressure disturbances at other points). Since the ocean has a finite memory, $A(\tau)$ should approach zero for large τ.

Suppose that both $s_k(t)$ and $p_k(t)$ are discretized to a time series of monthly values, $t=t_i$, and suppose $A(\tau)$ is negligible for $\tau > Lt_i$, where L is an integer. Then, (1) reduces to the discrete form:

$$s_k(t_j) = \sum_{i=-L}^{L} A(t_i)p_k(t_j-t_i) + \varepsilon_k(t_j) \qquad (2)$$

For an inverted barometer ocean, $A(0)$ would be -1.01 cm/mbar, and all other $A(t_i)$ would be zero.

Suppose $A(t_i)$ is independent of the tide gauge location. This is equivalent to assuming that the ocean responds to pressure in the same way at every location. Then, we can estimate $A(t_i)$ for each t_i by least squares fitting to all PSMSL tide gauge data simultaneously. We can least squares fit each $A(t_i)$ individually, or all the $A(t_i)$ simultaneously. As an example, if each data point is given equal weight, regardless of where or when it was taken, and if M_k is the total number of months in the time series for tide gauge k, then the least squares fit for an individual $A(t_i)$ gives:

$$A(t_i) = \frac{\sum_{k=\text{gauge}}\sum_{j=1}^{M_k} s_k(t_j)p_k(t_j-t_i)}{\sum_{k=\text{gauge}}\sum_{j=1}^{M_k} p_k(t_j-t_i)p_k(t_j-t_i)} \qquad (3)$$

Stacking

The procedure of stacking multi-station data to enhance small signals has been used by seismologists to improve their estimates of the Earth's free oscillation eigenfrequencies [Gilbert and Dziewonski, 1975], and to search for short period oceanic normal modes in the Pacific [Luther 1982]. Stacking is particularly useful in cases where the spatial dependence of the signal is known beforehand.

Suppose, for example, we want to test the hypothesis that the 18.6 year tide is equilibrium. The 18.6 year equilibrium amplitude at co-latitude θ, eastward longitude λ, and time t, has the form (using the corrected tables of tidal harmonics from Cartwright and Edden [1973]):

$$h(t,\theta,\lambda) = 2.794\ \zeta(\theta,\lambda) \cos(\omega t + \phi) \quad \text{cm} \qquad (4)$$

where ω is the frequency of the 18.6-year tide and ϕ is the phase at $t=0$. The function $\zeta(\theta,\lambda)$ represents the spatial dependence of the equilibrium tide, and would equal $(1+k-h)Y_2^0$ were it not for mass conservation of the oceans, sea floor loading, and gravitational self attraction (h and k are tidal Love numbers, and Y_2^0 is normalized so that the integral over the unit sphere of $|Y_2^0|^2$ is one). These latter effects cause the introduction of other spherical harmonics into $\zeta(\theta,\lambda)$ and increase the Y_2^0 component by about 10 percent. The function $\zeta(\theta,\lambda)$ can be found by iterative solution of eq. 102, Dahlen, [1976].

Once we have found $\zeta(\theta,\lambda)$, we least squares fit it to all the PSMSL tide gauge data at each time t. The average and secular trend is removed from each station prior to the fit. The resulting time series is referred to here as a stack. The stack is spectrally analyzed to search for a peak at the 18.6 year frequency, ω. If a peak is found, its amplitude and phase can be compared with the results expected for an equilibrium tide. To estimate the equilibrium amplitude and phase, an artificial data set is constructed by replacing the PSMSL sea level value at every gauge and at every time by the estimated equilibrium value, $h(t,\theta,\lambda)$, at that same position and time. The PSMSL and artificial data are then stacked against the equilibrium spatial dependence $\zeta(\theta,\lambda)$ and the spectral peak at 18.6 years is analyzed.

But, stacking against the equilibrium $\zeta(\theta,\lambda)$ is not enough. Even if there is good agreement between the PSMSL and artificial peaks, it does not guarantee that the tide is equilibrium. It is desirable to also stack against other $\zeta(\theta,\lambda)$. For example, the equilibrium $\zeta(\theta,\lambda)$ for the 18.6 year tide is nearly proportional to $Y_2^0(\theta,\lambda)$. We have stacked against other pure spherical harmonics as well, up to degree 6. These stacks serve two purposes. First, by stacking the artificial data we can assess the effectiveness of the stacking procedure. For instance, is the station coverage complete enough that different spherical harmonics are reasonably orthogonal? We find that it is. Second, by stacking the PSMSL data and looking at the spectra, we can further test the equilibrium assumption. For example, an equilibrium 18.6 year tide should show little power when stacked against pure spherical harmonics other than Y_2^0. Although we do not show results for these other stacks, below, we find no evidence of 18.6 year peaks that stand significantly above the noise continuum for any stacks against spherical harmonics other than Y_2^0. We compare these PSMSL results with Y_l^m stacks of the artificial data, and find that the 18.6 year peaks in the artificial stacks are also below the noise continuum of the PSMSL spectra. These results, then, are consistent with the equilibrium hypothesis. That hypothesis is addressed further, below, when we discuss the stacks against the equilibrium $\zeta(\theta,\lambda)$.

A similar procedure is used to investigate the 14 month pole tide. The equilibrium pole tide can be generated from polar motion data (in our case, we used polar motion data from the International Latitude Observatory, Mizusawa, Japan) using:

$$h(t) = \frac{-3\Omega^2 a^2}{2g} \text{Re}\left[\left[m_1(t) + im_2(t)\right]\zeta(\theta,\lambda)\right] \qquad (5)$$

Here, $m_1(t)$ and $m_2(t)$ are the x and y coordinates of the pole position, Ω and a are the Earth's mean rotation rate and radius, and Re denotes the real part. The spatial dependence, $\zeta(\theta,\lambda)$, in (5) is complex, and so when it is least squares fit to the PSMSL data it gives a complex stack. In the absence of sea floor loading, gravitational self attraction, and mass conservation, $\zeta(\theta,\lambda)$ would equal $(1+k-h)Y_2^1(\theta,\lambda)$. As in the case of the 18.6 year tide, $\zeta(\theta,\lambda)$ is modified by about 10% due to these additional effects, and it can be estimated as described by Dahlen [1976]. Stacks against the equilibrium $\zeta(\theta,\lambda)$ are described below. Stacks of the PSMSL data against spherical harmonics other than Y_2^1 do not exhibit significant peaks at the 14 month pole tide period, and are consistent with Y_l^m stacks of the artificial data.

As another application, we have stacked the PSMSL data to study the global rise in sea level. In that case, the expected signal is independent of position and has the form:

$$h(t,\theta,\lambda) = Dt \qquad (6)$$

where the constant D is unknown. To find D, we stack the data against $\zeta(\theta,\lambda) = 1$, which is equivalent to constructing a simple spatial average of the data at each time, and then solve, by least squares fit for the linear trend in the stack.

For any of these examples, the least squares fit to form a stack can either be weighted or unweighted. One weighting procedure we have found to be particularly useful involves stacking on a grid. The earth's surface is divided into grid elements, in our case either 10° latitude by 20° longitude or 15° latitude by 30° longitude. Sea level heights for each station (with secular trends and station averages removed) are multiplied by the equilibrium spatial dependence, $\zeta(\theta,\lambda)$, and the results are averaged over individual grid elements for a given time, t. Each grid average is then normalized by dividing by the grid average predicted for an equilibrium tide, and the results are added to form the stack.

For example, let $s_k(t)$ be the tide gauge reading for station k at time t, with the secular trend and station average removed. Let θ_k and λ_k be the co-latitude and eastward longitude of station k. Let $n_p(t)$ be the total number of stations in grid element p at time t, and suppose there are N grid points. Then, the gridded stack, $x(t)$, is defined as

$$x(t) = \frac{\sum_{p=1}^{N} \frac{1}{n_p(t)} \sum_{\substack{k=\text{gauges} \\ \text{in } p}} s_k(t) \, \zeta(\theta_k,\lambda_k)}{\sum_{p=1}^{N} \frac{1}{n_p(t)} \sum_{\substack{k=\text{gauges} \\ \text{in } p}} \zeta^*(\theta_k,\lambda_k) \, \zeta(\theta_k,\lambda_k)} \qquad (7)$$

Here, * denotes complex conjugation. This result, (7), is equivalent to a weighted least squares fit of ζ to the data at time t, and it reduces to an ungridded stack when the grid elements are so small that $n_p = 1$ for every p.

Individual station records exhibit considerable long period variability, most of it probably caused by forcing from surface winds. This makes it difficult to use a single tide gauge record to identify any of the spectral features we are considering here. Stacking the individual stations against the spatial dependence of the equilibrium solution amplifies any signal which has that spatial dependence, and this can significantly reduce the correlation with the wind-driven variability.

The use of a grid de-emphasizes areas with great concentrations of stations, such as northern Europe and Japan, by weighting them equally with areas of fewer stations. Gridding also tends to further reduce the correlation with long wavelength, wind-driven features. Roughly speaking, this correlation is reduced for wavelengths longer than the grid size, but is increased for shorter wavelengths. On the other hand, since isolated stations contribute more heavily to a gridded stack, that stack is more sensitive to noise in those individual station records. The sizes of the grids used here are chosen to be large enough to minimize the effects of wind-driven fluctuations and of gaps in the grid averages,

but small enough to avoid combining data from inland seas with data from open coastlines. The more sparse yearly data are stacked on a coarser grid than are the monthly data.

Subtracting the average tide gauge height for each station prior to a stack is necessary to establish a uniform benchmark for all stations. But, if a station does not report data over the entire time span of the stack, the computed station average is not the true average, and this can introduce systematic errors into the spectrum. The frequency content of these errors depends on the time spans of the individual station data. The errors are apt to be particularly important at periods close to and longer than the average of the station lengths. Although we have found no consistent way of eliminating these errors, subtracting station averages from the PSMSL and artificial data minimizes their impact on the estimated spectral amplitudes. Of the 722 stations in the PSMSL data set, only 7 cover the entire interval between 1900 and 1979. The pole tide results are much less sensitive to this problem since the average station length of 29 years is many times the period of the tide.

Fitting and removing secular trends from stations prior to stacking is found to remove another source of error, as power associated with these trends can leak into the spectral band of interest. In addition, before any spectral analysis was performed on the final stacks, any remaining secular trends are removed from the final time series.

The spectral methods employed here to analyze a stack include both untapered periodograms and a more sophisticated multi-taper method developed by Thomson [1982] and adapted for geophysical applications by Lindbergh et al. [1987]. For the multi-taper technique, the data are multiplied by one of six prolate spheroidal sequences, or tapers, to create six time series. A spectrum is derived from these series in which leakage from neighboring frequencies is minimized. A statistical F-test provides a confidence estimate that an apparent periodicity is truly sinusoidal.

Results

The response to pressure.

Figure 1 shows a plot of the cross-correlation function, $A(\tau)$ between sea level and atmospheric pressure, for lag times of up to 240

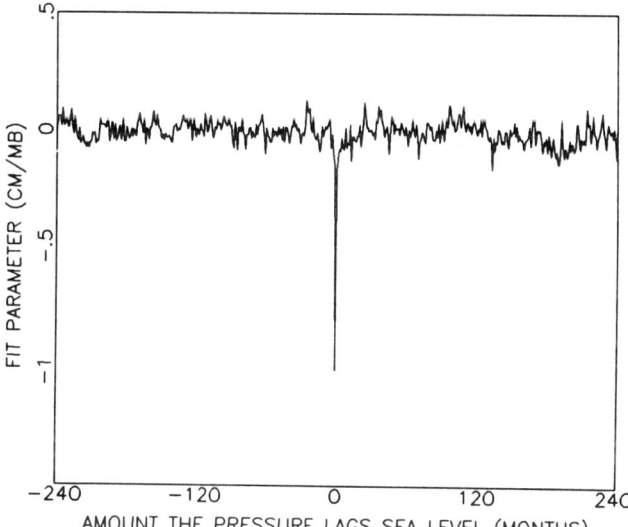

Fig. 1. The cross-correlation function, $A(\tau)$ between sea level and atmospheric pressure for all stations outside the North and Baltic seas. The response is inverted barometer for time lags of up to 240 months.

months. The pressure data are obtained from an objective analysis of global station pressure data for January, 1900 - April, 1973 [Wahr, 1983], and so we restrict the PSMSL data to this time period. All PSMSL stations during this time period are included in the fit, except for those in the North and Baltic seas and in the Gulf of Bothnia. Seasonal effects are removed from both sea level and pressure data sets prior to the fit. The result for $A(0)$ is -1.01 cm/mbar, and the results for all other $A(\tau)$ are close to zero. These results suggest that the response of the oceans to pressure is very nearly inverted barometer at periods of two months and longer. Peaks at 437 days in stacks of both the PSMSL and pressure data indicate that there is atmospheric and oceanic response to forcing at this period. With the exception of the data from the Baltic sea, where $A(0)= -1.25$ cm/mbar, and from the Gulf of Bothnia, where $A(0)= -1.6$ cm/mbar, data from all large oceanic areas give similar results. A fit to the southern hemisphere alone gives $A(0)=-1.04$ cm/mbar, with the other $A(\tau)$ close to zero.

The pole tide.

Monthly data were stacked, as described above, to study the pole tide. Before stacking the data, we removed the effects of atmospheric pressure, computed using the inverted barometer assumption. Again, we restrict our analysis to January, 1900 - April, 1973. We find that the removal of pressure affects the inferred pole tide amplitude by about 10 percent.

In this and succeeding analysis involving the pole tide, stations in the North and Baltic seas, and in the Gulf of Bothnia are removed from the data as there is an anomalously large pole tide at a period of 437 days in this region [see, for example, Miller and Wunsch, 1973]. A equilibrium stack of 26 stations in the North sea reveals the PSMSL power at this period to be 7 times the equilibrium power (the ratio of amplitudes is 2.6), in 42 Baltic sea stations the PSMSL power is 28 times equilibrium (the amplitude ratio is 5.3), and for the Gulf of Bothnia, the the PSMSL power exceeds the equilibrium power by a factor of 56 (the amplitude ratio is 7.5). The beat period between the pole tide period and one year is approximately 60 months. To minimize the correlation with the annual period, we choose to include in our analysis only those stations having greater than 60 months of data. This leaves us with 487 stations out of the original 721 in the full PSMSL data set.

Figure 2 shows a power spectrum of the equilibrium stack of these 487 stations. No grid has been used. Shown are the power spectra for the PSMSL data and for the artificial (ie. equilibrium) data. A band of frequencies centered on the annual frequency (.075-.09/month) is removed from both signals via least squares fit before plotting.

Note that in figure 2 the spectra for both the PSMSL data and the artificial data show a double peak for the pole tide, with one peak at 427 days and the other at 437 days. This is a well known feature of the polar motion spectrum. Most of the power contributing to this double peak is from data early in the time series. The two peaks are close enough together to require most of the 74 years of the data span to be adequately resolved. Figure 3 shows power spectra of the equilibrium stack of the 487 stations using a 10° by 20° grid. In both the gridded and non-gridded stacks the pole tide is clearly evident in the spectrum for the PSMSL data and there is good agreement with the equilibrium results. Note that the agreement is especially good for the gridded results.

Visual comparisons between the PSMSL and artificial spectra shown in the figures suggest that the global pole tide is close to equilibrium. To evaluate the agreement quantitatively, we compute amplitudes and phases for both the PSMSL and the artificial data sets. First, the amplitudes at the frequencies of the two apparent pole tide peaks are computed, and the results from the PSMSL data and artificial data

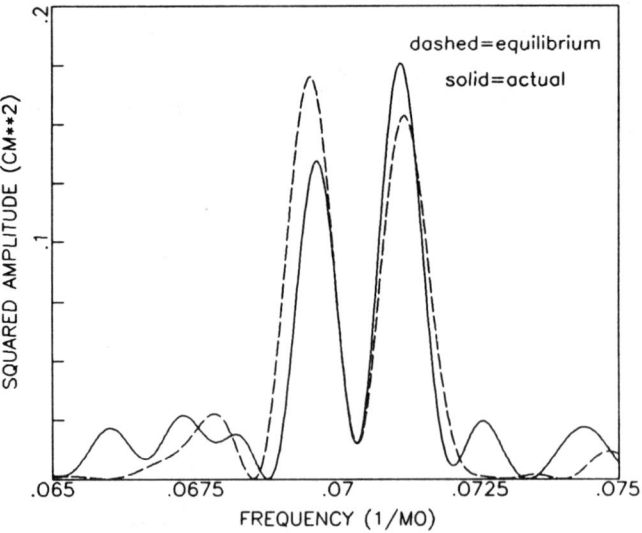

Fig. 2. Power spectra near the Chandler wobble frequency, for a Y_2^1 stack of 487 stations for ungridded stacks. Stations in the North and Baltic seas and in the Gulf of Bothnia are not included in the stacks. The Chandler wobble is clearly evident in the PSMSL results, represented by the sold line in all the plots (labeled 'actual') Annual periods have been removed, as have the effects of pressure.

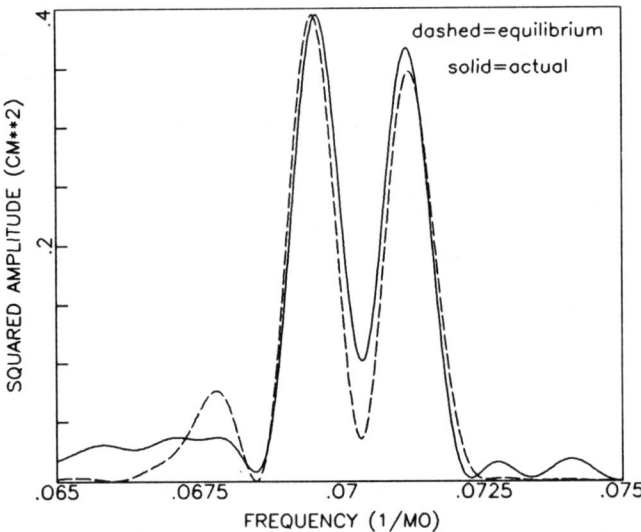

Fig. 3. Power spectra near the Chandler wobble frequency, for a Y_2^1 stack of 487 stations on a 10° by 20° grid. Stations in the North and Baltic seas and the Gulf of Bothnia are not included in the stacks. Annual periods have been removed, as have the effects of pressure. The PSMSL and artificial power are in excellent agreement.

are compared. We find, for the ungridded data shown in Figure 2, that the amplitude of the 427 day peak in the PSMSL data is 1.06 ± .28 times the artificial amplitude, and that of the 437 day peak is .86 ± .27 times the artificial.

We then repeat this procedure, but for the gridded stacks (figure 3). In this case, the 427 day and 437 day amplitudes are 1.02 ± .18 and

.98 ± .17 times the artificial, respectively. The uncertainties here are obtained from visual estimates of the background power in the vicinity of the pole tide. The background noise and the signal were assumed to be additive (in phase or 180° out of phase) in calculating the uncertainties. This background power is .03 cm² for gridded and ungridded plots.

To estimate the phase difference between the PSMSL and the artificial data, we fit the artificial data to the PSMSL data using a complex constant, after a filter is used to extract the power in a spectral band centered over the two pole tide frequencies (.069-.0715/mo). The phase difference between the PSMSL and the artificial data is roughly the same for both the gridded and ungridded stacks. The PSMSL leads the artificial data in the ungridded stack by 3° ± 50°. The phase lead found for the gridded stack is 0° ± 46°. The uncertainty in phase was obtained from the uncertainty in amplitude. Each degree of phase difference represents approximately 1.2 days.

If the artificial signal is fitted and removed from the PSMSL signal prior to the stack, the resulting power spectrum shows no peaks at the Chandler wobble frequencies for gridded and ungridded stacks. A spectrum is shown in figure 4 for the gridded stack.

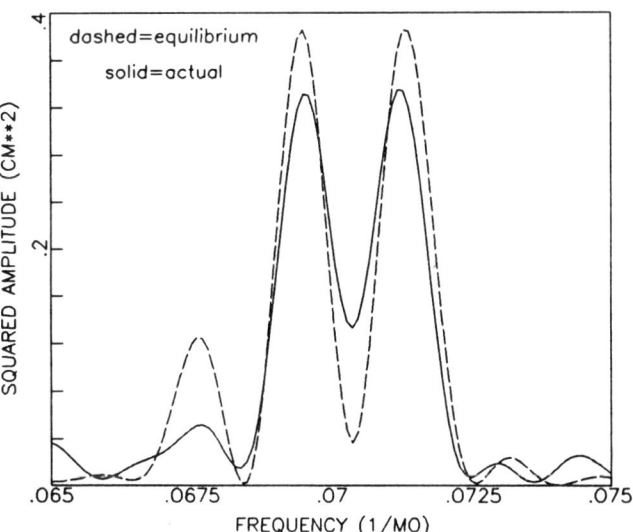

Fig. 5. A multi-taper spectrum near the Chandler wobble frequency, for a Y_2^1 stack of 487 stations on a 10° by 20° grid. North and Baltic sea stations are not included in the stack. A band of frequencies around the annual period have been removed. The stack is multiplied by each of a set of 6 eigentapers, and a spectrum is derived from the sum of these series in which leakage has been minimized.

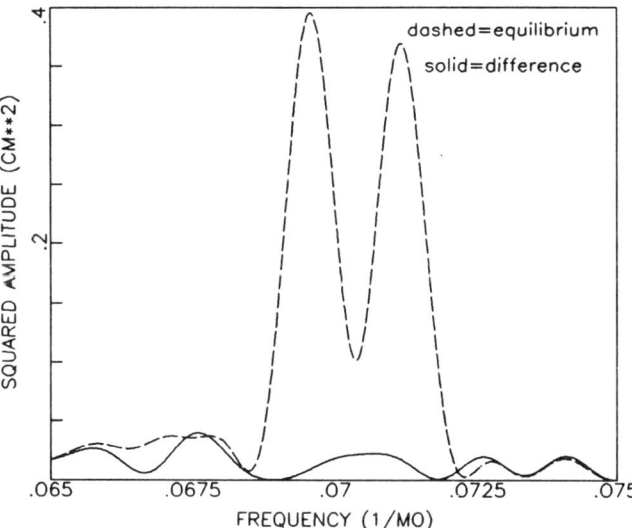

Fig. 4. Power spectra near the Chandler wobble frequency, for a Y_2^1 stack of 487 stations on a 10° by 20° grid. where the equilibrium tide has been fitted and subtracted from the final stack of the PSMSL data. The resulting spectrum shows no peaks at the Chandler wobble frequencies.

The multi-taper spectrum for the gridded equilibrium stack is shown in figure 5. The results are not significantly different from the conventional spectral results, probably because there is little spectral leakage into this band even in a conventional periodogram. The resolution of the peaks is slightly suppressed, as the 6 eigentapers used to weight the data for the Thomson spectrum each emphasize different portions of the data. Since these tapers are derived to minimize spectral leakage from neighboring frequencies, they weight the data independently of the number of stations contributing data to a given year.

The 18.6 year tide.

Yearly tide gauge data are stacked against the equilibrium 18.6 year tide for all 260 stations outside the Baltic sea that reported at least 19 years of data. The Baltic sea is the only geographic area where long period noise (22-23 years) so seriously masks the 18.6 year signal as to affect the estimated phase of a global stack. To identify these noisy stations, both the annual and 18.6 year periods are fitted and removed from the data and the standard deviation for the entire data set is calculated (σ=3.95 cm). The majority of stations with outliers greater than 3 σ are located in the Baltic sea. The σ for the all stations outside the Baltic sea is 3.48 cm. Data outliers are not removed from any of the 260 stations, as they do not affect the results for either the pole tide or the 18.6 year tide enough to justify biasing the data by their removal.

Figure 6 shows the time domain results for the PSMSL and artificial stacks against the equilibrium $\zeta(\theta,\lambda)$. An 18.6 variation is

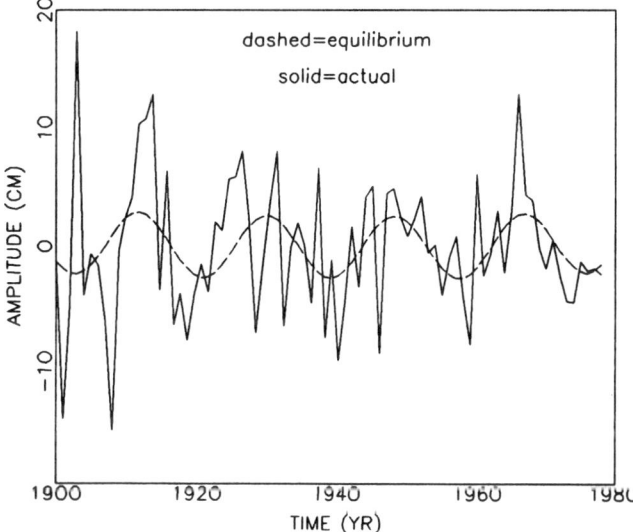

Fig. 6. The Y_2^0 stack of 260 stations as a function of time. The 18.6 year variation is evident, as is noise early in the time span.

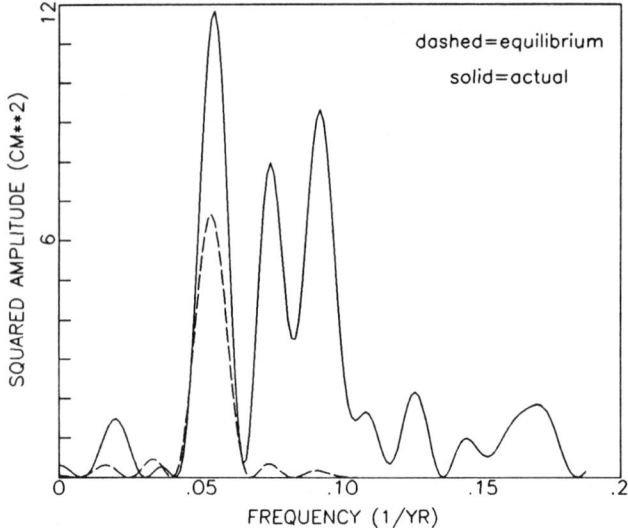

Fig. 7. Power spectra for a Y_2^0 stack of 260 stations outside the Baltic sea. The PSMSL amplitude is 1.2 times the artificial amplitude, but subtracting the stack of the artificial data from the PSMSL stack in the time domain (Figure 8) shows much of the difference to be attributed to uncorrelated noise at 18.6 years.

evident in the PSMSL result. Figure 7 shows the power spectra of ungridded equilibrium stacks. An 18.6 year peak stands out clearly above the noise in the PSMSL data, and has an amplitude 1.2 times the artificial amplitude. Figure 8 shows a power spectrum for the PSMSL data, where the final stack of the artificial data is subtracted from the final stack of the PSMSL data in the time domain. The spectrum shows that any discrepancy between the PSMSL and artificial data in Figure 7 may be attributed to noise, as the power at the 18.6 year period in Figure 8 is well below the noise continuum. Here the artificial data were not fitted before subtracting, so the low power at 18.6 years shows agreement in both amplitude and phase.

A multi-taper spectrum of the unweighted, ungridded data is shown in figure 9. The statistical F-test gives a 90 percent confidence that the 18.6 year signal of the PSMSL data is sinusoidal.

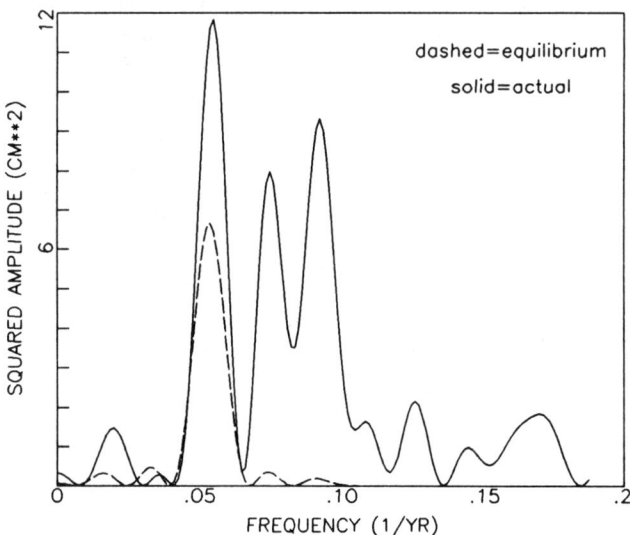

Fig. 9. Multi-taper spectra for the Y_2^0 stacks of the PSMSL and artificial signals. A statistical F-test of the PSMSL spectrum yields a 90 percent certainty that the 18.6 year peak is sinusoidal.

A conventional power spectrum for a stack against the equilibrium $\zeta(\theta,\lambda)$ on a 15° by 30° grid is shown in figure 10. The phase of difference between the time series is found by fitting an 18.6 year sinusoid to both the PSMSL and the artificial signals and comparing

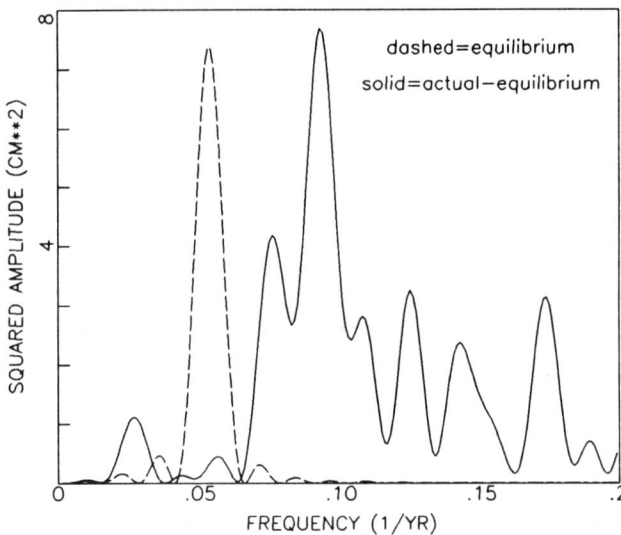

Fig. 8. Power spectra for a Y_2^0 stack of 260 stations where the equilibrium stack was subtracted from the final stack of PSMSL data. The power at 18.6 years is much lower than the noise continuum.

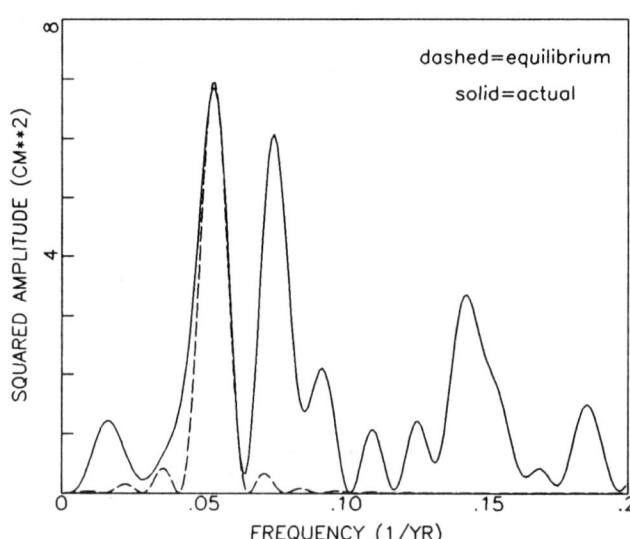

Fig. 10. Power spectra of an unweighted Y_2^0 stack of 260 stations done on a 15° by 30° grid. Baltic sea stations are not included in the stack.

their phases. The formal uncertainty for the PSMSL amplitude is calculated using an estimated noise level of 1.0 cm² for gridded and ungridded plots. The results are:

Ungridded PSMSL is 1.18 ± .22 times the artificial;
PSMSL leads by 1°±30°
Gridded PSMSL is 1.07 ± .22 times the artificial
PSMSL lags by 10°±30°

We combine these results to estimate an amplitude for the 18.6 year tide of 1.13 ± .22 times the equilibrium amplitude, and phase is equal to the equilibrium phase to within a 30° uncertainty.

The agreement between the PSMSL and artificial 18.6 year spectral peaks is not overly sensitive to the removal of blocks of stations (with the exception of the Baltic sea stations, which have already been removed) or to the use of time spans different from 1900-1979. For example, when all stations in other inland seas are removed from the data set, the power at 18.6 years is found to be little affected, although the frequency content of the noise did change.

The global rise in sea level

An observed secular sea level change at an individual station is not, by itself, evidence of a global rise in sea level. There could also be secular variations due to post-glacial rebound, local tectonic motion, or a shift in the wind-driven oceanic circulation pattern. The effects of these additional secular changes should be reduced in averages of global data. It is difficult, though, to adequately remove the effects of post-glacial rebound by averaging alone [Peltier, 1986]. Apparent changes in sea level at individual stations due to post-glacial rebound can be as large as 8.5 mm/yr, as is the case in the Gulf of Bothnia. And, a disproportionately large percentage of tide gauges are in the northern hemisphere, close to the centers of rebound. Large numbers of stations also lie in tectonically active areas, and no reliable model exists that allows us to remove the crustal motion from all these stations.

We digitized the rebound results from Peltier's [1986] post glacial rebound models for North America and northern Europe, and from Wagner and McAdoo [1986] for the remainder of the globe. In order to assess global sea level changes most accurately, we simultaneously fit the entire data set to a linear trend and to an artificial data set consisting of a set of trends predicted for each station by the combined rebound models as described above.

In order to establish a uniform benchmark for each station, a line is fit to each station record and the y-intercept of this line is subtracted from each station record before the simultaneous fit. The intercepts subtracted from each record in this way are not the true intercepts for those stations having less than 80 years of data. For those subsets of stations containing many short records, this introduces a systematic error into the final trend. For example, if a station record having only a few years of data contains power at periods greater than the record length, the truncated periodic signals are correlated with the secular variability, and any true secular trend over several decades could be masked or even reversed.

On the other hand, if we restrict the data set to only those stations having 80 years or more of data, there might not be enough stations to optimally average out the secular variability caused by sources other than the global sea level rise. To compromise, we first construct a subset of the data that includes all stations with at least N years of data, where N is greater than 1 but probably less than 80. We choose N so that when we simultaneously fit the rebound and sea level rise to the data, the post glacial rebound fit parameter is close to 1.

For an initial data set, we choose N=37. This leaves us with 120 stations, and the simultaneous fit results in a rebound coefficient of .9. The increase in sea level revealed by the fit is 1.2 ± .1 mm/yr. The uncertainty is the rms value of the time series of global, yearly averages of all 120 stations, after the rebound is removed. A post glacial coefficient of .9 is a good indication both that the post glacial models are giving reasonable results, and that the global tide gauge data are capable of resolving linear trends on the order of millimeters per year.

In fact, we have been able to improve the fit by further restricting our global data set. First, we exclude 23 additional stations from regions of the globe having significant tectonic activity (all stations on the west coast of North America and Japan). When the 97 remaining stations in this reduced data set are simultaneously fit to the rebound data and the global sea level rise, the rebound coefficient is .94 and the rise in sea level is 1.6 ± .12 mm/yr.

We further exclude all stations south of 30° north latitude. The post glacial rebound in this area is small and is reasonably the same everywhere (and so it is not easily separable from a global rise in sea level), and is strongly dependent on assumptions about the Pleistocene de-glaciation of Antarctica. This final data set consists of 84 stations, all with at least 37 years of data, situated north of 30° N latitude and away from Japan and the west coast of North America. The post glacial rebound data for all these stations are estimated from the results of Peltier's [1986] model. For these stations, the fit to the rebound model is especially good at .994 and the fit to the linear trend is 1.75 ± .13 mm/yr over the last 80 years. This trend is shown in figure 11.

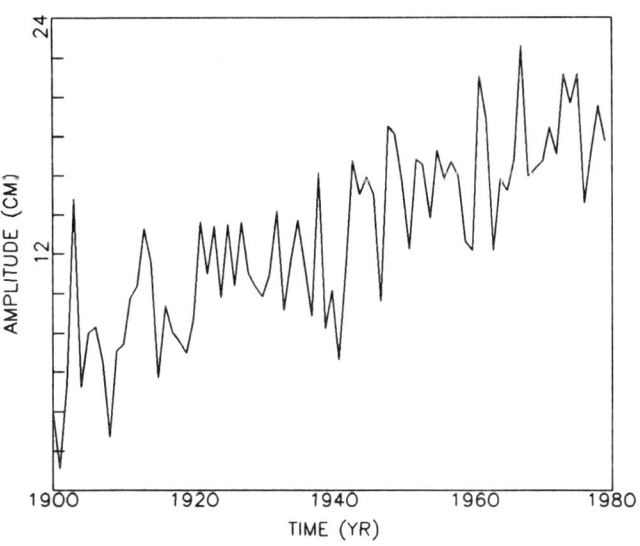

Fig. 11. Linear trend for sea level for 84 stations having greater than 37 years of data that lie north of 30° north latitude and are not in tectonically active areas (Japan and the west coast of North America are excluded). Vertical motion at the surface is simultaneously fit to these data using a model of post glacial rebound from Peltier [1986]. The fit to this model is excellent for these stations at .994, and the fitted rate of global increase of sea level is 1.75 ± .13 mm/yr.

From these three results, we estimate a global sea level rise during the last 8 decades of between 1.1 and 1.9 mm/yr, with a preferred value near 1.75 mm/yr. And, because the post glacial fit parameter is so close to 1, we tentatively conclude that the effects of post-glacial rebound are well described by current models, and that the tide gauge data are capable of resolving global linear trends at the millimeter per year level.

Summary

By stacking global tide gauge data, we can significantly improve the signal-to-noise ratio for long period tides, compared with results obtained from single station records. We find that the 18.6 year tide and the pole tide are clearly evident above the noise in the spectra for the equilibrium stacks. The peaks are not evident in stacks against other pure spherical harmonics. The observed amplitude and phase for the 18.6 year tide are consistent with the assumption of global equilibrium. The amplitude and phase of the pole tide are close to the equilibrium amplitude and phase for all regions outside the North Sea, Baltic Sea and Gulf of Bothnia. The 18.6 year result is somewhat less accurate than the pole tide result, partly because there are only 4 cycles of that tide in the 80 years of data.

We find that in all regions except the Baltic Sea and Gulf of Bothnia, the oceanic response to atmospheric pressure is very close to inverted barometer for periods longer than two months.

Finally, by simultaneously fitting the global the global sea level data to a linear trend and data prepared from a combined post-glacial rebound model, we conclude that the global rise in sea level over the last several decades was between 1.1 and 1.9 mm/year, with a preferred value of 1.75 mm/yr. Furthermore, the good agreement with the results from the post glacial models suggest that those results are reasonably representative of the true uplift, and that the tide gauge data are capable of resolving global changes in sea level at the millimeter per year level.

Acknowledgements.

We are grateful to Neamen Tewahade for his help in digitizing and interpolating the post-glacial rebound results. We thank the Permanent Service for Mean Sea Level for providing us with the tide gauge data. This work was supported in part by NASA (grant NAG5-485) and the Air Force Geophysical Laboratory (contract F19628-86-k-0011).

References

Barnett, T.P., Possible changes in global sea level and their causes, *Climate Change*, 5, 15-38, 1983.

Carton, J.A. and J.M. Wahr, Modeling the Pole tide and its Effect on the Earth's Rotation, *Geophys. J.R. astr. Soc.*, 84, 121-138, 1986.

Cartwright, D.E. and R.J.Tayler, New Computations of the Tide-generating Potential, *Geophys. J.R. astr. Soc.*, 23, 45-74, 1971.

Cartwright, D.E. and A.C.Edden, Corrected Tables of Tidal Harmonics, *Geophys. J.R. astr. Soc.*, 33, 253-264, 1973.

Chelton, D., D. Enfield, Ocean Signals in Tide Gauge Records, *J. Geophys. Res.*, 91, 9081-9098, 1986.

Currie, R.G., The Spectrum of Sea Level from 4 to 40 Years, *J. Geophys. Res.*, 83, 1837-1842, 1978.

Dahlen, F.A., The Passive Influence of the Oceans on the Rotation of the Earth, *Geophys. J.R. astr. Soc.*, 46, 363-406, 1976.

Dickman, S.R., The Damping of the Chandler Wobble and the Pole Tide, In press, 1988.

Emery, K.O., Relative Sea level from Tide-gauge records, *Proceedings of the National Academy of Sciences*, 77, 6968-6972, 1980.

Gilbert, F. and A.M. Dziewonski, An application of Normal Mode Theory to the Retrieval of Structural Parameters and Source mechanisms from Seismic Spectra, *Phil. Trans. R. Soc. A, 278*, 187-269, 1975.

Gornitz, V., Lebedeff, S., and Hansen, J., Global Sea Level Trend in the Past Century *Science*, 215, 1611-1614, 1982.

Hansen, J., D. Johnson, A. Lacis, S. Lebedeff, P. Lee, D. Rind, and G. Russel, Climate impact of increasing atmospheric carbon dioxide, *Science*, 213, 957-966, 1981.

Hosoyama, K., I. Naito, and N. Sato, Tidal Admittance of the Pole Tide, *J. Phys. Earth*, 24, 51-62, 1976.

Lambeck, K. and S.M. Nakiboglu, *Geophys. Res. Lett.*, 10, 857-860, 1983.

Lindberg, C.R., J. Park, and D.J. Thomson, Multiple Taper Analysis of Terrestrial Free Oscillations, parts I and II, *Geophys. J.R. astr. Soc.*, 91, 755-836, 1988.

Luther, D.S., Evidence of a 4-6 day Barotropic, Planetary Oscillation of the Pacific Ocean, *J. Phys. Ocean.*, 12, 644-657, 1982.

Merriam, J.B., Meteorological excitation of the annual polar motion, *Geophys. J. R. astr. Soc.*, 70, 41-56, 1982.

Miller, S. P., and C. Wunsch, The pole tide, *Nature Phys. Sci.*, 246, 98-102, 1973.

Munk, W.H., and D.E. Cartwright, Tidal Spectroscopy and Prediction, *Phil. Trans. R. Soc. Lond.*, A, 259, 533-581, 1966.

Munk, W.H., and E.M. Hassan, Atmospheric excitation of the earth's wobble, *Geophys. J. R. astr. Soc.*, 4, 339-358, 1961.

Munk, W.H. and G.J.F. Macdonald, *The Rotation of the Earth*, Appendix, Cambridge University Press, Cambridge, MA, 323 pp, 1975.

O'Connor, W.P. and T.B. Starr, Approximate Particular Solutions for the Pole Tide in a Global Ocean, *Geophys. J. R. astr. Soc.*,75, 397-405, 1983.

O'Connor, W.P., The 14 Month Wind Stressed Residual Circulation (Pole Tide) in the North Sea, *NASA Technical Memorandum 87800*, 1986.

Peltier, W.R., Deglaciation-Induced Vertical Motion of the North American Continent and Transient Lower Mantle Rheology, *J. Geophys. Res.*, 91, B9, 9099-9123, 1986.

Pugh, D.T. and H.E.Faull, *Monthly and Annual Mean Heights of sea Level, Permanent Service for Mean Sea-Level*, Institute of Oceanographic Sciences, Bidston Observatory, Birkenhead, Merseyside L43 7RA, England, 1983.

Rabbel, W., and J. Zschau, Static deformation and gravity changes at the earth's surface due to atmospheric loading. *J. Geophys.*, 56, 81-99, 1985.

Rubincam, D.P., Post Glacial Rebound Observed by Lageos and the effective viscosity of the lower mantle. *J. Geophys. Res.*, 89, 1077-1088, 1984.

Smith, M.L. and F.A. Dahlen, The Period and Q of the Chandler Wobble, *Geophys. J.R. astr. Soc.*, 64, 223-282, 1981.

Thomson, D.J., Spectrum Estimation and Harmonic Analysis, *Proceedings of the IEEE*, 70, 9, 1055-1096, 1982.

Van Dam, T. and J.M. Wahr, Displacements of the earth's surface due to atmospheric loading: effects on gravity and baseline measurements. *J. Geophys. Res.*, 92, 1281-1286, 1987.

Wagner, C.A. and D.C. McAdoo, Time Variations in the Earth's Gravity Field Detectable With Geopotential Research Mission Intersatellite Tracking, *J. Geophys. Res.*, 91, B8, 8373-8386, 1986.

Wahr, J.M. The effects of the atmosphere and oceans on the earth's wobble and on the seasonal variations in the length of day - 2. Results. *Geophys. J. R. astr. Soc.* 74, 451-487, 1983.

Wilson, C.R., Haubrich, R., Meteorological excitation of the earth's wobble. *Geophys. J. R. astr. Soc.*, 46, 707-743, 1976.

Wunsch, C., Bermuda Sea Level in Relation to Tides, Weather, and Baroclinic Fluctuations, *Reviews of Geophys. and Space Phys.*, 10, 1-49, 1972.

Wunsch, C., Dynamics of the North Sea Pole Tide Reconsidered, in press, 1986.

Yumi, S. and K. Yokoyama, *Yumi, S. and K. Yokoyama, Results of the International Latitude Service In a Homogeneous System*, Central Bureau of The International Polar Motion Service, International Latitude Observatory of Mizusawa, Mizusawa, Japan, 1980.

ATMOSPHERIC EXCITATION OF THE EARTH'S ROTATION RATE

J.B. Merriam

University of Saskatchewan, Saskatoon, Saskatchewan S7N-0W0

Abstract. Modern techniques for the determination of the Earth's rotation rate: long-baseline interferometry, satellite laser ranging, and lunar laser ranging, now permit the orientation of the Earth to be determined with an accuracy of 5 cm, which corresponds to about 10^{-4} sec in Universal Time. This nearly order-of-magnitude improvement over what was available ten years ago makes it feasible to look at variations in the length-of-day on much shorter timescales. At the same time, the requirements of operational weather forecasting have resulted in more detailed knowledge of the variations of the angular momentum of the atmosphere. The result has been a convincing demonstration over the last several years that virtually all of the random variations in the length-of-day, at periods between a few years and a day, are due to atmospheric variations. Geophysicists and meteorologists have both exploited this discovery. Removal of the atmospheric signal from the length of day, results in a data set in which other interesting phenomena of geophysical interest can be studied. Meteorologists have had some success in using the rotation data to deduce the angular momentum of the atmosphere at times in the past when sufficient global coverage was not available to do this directly. Outstanding problems are: the low frequency variations in atmospheric angular momentum, which the passage of time will correct, and the details of the mechanism by which angular momentum is exchanged with the mantle.

Introduction

Most people would find it very difficult to believe that winds in our diffuse atmosphere could alter the rotation rate of such a massive flywheel as the Earth. This incredulity stems from a lack of appreciation for the sensitivity of the instruments which monitor the rotation of the Earth more than a misconception of how strong the winds really are. However, we should not underestimate how tightly the atmosphere and solid earth shell are coupled. The strongest axial torques on the shell are from the exchange of atmospheric angular momentum (AAM); they are typically a hundred times larger than the secular tidal torque, and ten times larger than the core-mantle coupling torque.

Changes in the length-of-day (LOD) occur across a broad spectrum. Figure 1 shows a schematic amplitude spectrum of the LOD from days to a century. The principal tidal components: fortnightly, monthly, terannual, semiannual, annual, and lunar nodal are shown as vertical lines. The broad band atmospheric spectrum, which includes winds and a small (10 percent) contribution from pressure is shown as a solid line, broken by short dashes where the spectrum is uncertain. This occurs at less than about 2 weeks, and at more than a few years. A terannual component of 0.05 ms in LOD has been reported by Morgan et al. [1985], but as this is not supported by the lod observations, it may be the result of the omission of stratospheric winds. There may as well be a monthly oscillation in AAM with uncertain origin [Merriam, 1984, Morgan et al., 1985]. If we can extrapolate the short period fluctuations in AAM to periods less than 2 weeks, then amplitude should fall off roughly linearly with period; the atmospheric torque seems to be nearly independent of period from about a year to 2 weeks [Eubanks et al., 1985]. At the long period end of the spectrum there is even greater uncertainty. The long dashed line represents the amplitude spectrum of the decade fluctuations, which amount to a few milliseconds. The atmosphere has enough angular momentum to explain a large part of the decade fluctuations in LOD, if a substantial fraction of this angular momentum can be transferred to the shell, but the last 10 years of data suggest that this does not happen.

The statistical uncertainty in the LOD observations is about 0.05 ms at 2 week periods and decreases at longer periods. Variations in LOD from sources not represented here: ocean currents, groundwater, and sea-level, probably amount to less that 0.2 ms at any one time, but the amplitude spectrum from these sources might reach only 0.05 ms. The exchange of angular momentum with the atmosphere dominates the LOD spectrum at all periods less than decadal.

The connection between the winds and the LOD is made because the Earth plus atmosphere must conserve angular momentum (or nearly so). This means that if the net angular momentum of the atmosphere increases, the angular momentum of the solid Earth must decrease, making the day longer. Specifically, for every $10^{26} kg\, m^2\, s^{-1}$ that the atmosphere loses to the solid Earth the LOD decreases by 1.68 ms. Before wind data were available with sufficient altitude coverage to compute the angular momentum of the atmosphere reliably, several efforts were made to use the surface wind data and a model of friction to derive the torque on the shell, but this approach is unreliable because the exact nature of the coupling is so poorly understood. Indeed, there is some potential in the interdisciplinary interaction between geodesists and meteorologists to study the coupling mechanism.

Instrumentation has been so important to the study of the effect of winds on the LOD that the history of the subject could be neatly divided into old and recent, with the demarcation set around 1980, when very long baseline interferometry (VLBI), satellite laser ranging (SLR) and lunar laser ranging (LLR) begin to be heavily weighted into the rotation observations, and with the commencement of a world-wide program to monitor the global winds twice daily.

Copyright 1990 by
International Union of Geodesy and Geophysics
and American Geophysical Union.

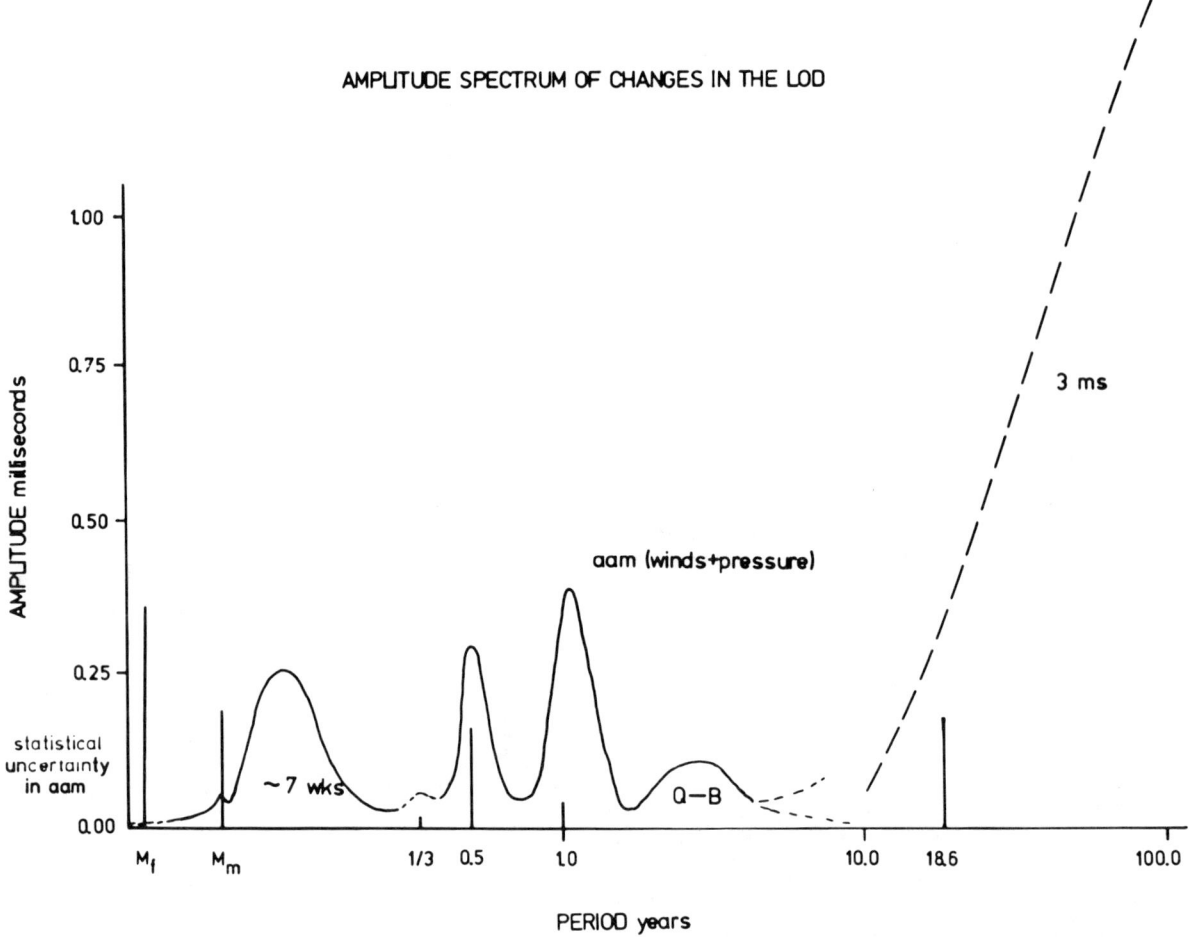

Fig. 1. A schematic amplitude spectrum of changes in the length of day. The vertical lines are the tidal terms and the curved line is the contribution of atmospheric angular momentum, broken by short dashes where the spectrum is uncertain. The long dashed line represents the decade variations.

In this relatively new subject, ancient history begins in 1950 with work by Munk and Miller, who showed that winds could produce a measurable seasonal oscillation in the LOD. In fact, a seasonal oscillation in the LOD of the same amplitude, discovered by Stoyko[1936], had already gone unexplained for 14 years. Some years later, recompilations of the wind data, which included more of the atmosphere, required that this figure be revised downward by about a third to again bring it into agreement with the seasonal term in the LOD which in the meantime also had to be reduced by a factor of three.

The introduction of quartz oscillators strengthened the observed connection between the winds and the seasonal LOD, and the introduction of atomic clocks probably contributed to the discovery of a quasi-bienniel oscillation in the LOD [Iijima and Okazaki,1966], which has a counterpart in the atmospheric circulation. Lambeck and Cazenave[1973,1974] confirmed the quantitative connection between the quasi-biennial terms in atmosphere and rotation and made the first study of the broadband excitation of changes in rotation rate by winds. Although their compilation of wind data only permitted a study of fluctuations with periods greater than 2 months, they concluded that nearly all of the observed irregular fluctuations in the LOD were of meteorological origin and correctly predicted that meteorological influences would be found in the rotation data at periods shorter than 2 months. They also hypothesized that exchange of angular momentum between the atmosphere and solid Earth could explain at least part of the decade fluctuations which were, and still are, widely attributed to core-mantle coupling. Lambeck and Hopgood[1981] updated this work and, while they did not investigate higher frequencies, they did confirm the earlier findings that the atmosphere dominates the excitation at periods less than a few years and that up to 20 percent of the decade fluctuations may be due to the atmosphere. More importantly, Lambeck and Hopgood also provided the first investigation of the *distribution* of angular momentum in the atmosphere.

As has already been mentioned, the growing interest in this field is a direct result of improvements in monitoring the LOD and the winds. The latter owes much to the inception of the Global Atmospheric Research Program (GARP) and the Special Observing Period of the First GARP Global Experiment in 1979. Hide et al. [1980] used this data and showed that the fluctuations in the angular momentum of the atmosphere were mirrored in the excess LOD data of the Bureau International de l'Heure (BIH) at periods down to 2 weeks. The earlier predictions that the atmosphere drives changes in the LOD at these short periods were vindicated, and a series of papers published between then and now vastly increased our awareness of at least the kinematics of atmospheric angular momentum.

The Spatial Structure of Atmospheric Angular Momentum

Winds are driven by pressure anomalies that are produced by solar heating and modified by rotation and the distribution of continents and oceans. Neglecting the latter, we see that rising warm air and low pressures are generated in a narrow belt around the equator and cool sinking air and high pressure over the poles. There is another belt of dry descending air around 30 degrees latitude. The return flow at the surface is deflected to the right in the northern hemisphere, producing the observed pattern of polar easterlies, mid-latitude westerlies, easterly trade winds, the relatively windless doldrums around the equator and the so-called horse latitudes near 30 degrees. This pattern is rather variable but the major changes with latitude, altitude, and season that have an important influence on the lod can be appreciated by looking at east-west winds in an idealized cross-section through the atmosphere.

Figure 2 shows idealized profiles of wind speeds, in $m s^{-1}$ for January and July. Zones of easterly winds are ruled. The surface pattern evident in the previous figure reveals itself as two cores of westerlies centered just above the tropopause (about 11 km height or 200 mb) and a trade wind belt which narrows with altitude throughout the troposphere and then broadens asymmetrically through the stratosphere. The westerly cores strengthen in the hemispheric winter and shift slightly towards the equator. The asymmetry of the easterly trade winds in the stratosphere is accompanied by a polar winter jet which is more prominent in the Southern Hemisphere than in the Northern Hemisphere. Most of the variability in these patterns is in the westerly cores, with the Northern Hemisphere being much more variable than the Southern Hemisphere. Wind speeds alone do not tell the whole story, and it is really axial angular momentum which is of interest; thus the polar regions have relatively little influence on the axial angular momentum, by virtue of location as well as wind speed, while the tropics and especially mid-latitudes are dominant.

Figure 3 shows a comparison of the LOD, from atmospheric forcing, as determined by the European Center For Medium Range Weather forcasting (EC) and modern space geodetic techniques [Dickey and Eubanks, 1986]. The rms difference, after removing a bias between the two, is only 0.07 ms. There are numerous episodes of misfit of about 0.2 ms, that persist for a month or more, and there may as well be larger misfits, at longer periods, that have been absorbed into the bias. The rms difference can be explained by a combination of errors in the observed LOD and errors in the AAM. The compilations of AAM by both the EC and the National Meteorological Center (NMC) show rms differences of about 0.07 ms. Since these are compiled from essentially the same wind data, the rms difference is probably a measure of the error introduced by the reduction procedure. The more prominent episodes of misfit are small enough to be accounted for by the neglect of stratospheric winds, and possibly the oceanic buffering of angular momentum exchange between the atmosphere and Earth. Core-mantle coupling and groundwater probably make insubstantial contributions at these periods.

Temporal Variations in the Angular Momentum of the Atmosphere

Figure 4 shows the angular momentum of the atmosphere from 1976 to 1985, and the changes in the LOD that it produced. Many of the important and interesting features of the angular momentum of the atmosphere are evident upon even a casual inspection of this figure.

The annual term is dominant with a variable amplitude of between 0.3 and 0.5 ms in the LOD and accounts for approximately three-fourths of the total variance. It is evidently not a sinusoid but has peaks that are distinctly broader than the valleys. The atmospheric angular momentum seems to linger in a boreal winter mode between rather precipitous excursions to a summer mode. This feature is also evident in the series of subharmonics of the annual period [Eubanks et al.,1985], including a terannual component (0.05 ms in LOD) which Morgan et al. [1985] find is much larger than the LOD data admit. There is also a semiannual component of about 0.2 ms in LOD.

There has been much discussion of the annual and semiannual signals in both the AAM and LOD. Early compilations of seasonal variations in AAM seemed to indicate that they could not explain all of the seasonal variations in LOD [Lambeck and Cazenave 1973, Lambeck and Hopgood, 1981, Eubanks et al., 1985 and Morgan et al., 1985] and so a variety of other meteorological, hydrological and oceanographical causes for the discrepancy were suggested. Later work by Rosen and Salstein [1985] showed that the imbalance was due to the neglect of stratospheric winds, which contain about 16 percent of the annual variation and 40 percent of the semiannual variation in AAM. This is about 0.1 ms for both. The unbalanced terannual component [Morgan et al., 1985] might be similarly brought into balance by the inclusion of stratospheric data. There now appears

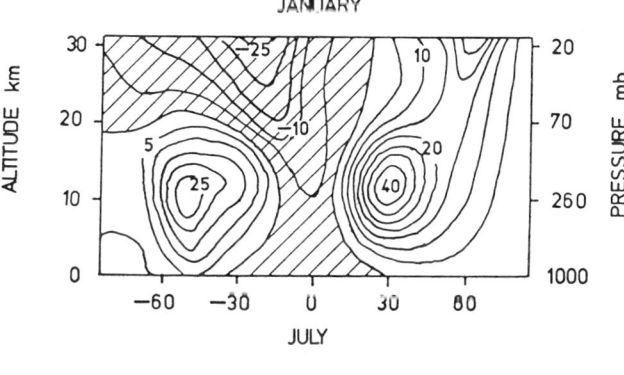
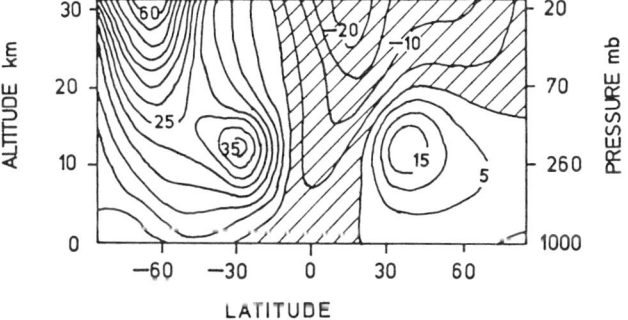

Fig. 2. A cross section of zonal wind speeds through the atmosphere, from the surface to 30 km height and from pole to pole, for January and July. Zones of easterly winds are ruled and wind speeds are in m/sec. Adapted from Lambeck and Hopgood (1981).

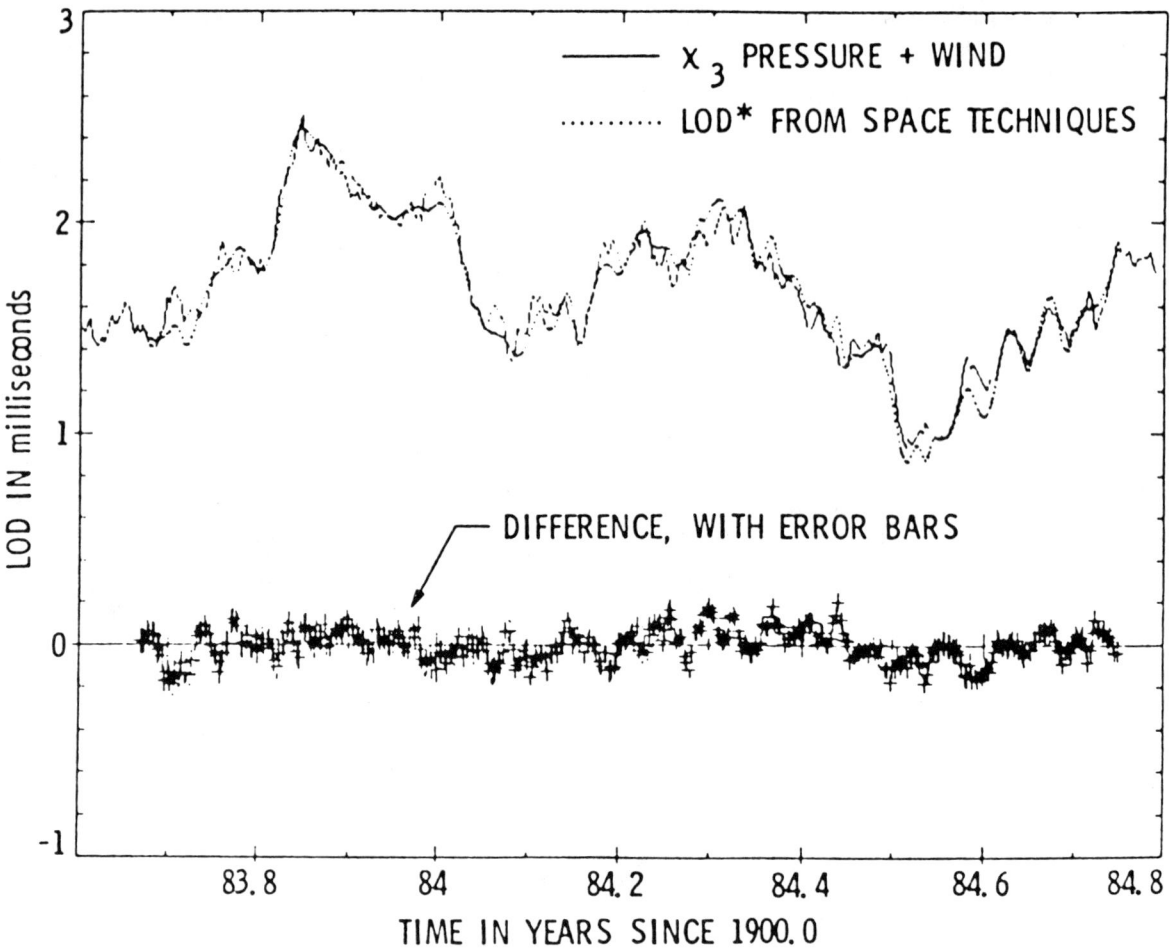

Figure 3 A comparison of the changes in the length of day, as observed by space geodetic techniques (corrected for tidal variations), and as inferred from changes in atmospheric angular momentum. From Dickey and Eubanks (1986).

to be no reason to invoke anything but the atmosphere to explain seasonal changes in LOD.

The structure of the annual signal has been known for some time to be dominated by the asymmetry between the Northern and Southern Hemispheres (Figure 5). Most of the total annual angular momentum variations of the atmosphere are from the Northern Hemisphere, with the Southern Hemisphere variations about 180 degrees out of phase and oscillating by half as much despite containing nearly twice as much angular momentum on average. So great are the fluctuations in the Northern Hemisphere that its angular momentum can apparently become negative for short periods in July and August, although this is always so small that it could as easily be error.

Much of the remaining short period variance evident in Figure 4 is in a band between 3 to 9 weeks, which seems to consist of a family of pulsed oscillations. Its amplitude is highly variable but averages about 0.2 ms in LOD. This signal was found in the LOD data [Feissel and Gambis, 1980], before it was known to be a global atmospheric phenomenon (in the sense of net angular momentum) although there were scattered and local reports of oscillations in high altitude winds at this period [Madden and Julian, 1972]. In contrast to the annual variations, this "mode" is in phase in the Northern and Southern Hemispheres, and largely a tropical or subtropical phenomenon, centered near the tropopause [Anderson and Rosen, 1985]. Feissel and Nitschelm [1985] have presented evidence that the oscillations occur in pulses which last for 160 to 445 days with a relatively short recurrence time, of 20-60 days, between pulses.

Despite its equatorial confinement, inconspicuous presence in the atmosphere, and recent discovery, we should not underestimate the importance of this oscillation to the coupling of the shell and atmosphere. It frequently involves the transfer of $0.5 \times 10^{26} \, kg \, m^2 \, s^{-1}$ in about 30 days, implying a torque of $2 \times 10^{19} \, Nt \, m$. This may be the largest axial torque exerted on the Earth. For comparison, the secular tidal torque is only $0.5 \times 10^{17} \, Nt \, m$, the annual atmospheric torque is about $4 \times 10^{18} \, Nt \, m$, and the core-mantle coupling torque has reached a maximum of only $10^{18} \, Nt \, m$ this century [Morrison, 1979].

This oscillation, or family of oscillations, is visually observable as a cycle of cloudiness that nucleates in the Indian ocean and intensifies as it drifts eastward into the Pacific to eventually dissipate in the eastern Pacific. Suspected teleconnections with mid-latitude jets [Anderson and Rosen, 1985] and the modulation of the Indian monsoon are being investigated [Weickmann, 1983]. Uncovering the dynamics of

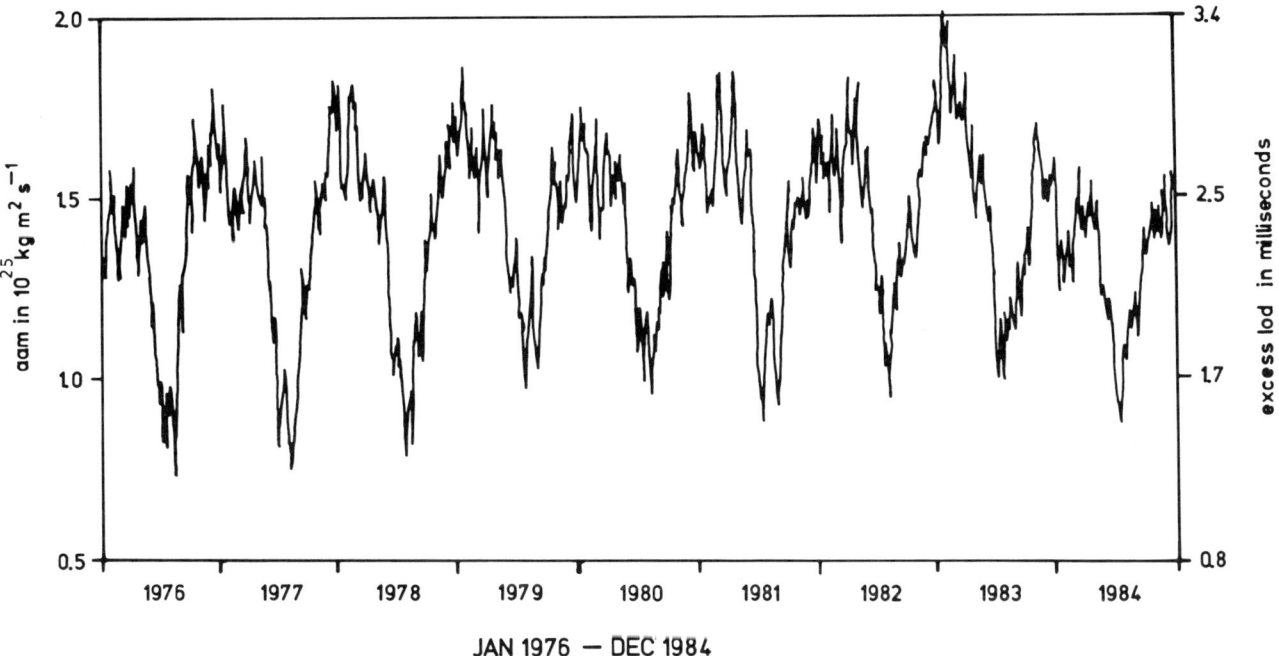

Figure 4 The angular momentum of the atmosphere from 1976 to 1985.

this system may be the key to further advances in understanding the general circulation of the atmosphere and in extending weather forecasts beyond the current 2 week limit [Hide,1984].

Despite its possible importance to weather prediction and its predominance in the angular momentum budget, we do not seem to have settled on a name for this phenomenon. It has been variously described as: the $\sim$ 50 day oscillation, [Langley et al, 1981] the 7 week oscillation, [Hide et al., 1980], the 30-70 day oscillation, [Morgan et al., 1985], the 60-40 day oscillation, [Eubanks et al., 1985], the 40-50 day oscillation, [Madden and Julian, 1972], and the 50-55 day oscillation, [Feissel and Nitschelm, 1985].

Also apparent in Figure 4 is a positive dc value of angular momentum of about $1.4 \times 10^{26}\ kg\,m^2\,s^{-1}$. This represents a super rotation of the whole atmosphere at a rigid body rate of about $6m\,s^{-1}$ at the equator. How this is maintained, against the tendancy for friction to bring the atmosphere into co-rotation with the solid Earth, is still not understood. If all of this angular momentum were dumped to the rotation of the Earth, the lod would decrease by about 2.3 ms.

Since the decade variations in the LOD have a maximum of only 7 ms over the last century [Morrison, 1979], and typically less, the atmosphere can potentially explain a large part of the decade variations. However, over the decade for which we have good data on the AAM it has not shown a long-period trend of the same magnitude as the LOD data. Eubanks et al. [1985] find a linear rate in the angular momentum of the atmosphere (below 100 mb) of $0.004\ ms\,yr^{-1}$ between 1976 and 1982, while the regularized LOD (the LOD with tidal effects removed) decreased at $-0.19\ ms\,yr^{-1}$ during the same period. If this linear decrease in LOD were to be explained by a change in AAM, that was not accounted for by their data, then they would have to have missed twenty percent of the angular momentum of the atmosphere, and so they conclude that this linear rate in LOD could not have been caused by the atmosphere. The work of Taylor and Mayr [1985] suggests more caution. In a limited span of time [1979] the angular momentum of the stratosphere above 100 mb accounted for 10 percent of the total. If this amount is variable year-to-year it is thus conceivable that the secular trend in LOD, found by Eubanks et al, has an atmospheric origin. Omitting the large changes in LOD between 1900 and 1920, then the decade variations in LOD are characterized by an amplitude of perhaps a millisecond. This is entirely within the scope of AAM and it is conceivable that any but the largest decade changes in LOD could have an atmospheric origin. It is premature to judge the power in the decade changes in AAM.

The definition of the decade changes in AAM will be awaited by meteorologists and geodesists alike. Meteorologists will be interested in just how stable the super rotation of the atmosphere is and geophysicists will want to remove the long term AAM signal from the LOD and study the residual for the messages it must contain on core dynamics. Once we have such a residual LOD we may be able to learn more about the spectrum of core oscillations and the frequency dependance of core- mantle coupling, if not details of the coupling itself.

Not so evident on the scale of Figure 4 is the quasi-bienniel oscillation, which actually refers to a band from just under 2 years

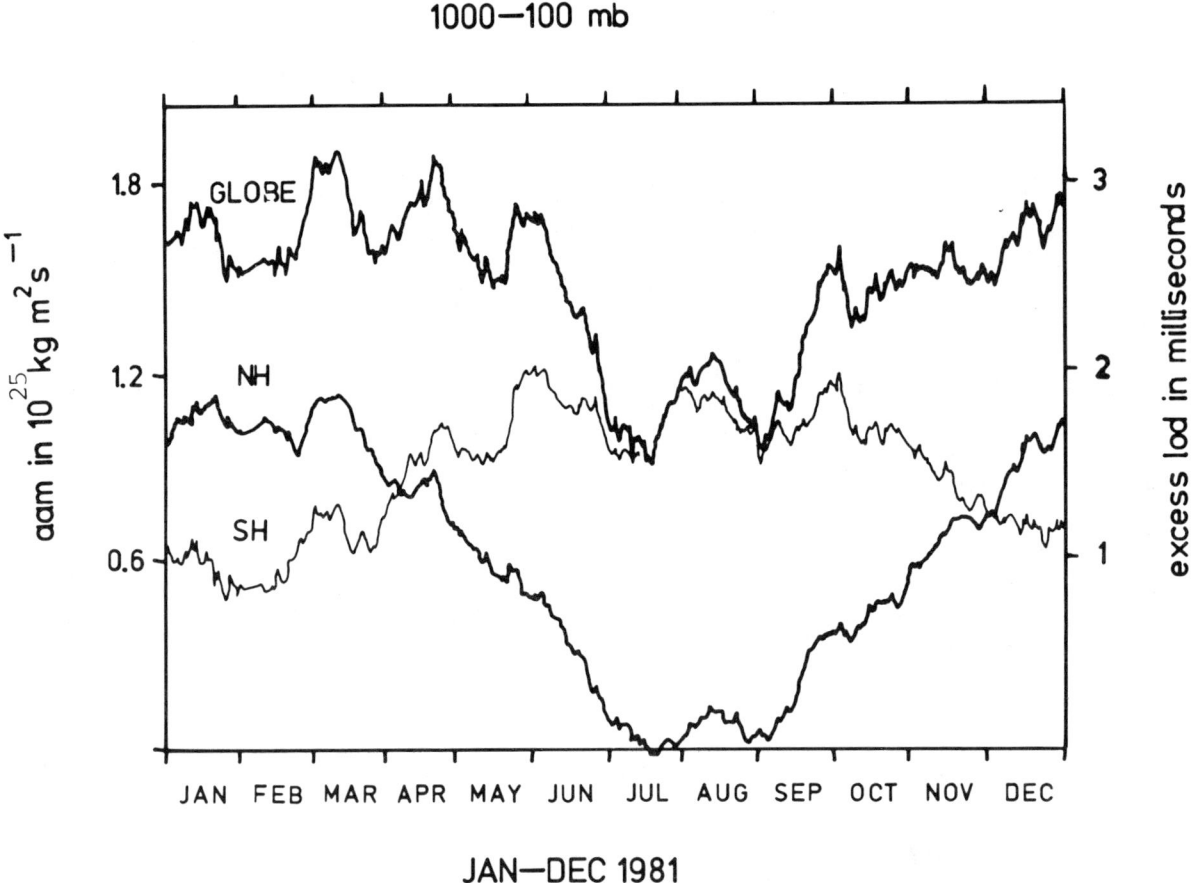

Figure 5 The atmospheric angular momentum in the Northern and Southern hemispheres during 1981. The Northern and Southern hemispheres are 180° out of phase and the Northern hemisphere is much more variable than the Southern hemisphere and on average contains less angular momentum. Adapted from Rosen and Salstein 1983.

to about 4 years. It has a variable amplitude of less than 0.1 ms. It is a well-known, though not very well understood, feature of the atmospheric circulation that we may eventually come to identify more closely with the El Nino-Southern Oscillation. Chao [1984] has found that the Southern Oscillation Index (roughly the mean monthly difference in atmospheric pressure between Tahiti and Darwin) is correlated with the interannual (between 1 year and 10 years) variations in the LOD. Most of the variations in winds which are quasi-biennial are confined to a narrow band around the equator and to greater heights (perhaps 20 to 30 km or 50 to 10 mb) than the annual wind cycle. The usual compilations of AAM, extending to only 100 mb or even 50 mb, therefore exclude much of the region in which the bienniel oscillation is prominent. A proper study of the correlation between the Southern Oscillation Index and LOD should examine the angular momentum of the atmosphere to at least 10 mb. Lambeck and Hopgood [1981] have presented evidence that the equatorial forcing is somewhat rectangular, with a normal cycle of easterly winds in the stratosphere lasting for about 2 years, interrupted by a counter-cycle of westerly winds lasting less than

a year. Interestingly, the four episodes of easterly winds in their data correspond to episodes of large negative values of the Southern Oscillation Index.

Future Directions

The BIH has for many years published a regularized Universal Time which has the relatively well known tidal terms removed. We are approaching the point where we should consider the publication of an atmospherically regularized LOD time series which has been corrected for variations in atmospheric angular momentum. This will permit a closer look at other variations in the LOD. Merriam [1984] has shown that the tidal love numbers can be more accurately measured in LOD data once the "noise" from atmospheric fluctuations has been removed. While the fortnightly (M_f) term in LOD may be dominated by observational noise which is larger than the "noise" from the winds it is more certain that the monthly term (M_m) can be more accurately measured once the atmospheric influence is removed. Moreover, the fortnightly and monthly terms in AAM are larger than can easily be explained.

We are now certain that virtually all of the irregular variations in the LOD, between a few years and a week, are due to atmospheric influences. As more data accumulates we should be able to examine the critical band where the exchange of angular momentum with the core begins to influence the LOD. Preliminary work suggests that 5 years may be an appropriate cut-off but more time, and more complete stratospheric data, are required. Morrison's [1979] work on the decade fluctuations in the LOD indicates a minimum in power at about the 10 year period. Perhaps this minimum in the spectrum is the cusp between the tail-off of atmospheric influence and the ascendancy of core-mantle coupling. Does the vigour of core motions, as seen through the frequency dependent filter of core mantle coupling, drop off sharply at periods less than 5 years? I am sure that there is information in this critical band that will help us to understand core motions better.

While we have assured ourselves that conservation of angular momentum holds in the atmosphere-shell system we are relatively ignorant as to how this is achieved. Numerous attempts have been made to compute the coupling torque between the atmosphere and the shell [Wahr, 1983, Lambeck, 1980, Wahr and Oort, 1984, Swinbank, 1985] but results do not agree with the computed angular momentum transfers. Intuitively, it makes sense that the coupling should be dominated by the montain torque originating in the difference in East-West pressure across mountain chains, but it seems that surface friction may be larger, and perhaps steadier. However, results are depressingly erratic with some studies indicating a complete insensitivity of atmospheric angular momentum to the presence of mountains; if mountains are made bigger in a model of atmospheric circulation the consequent rearrangement of winds reduces the surface friction, so that the net transfer of angular momentum is unaffected. This has lead Lambeck [1980] to suggest that angular momentum is conserved in the atmosphere in the first instance and only a small residual is transmitted to the Earth. This appears to be a natural consequence of the existence of a layer in the lowest atmosphere which is tightly coupled to the Earth. One should bear in mind that the orographic part of the atmosphere (ninety-five percent of the surface of the Earth is no more than 3 km above sea level) contains only four percent of the super-rotation angular momentum of the atmosphere and moreover has the least variablity in angular momentum. Hence, this layer almost rotates with the Earth (its rigid body equatorial velocity with respect to the solid Earth is less than $1 \ m \ s^{-1}$) and it is mostly a medium for transfers of angular momentum between the stratosphere and Earth; it contributes very little to the angular momentum budget of the atmosphere. This suggests that the important level for angular momentum transfers is at the top of the orographic layer. As long as the orographic shell is closely coupled to the Earth any injection of angular momentum across its upper boundary must be balanced by an opposite torque from the Earth, and any distribution of mountains and surface friction which can produce such a tightly coupled layer must inevitably produce the same torque. A more worthwhile approach to the coupling problem may be to examine how angular momentum is transferred within the atmosphere, rather than how the final transfer is effected at the Earth's surface. This will inevitably lead to a better description of the variation of the skin friction coefficient with wind speed and terrain roughness, and this in turn will improve long range weather prediction.

There has been ample demonstration that the stratosphere above 100 mb, and even above 50 mb, contains significant and variable angular momentum. The bienniel oscillation is largely confined above these levels, the residence for longer period oscillations of AAM may be here, and there is significant seasonal and short period power as well. If observations are too few and unreliable to permit a routine computation of the angular momentum at these heights, then a quasi-geostrophic extrapolation of lower winds should be developed and tested. This has already been done by Taylor et al [1985] and Rosen and Salstein [1985], but problems with geostrophy in the important equatorial regions need to be addressed.

Geodesists are of course mainly concerned with the net angular momentum of the atmosphere, but I would like to make an appeal for a better understanding of the angular momentum structure of the atmosphere. Are there distinct reservoirs for the annual, semi-annual, seven weeks and decade variations in angular momentum? Rosen and Salstein [1985] have have shown that 40 percent of the semi-annual and 16 percent of the annual fluctuation is from levels above 100 km (less than 16 mb) and the bienniel oscillation is confined to pressures below 100 mb, ie., above 20 km. How does the atmosphere exchange angular momentum with itself, both meridionally and vertically? Anderson and Rosen (1983), Whysall et al [1985] and Taylor et al [1985] have studied some of these questions, but I think that much more could be done.

I have already mentioned the super rotation of the atmosphere as a phenomenon whose cause is completely unknown. It makes intuitive sense that without the constant input of angular momentum to the atmosphere the atmosphere would eventually co-rotate with the shell, but this may be a misconception. Perhaps its makes more sense to hypothesize that the circulation of the atmosphere is such that it does in fact result in a minimum value for the torque between the atmosphere and shell, given that the Sun will always force some circulation.

There is in fact a steady input of angular momentum to the atmosphere from the solar diurnal tide. This is $2 \times 10^{15} \ kg \ m^2 \ s^{-2}$, enough to produce the super-rotation in 2000 years, but this would imply a similar time constant for the injection of excess angular momentum into the atmosphere to be coupled to the rotation of the shell, and that is clearly too long. This raises an interesting question in that, if the observed super-rotation proves to be stable on longer time periods than we have been able to examine closely, then what is the spin down time for the removal of any excess above the stable value?

Numerous attempts are being made to forecast the LOD. This has important practical applications to navigation because fewer resources need to be commited to Earth orientation measurements at critical times if we can predict the LOD in advance by using circulation models to extrapolate the AAM. Rosen et al [1987] report some success at forecasting the LOD at lead times of 1-10 days.

The excellent match of AAM derived LOD and observed LOD means that the residence time for any angular momentum dumped from the atmosphere to the oceans must be short. A 0.2 ms discrepancy between observed LOD and AAM derived LOD involves about $10^{25} kg \ m^2 \ s^{-1}$ of angular momentum. Whitworth and Petersen [1985] estimate the mean transport of the Antarctic Circumpolar Current, which probably contains most of the oceans axial angular momentum, to be $10^8 \ m^3 \ s^{-1}$, with fluctuations of comparable magnitude. The relative angular momentum of the current is then about $10^{25} \ kg \ m^2 \ s^{-1}$, in line with the budget imbalance. We need a more through study of the coupling of the ocean and atmosphere.

Acknowledgements. This work was supported by the Natural Sciences and Engineering Research Council of Canada, through operating grant A1084.

References

Anderson, J. R., and R. D. Rosen, The latitude-height structure of the 40-50 day variations in atmospheric angular momentum, Jour. Atmos. Sci., 40, 1584-1591, 1983.

Chao, B. F., Interannual length-of-day variation with relation to the Southern Oscillation/El Nino, Geophys. Res. Let., 11, No. 5, 541-544, 1984.

Dickey, J. O., and T. M. Eubanks, Atmospheric excitation of the earth's rotation: progress and prospects, Paper presented at the international symposium Figure and Dynamics of the Earth, Moon and Planets, Prague, Czechoslovakia. JPL geodesey and geophysics preprint no. 149, 1986.

Eubanks, T. M., J. A. Steppe, J. O. Dickey and P. S. Callahan, A spectral analysis of the earth's angular momentum budget, Jour. Geophys. Res., 90, No. B7, 5385-5404, 1985.

Feissel, M., and D. Gambis, La mise en évidence de variations rapides de la dureé du jour, Comptes Rendus Acad. Sci. Paris, 291, 271-273, 1980.

Feissel, M., and C. Nitschelm, Time-dependent aspects of the atmospheric driven fluctuations in the duration of the day, Annales Geophysiques, 3, 2, 181-186, 1985.

Hide, R., Rotation of the atmospheres of the earth and planets, Phil. Trans. R. Soc. Lond., 313, 107-121, 1984.

Hide, R., N. T. Birch, L. V. Morrison, D. J. Shea, and A. A. White, Atmospheric angular momentum fluctuations and changes in the length of day, Nature, 286, 114-117, 1980.

Iijima, S., and S. Okazaki, On the bienniel component in the rate of rotation of the earth, Jour. Geod. Soc. Jap., 12, 91, 1966.

Lambeck, K., The Earth's Variable Rotation: Geophysical Causes and Consequences, Cambridge Univ. Press., Cambridge, 449pp., 1980.

Lambeck, K., and A. Cazenave, The earth's rotation and atmospheric circulation- I. seasonal variations, Geophys. Jour. Roy. Astron. Soc., 32, 79-93, 1973.

Lambeck, K. and A. Cazenave, The earth's rotation and atmospheric circulation- II. The continuum, Geophys. Jour. Roy. Astron.Soc., 38, 49-61, 1974.

Lambeck, K., and P. Hopgood, The earth's rotation and atmospheric circulation, from 1963 to 1973, Geophys. Jour. Roy. Astron. Soc. 64, 67-89, 1981.

Langley, R. B., R. W. King, I. I. Shapiro, R. D. Rosen, and D. A. Salstein, Atmospheric angular momentum and the length of day: a common fluctuation with a period near 50 days, Nature, 294, 730-732, 1981.

Madden, R. A., and P. R. Julian, Description of global-scale circulation cells in the tropics with a 40-50 day period, Jour. Atmos. Sci., 29, 1109-1123, 1972.

Merriam, J. B., Tidal terms in universal time: effects of winds and mantle Q, Jour. Geophys. Res. 89, No. B12, 10,109-10114, 1984.

Morrison, L. V., Re-determination of the decade fluctuations in the rotation of the Earth in the period 1861-1978, Geophys. Jour. Roy. Astron. Soc., 58, 349-360, 1979.

Morgan, P. J., R. W. King and I. I. Shapiro, Length of day and atmospheric angular momentum: a comparison for 1981-1983, Jour. Geophys. Res., 90, B14, 12645-12,652, 1985.

Munk, W. H., and R. Miller, Variations in the earth's angular velocity resulting from fluctuations in atmospheric oceanic circulation, Tellus, 2, 93-101, 1950.

Rosen, R. D., and D. A. Salstein, Variations in atmospheric angular momentum on global and regional scales an the length of day, Jour, Geophys. Res., 88, No. C9, 5541-5470, 1983.

Rosen, R. D. and D. A. Salstein, Contribution of stratospheric winds to annual and semiannual fluctuations in atmospheric angular momentum and the length of day, Jour. Geophys. Res., 90, No. D5, 8033-8041, 1985.

Rosen, R. D., D. A. Salstein, T. Nehrkorn, M. R. P. McCalla, A. J. Miller, J. O. Dickey, T. M. Eubanks, and J. A. Steppe, Medium range numerical forecasts of atmospheric angular momentum, JPL Geodesy and Geophysics preprint No. 159, 1987.

Stoyko, N., Sur l'irregularité de la rotation de la terre, Comptes Rendus Acad. Sci. Paris, 203, 29, 1936.

Swinbank, R., The global atmospheric angular momentum balance inferred from analyses made during fgge. Quart. Jour. Roy. Met. Soc.,111, 977-992, 1985.

Taylor, H. A., and H. G. Mayr, Contributions of high-altitude winds and atmospheric moment of inertia to the angular momentum-earth rotation relationship, Jour. Geophys. Res., 90, No. D2, 3889-3896, 1985.

Wahr, J. M., and A. Oort, Friction and mountain torque estimates from global atmospheric data, Jour. Atmos. Sci., 41, No. 2, 190-204, 1984.

Weickmann, K. M., Intraseasonal circulation and outgoing longwave radiation modes during northern hemisphere winter, Mon. Weather Rev., 111, 1838-1858, 1983.

Whitworth, T. and R. G. Peterson, Volume transport of the Antarctic Circumpolar Current from bottom pressure measurements, Jour. Phys. Oceanography, 15, No. 6, 810-816, 1985.

Whysall, K. M., R. Hide, and M. J. Bell, 1985. Current work on the earth's rotation at the United Kingdom Meteorological Office, in Proceedings of the International Conference on the Earth Rotation and the Terrestrial Reference Frame, 2, July 31-Aug. 2 1985 Columbus, Ohio, 771 pp., 1985.

EARTH ROTATION AND CLIMATIC PERIODICITIES

J.P. ROZELOT

Observatoire du Pic du Midi-Toulouse
Present address: CERGA, Avenue Copernic, 06130 GRASSE, France

D. SPAUTE

Planetary Science Institute, Speedway, 2040 Tucson, U.S.A

Abstract. Earth rotation and polar motion studies are embarking on a new area with, on one hand, the advent of highly accurate space geodesic techniques such as S.L.R., L.L.R. or V.L.B.I., and on the other hand, the availability of good climatic data such as atmospheric angular momentum measurements. It is now possible to examine the long period variations of the both series of data: we will show here the strong correlation existing between the periodicities deduced from U.T.1 measurements and those of climatic series.

Introduction

Since Hide et al. [1980] first posed the problem of jointly improving the changes in the global atmospheric angular momentum and short period fluctuations of Earth rotation and polar motion, progress has been slow. Recent work by Dicke et al. [1986] and Eubanks et al. [1986] indicate for instance that "fluctuations in Earth rotation over the time scale of a year or less are dominated by atmospheric effects with the wind term dominating" and "that there is a relationship between the Southern Oscillation and variations in the duration of the day". The basic idea contained in such works is to try to prove the physical reality of such relationships, that would mean, in return, that the Earth's rotation might be able to predict climatic paroxysms, such as El Niño events. This would be of importance either to determinate long range weather forecasting, by extrapolation of accurate measurements of Earth rotation parameters, or to attempt to reconstruct past climate by considering long-term changes in the rotation of the Earth.

Methodology

We simply made a comparative spectral analysis of the Earth rotation on one hand and climate variability on the other one, to facilitate a more careful study of their possible physical interpretation.

a/ **Concerning the Earth's rotation** data has been provided by the "Bureau International de l' Heure à Paris" for the period going from 1880 to 1984. Note that the LOD gradually increased from about 1928 until 1972, and then began a period of decline that has lasted until the present.

The method used is this of Sneyers [1976] (see also Ulrich and Clayton [1976]), which is based upon an autoregressive least squares analysis and is satisfactory for frequencies which are low in comparison with the record length of the data. This method is more suitable for our purpose (avoiding spurious periodicities) than those allowed by classical Fourier transforms analysis.

Results obtained on $(UT1-TA)_{BIH}$ data are given in Table 1, for the 2-12 year range.

A great number of the other authors have already made such spectral analysis; it can be quoted

TABLE 1. Earth's rotation periodic terms in the 2-12 year range.

2.19	7.03	(in years)
2.83	8.56	
3.75	10.2	
4.9	12.2	

(Korsun and Sidorenko [1971], Lambeck and Cazenave [1973], Okasaki [1977], Vondrak [1977], Emetz and Korsun [1979], Zeng-Da-Wei [1980], Pomia and Proverbio [1980], Djurovic [1981], Carta et al. [1982]. Results are in good agreement.

b/Concerning climatic data good times series are now available. They last for about 200 years; there is a 3 centuries temperature record for Central England and an estimation of dryness/wetness index of 5 centuries for China.

Climatological records used for this study are given in Table 2.

The computational procedure is described in Pecker [1982], Vines [1981], Galindo and Otoala [1981]. Results are given in Figures 1 to 7 for the stations involved. In recent years, several groups have published time series of large scale average surface air temperature, which are analysed in Wigley et al. [1985]. The same method of least

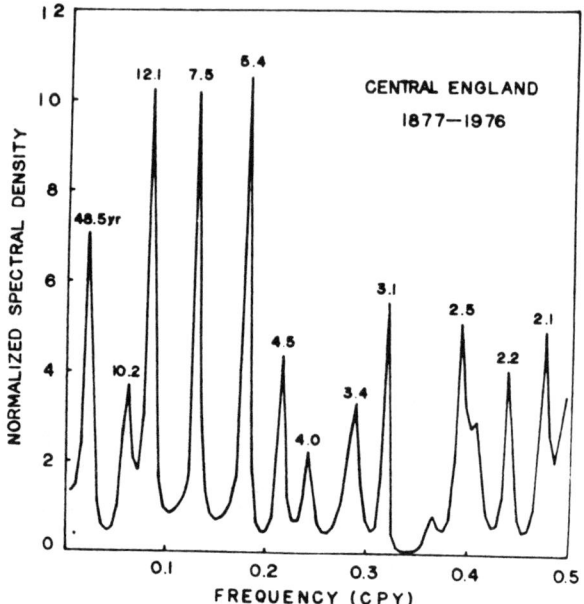

Fig.1. Annual mean air temperatures spectrum for Central England (1877-1976), as estimated by the non-integer spectral technique. Confidence levels according to the F test are indicated (see also Galindo et al., 1981).

TABLE 2. Climatological records according to various cities lasting for four centuries.

City or Geographical area	Period	Length
Mexico city	100 yr	1876-1976
Lisbon	100 yr	1876-1976
Central England	100 yr	1876-1976
and also	314 yr	1659-1953
New Zealand (20 stations)	60 yr	1900-1960
South Africa (20 stations)	63 yr	1910-1973
Paris	88 yr	1885-1973
and also	216 yr	1757-1973
Upernivik	88 yr	1885-1973
Quixeramobim	64 yr	1896-1960
Accra	72 yr	1888-1960
Entebe	55 yr	1905-1960
Batavia	94 yr	1866-1960
Singapour	49 yr	1911-1960
Shangai	509 yr	1770-1979

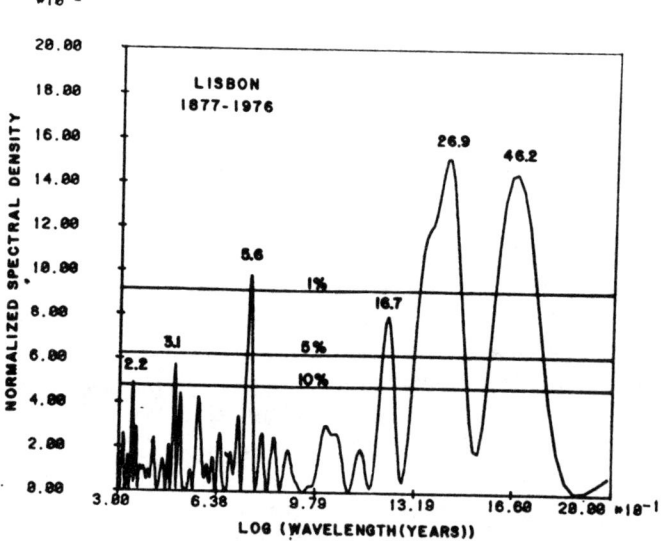

Fig.2. Annual mean air temperatures spectrum for Lisbon, Portugal (1900-1900), as estimated by the non-integer spectral technique. Confidence levels according to the F test are indicated (see also Galindo et al.,1981).

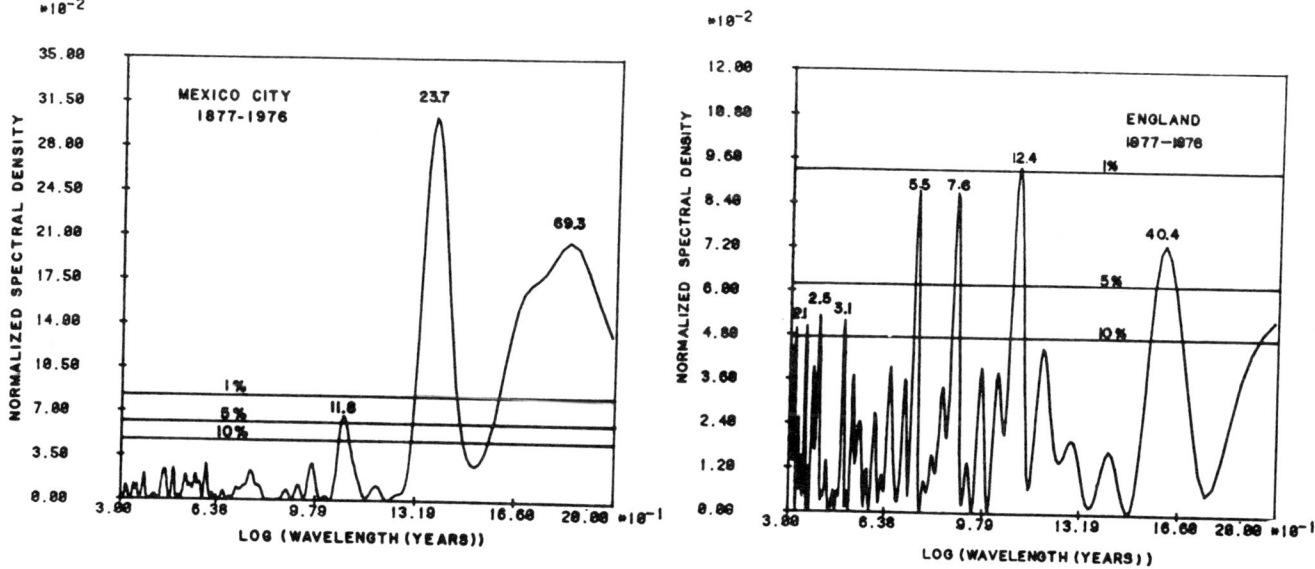

Fig.3. Annual mean air temperatures spectrum for Mexico City (1877-1976) as estimated by the non-integer spectral technique. Confidence levels according to the F test are indicated (see also Galindo et al.,1981).

Fig.4. Annual mean air temperatures spectrum for Central England (1877-1976), as estimated by the Maximum Entropy method of spectral analysis (see also Galindo et al.,1981).

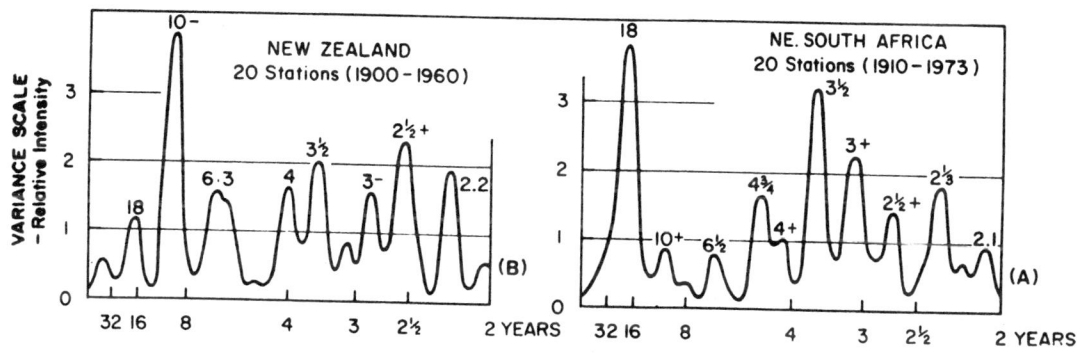

Fig. 5. Power spectra from analysis of 20 climatological stations in New Zealand (period 1900-1960) and in South Africa (period 1910 -1973); see also Vines, 1981.

square analysis is in progress within this data. We get a lot of 67 periodicities, which can be grouped, as there is no physical reason for a spatial stationarity, in eight classes (see Figure 8). These frequency bands are as follows:

2.1 - 3.5	years	18 - 24	years
4.0 - 6.5	"	25 - 40	"
7.3 - 8.3	"	50 - 70	"
10.0 - 17.0	"	200 - 250	"

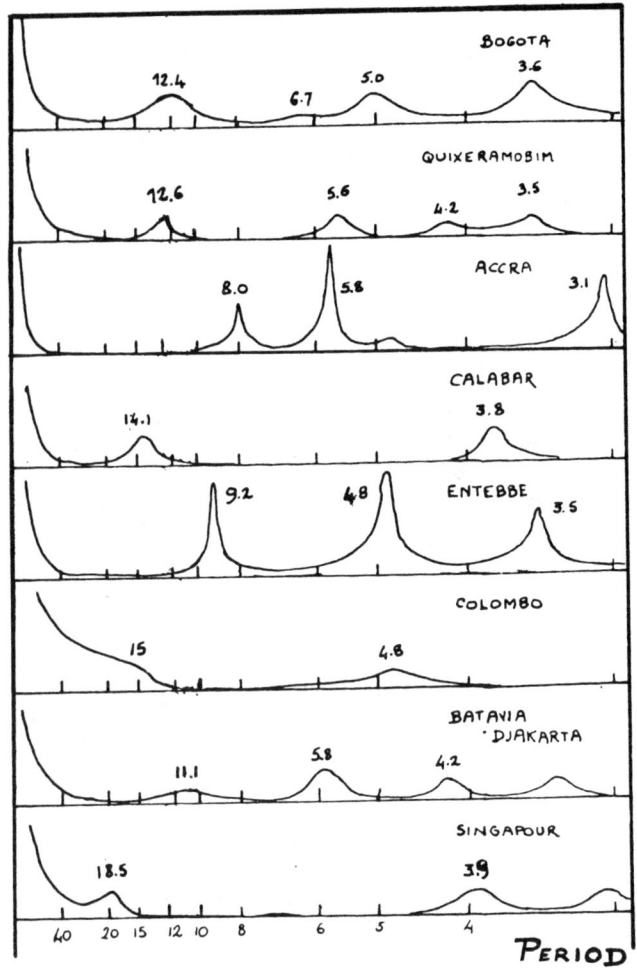

Fig. 6. Annual mean air temperatures spectrum for eight equatorial climatological stations (see also Pecker, 1982).

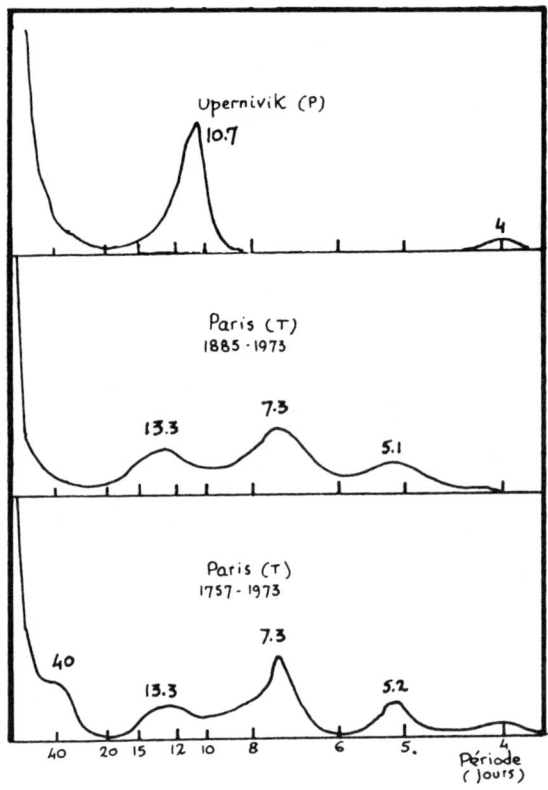

Fig. 7. Annual mean air temperatures spectrum for three climatological stations of northern latitudes (see also Pecker, 1982).

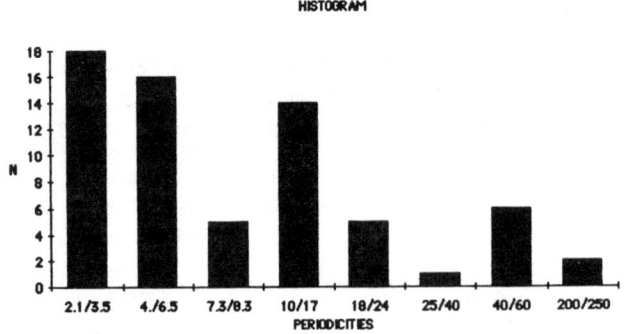

Fig. 8. Histogram of the 67 periodicidies deduced from the seven annual mean air temperatures spectrum given in Fig.1 to 7.

In each class, we computed the mean, and we plotted the climatic periodicities versus the Earth's rotation periodicities. Results (see Figure 9) give a correlation coefficient between the two series of 0.9.

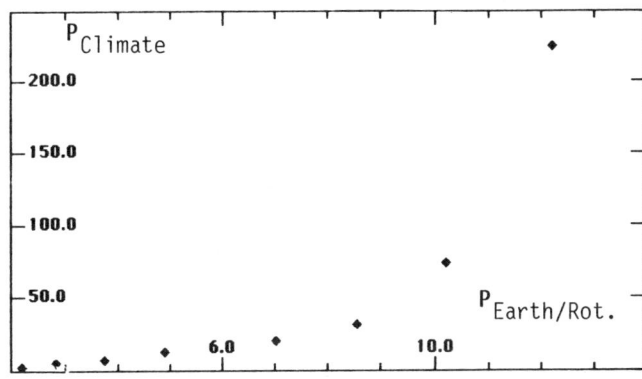

Fig. 9. Graph showing the relationship between the climatic periodicities and the Earth's rotation periodicities.

Conclusion

First of all, it may be argued that the relationship presented here is merely accidental. This objection may arise for at least two reasons:

a/ All climatic periodicities found by analysing records are regularly critised, and we are still in a position that nothing can be said as to whether they exist or not, and mainly because a physical explanation cannot be given.

b/ One problem in relating the free oscillation of the atmosphere to geodetic measurements is their irregular nature. Statistical analysis, although useful, be employed and interpreted with caution in climatology. Further work is required to assess the validity of the results.

However, when our solar system is realized to be a complete whole, it can be suggested that the short lived variabilities (periods less than 10 years) are related more to the general circulation and the longer ones to external forces. Briefly speaking and presented here as working hypotheses, it can be suggested that :

a/ at low frequencies, climate is influenced by astronomical cycles, and mainly by variations in the elements of the Earth orbit;

b/ at high frequencies, fluctuations in Earth rotation are dominated by atmospheric effects;

c/ between the two fields, it may exist external forces which must be investigated.

References

Hide, R., Birch, N.T., Morrisson, L.V., Shea, D.J., and White, A.A., Atmospheric angular momentum fluctuations and changes in the length of the day, Nature, 286,114,1980.

Eubanks, T.M., Steppe, J.A., and Dicke, J.O., The Niño, the southern oscillations and the Earth rotation, Jet Propulsion Laboratory reprint, N° 143,1986.

Dicke, J.O., Eubanks, T.M., and Steppe, J.A., High accuracy Earth rotation and atmospheric angular momentum, Jet Propulsion Laboratory reprint, N° 144, 1986.

Sneyers, R., Application of least squares to the search for periodicities, Journal of Applied Meteo. 15, 387-393,1976.

Ulrich, J.T., and Clayton, R.W., Phys. Earth Planet. Inter., 12,188,1876.

Korsun, A.A., and Sidorenkov, Soviet Astro , 14, 896,1971.

Lambeck, K., and Cazenave A., Geophys. Journal Roy. Astron. Soc., 32, 79,1973.

Okasaki, S., Pub. Astron. Soc. Japan., 29, 619,1977.

Vondrak, J., Studia Geophys.Geod., 21,107,1977.

Emetz, A.I., and Korsun, A.A. , Time and Earth's rotation, I.A.U. Symposium, 82, 59, 1979.

Zeng Da Wei, Chinese Astronomy, 4, 298, 1980.

Poma, A. and Proverbio, C., Voroff Zentral Inst. Phys. der Erde, 1980.

Djurovic, D., Solar activity and Earth's rotation, Astron. Astrophys., 100,156,1981.

Carta, F., Chlistovsky, F., Manara, A., and Mazoleni, F., A comparative spectral analysis of the Earth's rotation and the solar activity, Astron. Astrophys., 114, 388, 1982.

Pecker, J.C. , in "Compendium in Astronomy", ed. Mariapoulos E.G., Reidel D. pub. Co., Dordrecht, p. 51, 1982.

Vines, R.G., in "Sun and climate", ed. C.N.E.S., Toulouse, p. 55,1981.

Galindo, I. and Olaola, J.A. in "Sun and climate", ed. C.N.E.S.,Toulouse, p. 67,1981.

Wigley, T.M.L., Angell, J.K., and Jones, P.D., Analysis of the temperature record, DOE/ER- 0235 report, U.S. Department of Energy, p. 57,1985.

ENSO-RELATED SIGNALS IN EARTH ROTATION, 1962-87

Martine Feissel

Bureau International de l'Heure, Observatoire de Paris
61, avenue de l'Observatoire F-75014 Paris

Jean Gavoret

Institut de Physique du Globe de Paris
4, Place Jussieu, F-75005 Paris

Abstract. A filter derived from the CENSUS X-11 Seasonal Adjustment algorithm is applied to a 1962-87 time series of the duration of the day. Features in the seasonal and irregular parts of the time series that would be associated with the occurence of an El Niño Southern Oscillation (ENSO) event are looked for. ENSO Northern Hemisphere winters and the preceding Southern Hemisphere winter tend to exhibit westerlies seasonal anomalies, with an increase in amplitude and frequency of the short term oscillations. The seasonal oscillation, except in the case of the strong 1982/83 event and, less importantly, in 1976-77, is not highly perturbed.

Introduction

The ENSO phenomenon is known as an association of large seasonal anomalies in sea surface temperature and in zonal winds which affect every two to six years an important part of the Earth's hydrosphere and atmosphere. Through the conservation of the total angular momentum of the Earth, the corresponding zonal wind anomalies are expected to reflect themselves in the rotational velocity of the Earth, the variations in the duration of the day (LOD) being proportional to the variations in the axial angular momentum of the atmosphere [Barnes et al., 1983]. The strong ENSO event of 1982/83 has affected the variation in the duration of the day to a large extent [Eubanks et al., 1985].

On the other hand, in addition to the dominant seasonal variations, the Earth rotation is permanently subject to quasi-periodic oscillations in LOD in the range of 20 to 100 days, with peak-to-peak amplitudes of 0.2 to 0.6ms. These oscillations are totally correlated with those in the atmospheric angular momentum (AAM). They can start at any time, with a slight preference for the Southern Hemisphere winters, and persist for variable durations, ranging from 170 to 450 days [Feissel and Nitschelm, 1985].

The mechanisms which give rise to ENSO events as well as those which could explain the 20-100 days oscillations are still largely hypothetical. These two phenomena might be related to one another [M.Ghill, personal communication 1987]. The aim of the present analysis is to isolate particular patterns in the time series of the duration of the day which might be related to the occurence of El Niño events over 1962-87. During this period, six of these events are recorded, in 1963/64, 1965/66, 1969/70, 1972/73, 1976/77, 1982/83. A weak El Niño event is considered to have taken place in 1986/87.

Data analysis

The data analyzed is the series of universal time, UT1-TAI, in the latest BIH combined solution, ERP(BIH) 87 C 02 [Feissel and Guinot, 1988]. It is available at 5-day interval over 1962-87. Starting with 1967 the individual 5-day values are statistically independent; before 1967 they are based on a smoothed series with a frequency cutoff at 10c/year. In order to recover the short-term variations, the series ERP(BIHSO)84 A 02 [Li, 1984], based on optical astrometry, is used for the years 1962-66.

The series of the duration of the day is obtained by taking the time derivative of the series of UT1-TAI, corrected for the effect of zonal tides according to [Yoder et al. 1981]. The pre-

Copyright 1990 by
International Union of Geodesy and Geophysics
and American Geophysical Union.

cision of an individual determination improves with time, evolving from 0.20ms over 1962-1966 and 0.10ms over 1978-82 to 0.01ms starting with 1984.

The multifrequency time series of the duration of the day is split into three parts, the low frequency trend, the seasonal oscillation and the remaining irregularities in the range of a few days to a couple of years. For this purpose, a filter derived from the CENSUS X-11 algorithm [Shiskin et al.,1965] is applied. This filter is in wide use in time series analysis in the fields of economy.

CENSUS X-11 consists in a standard sequence of repeated implementations of three basic operations:

1- N-years running mean (rectangular filter) to derive a trend; in the present study, N=3.
2- Running mean over five years (non rectangular filter) of data for the same dates inside a year.
3- Severe intermediate filtering of irregularities, 30 to 50 percent of the individual points being rejected or down-weighted prior to the re-implementation of the operations 1 or 2.

The original time series as well as the three time series obtained by applying the filter to it are plotted on figure 1.

Interseasonal patterns

The most dramatic feature in the irregular signal is a triangular pattern associated with the 1982/83 ENSO event. Starting with 1982.0 there is a steady increase of the westerlies anomaly (increase in the duration of the day) which culminates in January-February, then a steady decrease until the end of 1983. Large short term oscillations are superimposed on this interannual pattern. Smaller scale triangular patterns similar to the 1982/83 one can be recognized for 1963/64 (a weak El Niño), 1977/78 (non-El Niño) and 1986/87 (weak El Niño). A second group, 1969/70 and 1976/77 is characterized by a westerlies anomaly in the first year and an easterlies anomaly in the second year. The third group, 1965/66 and 1972/73, shows a westerlies anomaly spreading over the two years.

Seasonal characteristics

Figure 2 shows the amplitudes and phases of the annual and semi-annual oscillations deduced from the yearly least-squares analysis of the detrended series. The formal uncertainties are approximately 0.03ms on the amplitudes and 10 days on the phases.

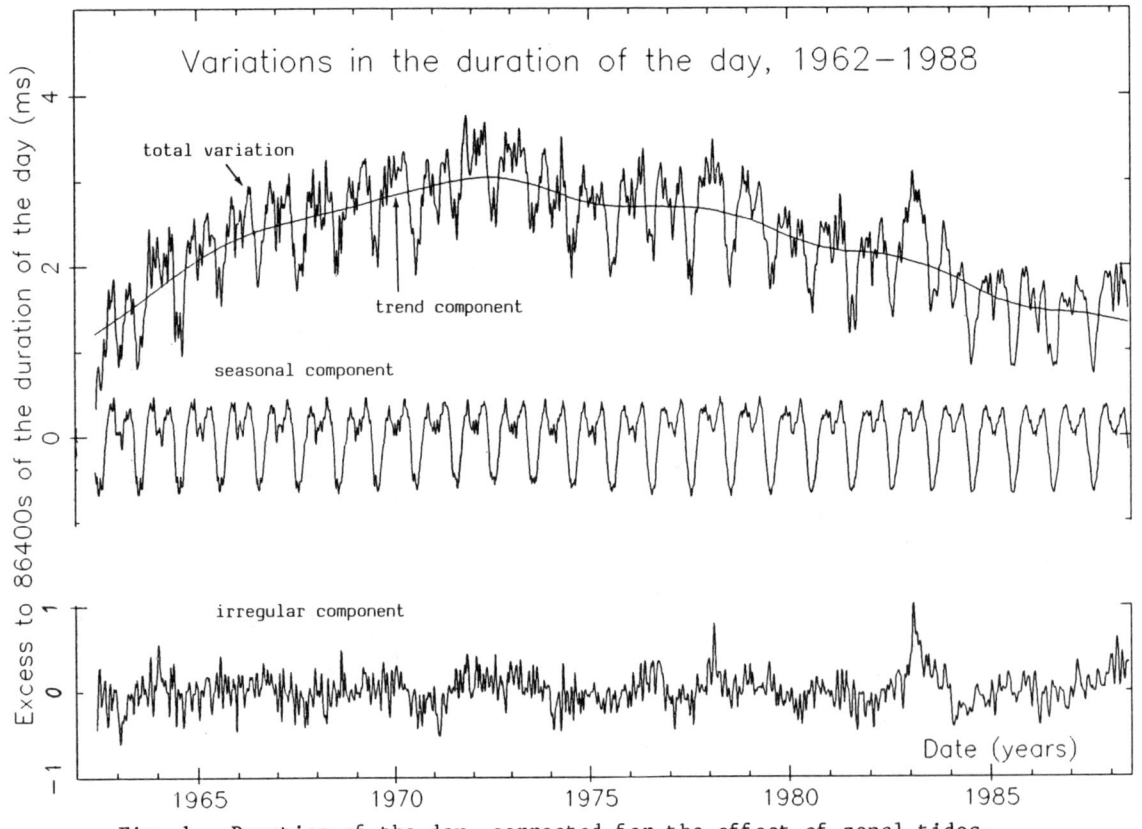

Fig. 1. Duration of the day, corrected for the effect of zonal tides. Total variation and decomposition in trend, seasonal and irregular components.

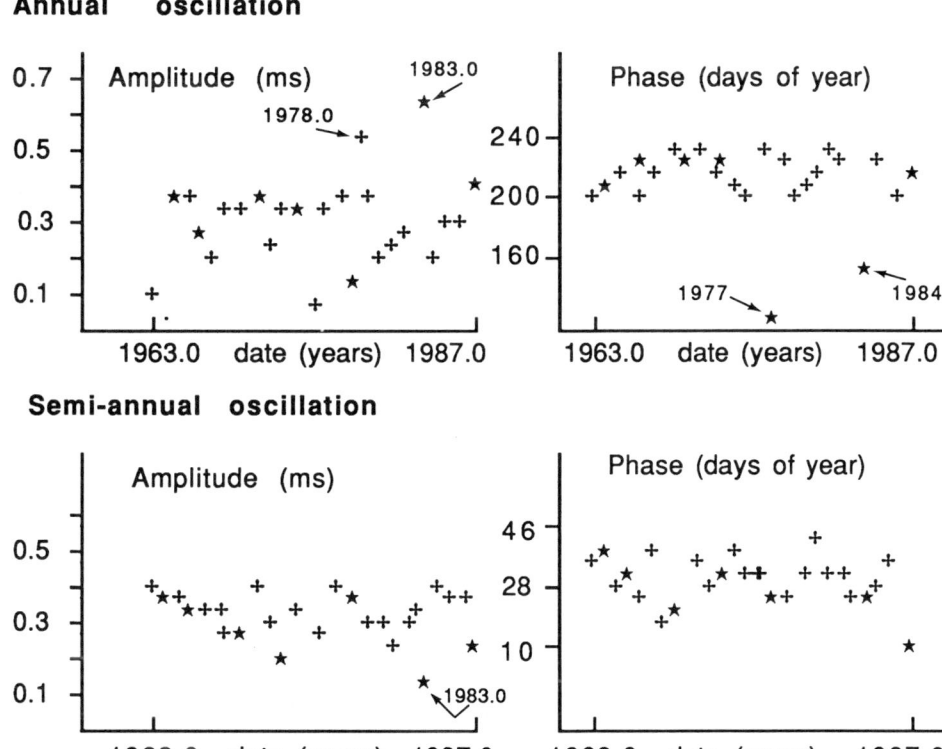

Fig. 2. Annual and semi-annual oscillations. The values for the Northern Hemisphere Winters with an ENSO event are marked by *.

The amplitudes are remarkably stable; in particular the data of the first years, which have a lower precision in the short term, have a stability in the seasonal frequencies which is comparable to the more recent data, based on VLBI. The only remarquable anomalies are the high annual and low semi-annual terms for 1983.0, and the high annual term for 1978.0.

The phases of the annual terms indicate the date of the minimum of the duration of the day, when the global westerlies start to increase. Except in two cases, this change takes place within 20 days of August 7. Two anomalous phases are found in 1976/77 (early April 1977) and 1983/84 (mid-June 1984). The phase of the semiannual term, defined also as the date when the westerlies start to increase, is within 20 days of 30 Jan/30 July. The values of the phases for ENSO events years do not deviate significantly from the other ones.

Higher Frequency Oscillations

In this section, the irregular series of figure 1 is treated by a band pass filter (difference of two Vondrak smoothings [Vondrak,1977]) which retains the oscillations between 20 and 100 days. The series is then organised in 48 segments of a half-year duration, centered on the start and the middle of each year, from 1963.0 to 1987.0. Two of the half-year segments correspond to the Northern Hemisphere Winter (NHW), among which six had an ENSO event. Similarly the other 24 half-year intervals comprise 11 "normal" Southern Hemisphere winters (SHW), seven SHW preceding an ENSO event (SHW-1), and six SHW following it (SHW+1). The average rms LOD in the 20-100d span is listed in Table 1 for the five types of half-years.

In each of the 48 intervals we select separately by least squares the periods between 20 and

TABLE 1. RMS signal and Number of component per half-year in the 20-100 d irregularities

half-year	rms LOD (ms)	mean number of components 0.10 to 0.15ms	0.15 to 0.30ms	total
normal NHW	0.20	1.1	0.5	1.6
normal SHW	0.21	1.3	0.3	1.6
SHW-1	0.22	1.1	1.0	2.1
ENSO	0.24	1.2	1.2	2.4
SHW+1	0.22	1.6	0.2	1.8

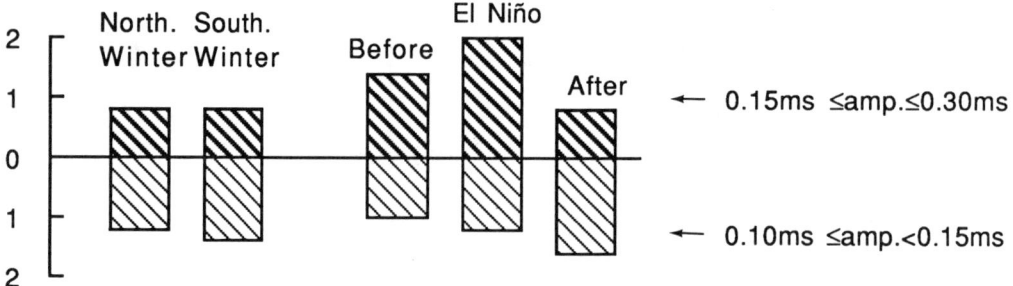

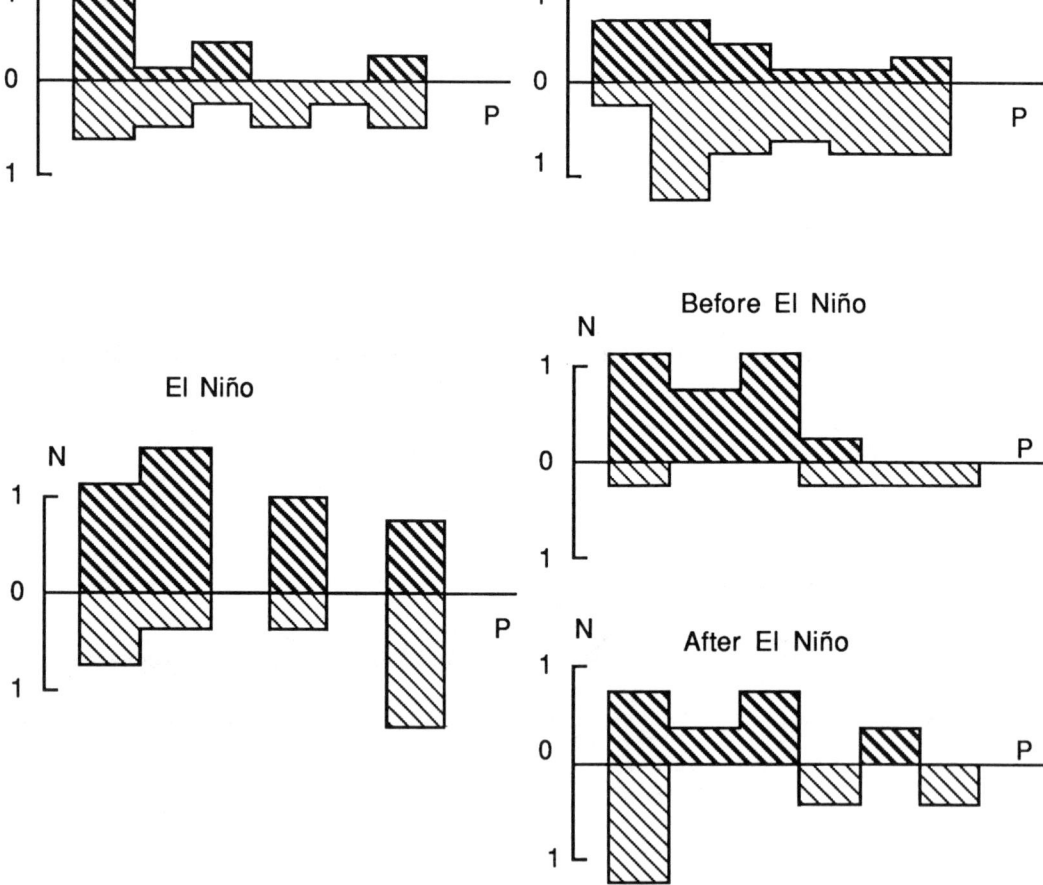

Fig. 3. 20-100d oscillations : interseasonal variability. The period slots have a width of 14d ; they are centered on 26d, 40d, 54d, 68d, 81d, and 94d (from left to right).

100 days for which statistically significant amplitudes over 0.1ms are adjusted. Between zero and five components with amplitudes larger than 0.1ms can be selected in each of the half-years. The mean numbers of such components for the five classes of half-years are listed in Table 1. The ENSO NHW and the preceding SHW are characterized by a larger number of high amplitude components; the SHW following ENSO have a larger number of low amplitude components.

The selected components are not evenly distributed in the range of periods from 20 and 100 days. Six period slots approximately 14 days wide are considered. The corresponding number of components are plotted in Figure 3. The normal NHW has a slight excess in high amplitude components in the period slot 20-33 days and the SHW a slight excess in low amplitude components in the slot 34-47 days; otherwise, both have a rather even distribution relative to periods. The distributions for the ENSO NHW and the related SHW-1, SHW+1 are more irregular. They suggest a tendency to have large amplitude components with periods under 60 days before an ENSO event, similar tendency after it, but with lower amplitudes; the distribution for ENSO NHW has several peculiarities : deficiency of components in the period slots 48-61d and 76-88d, excess of large amplitudes for periods under 47 days, excess of low amplitudes for periods over 89 days.

Summary

Our analysis of the Earth rotation irregularities over 1962-87 leads to several remarks in relationship with ENSO events.

The 1982/83 episode corresponds to the most striking large scale irregularity that is visible in the duration of the day series, with a large triangular pattern spreading over two years. The peak-to-peak variation is 1.6ms, superimposed to the mean seasonal variation of total amplitude 0.96 ms. This large scale irregularity reflects itself in the annual and semi annual terms, the first one having an amplitude twice as large as the one of the other years and the second one only half. The phase of the annual term over 1983.5-1984.5 indicates that the start of the seasonal increase of the westerlies after ENSO was earlier by 1.5 month than the usual date.

The second remarkable seasonal/interseasonal anomaly covers the years 1976 through 1978, which include the 1976/77 El Niño event. Over these years, the seasonal cycle is perturbed by a relatively high bi-annual oscillation (0.6 ms peak-to-peak). A sudden increase of 0.8ms in the duration of the day, spreading over one month, is clearly visible in early 1978.

In the 20-100d range of periods, the non-ENSO Northern and Southern Hemisphere Winters have similar spectra, with an average of 1.6 component with amplitudes generally under 0.15ms. The ENSO Northern Hemisphere Winters and the preceding half-years are characterized by an average of 2.2 components with relatively large amplitudes. The three half-years before, during and after an ENSO events show and excess of large amplitude (over 0.15ms) and high frequency (periods under 50-60d) components.

References

Barnes,R., Hide,R., White,A., Wilson,C.:1983. Atmospheric angular momentum fluctuations, length-of-day changes and polar motion, Proc. Roy. Soc. Lond.,A 387 , 31-73.

Eubanks,T.M., Steppe,J.A., Dickey,J.O.:1986. The El Niño, the Southern Oscillation and the Earth Rotation, Proc.NATO Advanced Research Workshop on Earth rotation:solved and unsolved problems , Reidel, 163-186.

Feissel,M. and Nitschelm,C.:1985. Time dependent aspects of the atmosphere driven fluctuations in the duration of the day. Annales Geophys.3 , 181-186.

Feissel,M. and Guinot,B.:1988. A new homogeneous series of the Earth Rotation Parameters, 1962-1987 Annual Report of the BIH for 1987, D-79-84, Observatoire de Paris.

Li,Z.X.:1984. Nouvelle évaluation des paramètres de la rotation de la Terre d'après les mesures optiques, 1962.0-1982.0. Thèse d'Etat, Observatoire de Paris.

Shiskin,J., Young,A.H., and Musgrave,J.C.:1965. The X-11 variant of the Census Method II seasonal adjustment program. U.S.Dept.of Commerce, Bureau of the Census, Technical Paper nr 15 .

Vondrak,J.:1977. Problem of smoothing observational data II. Bull.Astron.Inst.Czech.28, 84-89.

Yoder,C.F., Williams,J.G., and Parke,M.E.:1981. Tidal Variations of Earth Rotation, J. of Geophys. Res. 86 , 881-891.

FORECASTING ATMOSPHERIC ANGULAR MOMENTUM AND LENGTH-OF-DAY
USING OPERATIONAL METEOROLOGICAL MODELS

R. D. Rosen, D. A. Salstein and T. Nehrkorn

Atmospheric and Environmental Research, Inc.
840 Memorial Drive, Cambridge, MA 02139

J. O. Dickey, T. M. Eubanks and J. A. Steppe

Jet Propulsion Laboratory, California Institute of Technology
4800 Oak Grove Drive, Pasadena, CA 91109

M. R. P. McCalla and A. J. Miller

Climate Analysis Center/NMC
W/NMC53, WWB, Washington, D.C. 20233

Abstract. Forecasts of zonal wind fields produced by the medium-range forecast model of the U.S. National Meteorological Center are used to create predictions of the atmosphere's angular momentum at lead times of 1-10 days. The skill of these forecasts, which are of interest to those concerned with monitoring changes in the length-of-day for navigational purposes, is assessed, and the regions in the atmosphere that contribute most importantly to forecast errors are identified.

Introduction

Forecasts of parameters related to the Earth's orientation in space are used in precise navigation, both for terrestrial activities and for deep space missions. Techniques currently used by the geodesy community to produce such forecasts on 1-10 day time scales are statistical in nature, involving extrapolating time series of observed values of these parameters. We have been examining a new approach to forecasting changes in length-of-day (Δl.o.d.) on these time scales by using the output from operational numerical weather prediction models. This approach is based on the high correlation shown to exist generally between Δl.o.d. and changes in the atmosphere's angular momentum, M [Hide et al., 1980; Rosen and Salstein, 1983, 1985;

Copyright 1990 by
International Union of Geodesy and Geophysics
and American Geophysical Union.

Eubanks et al., 1985]. Hence dynamically based atmospheric forecasts of M might serve as proxies for useful forecasts of Δl.o.d.

A necessary requirement for the success of this new approach is that the dynamical forecasts of M be more skillful than those of statistical competitors. We have chosen persistence as the statistical model against which to judge the quality of the dynamical forecasts, because there is some evidence to suggest that fluctuations in M on the 1-10 day time scales obey a power law spectrum that is consistent with a random walk process [Eubanks et al., 1985]. In Rosen et al. [1987], we demonstrated that forecasts of M derived from the medium-range forecast (MRF) model of the U.S. National Meteorological Center (NMC) were on average more skillful than persistence at all lead times from 1 to 10 days for the period December 1985 - November 1986. Here we extend the results of Rosen et al. by incorporating data through June 1987 and by examining in more detail the sources of the errors that are present in the MRF forecasts of M. Further discussion of the subject addressed here may be found in Rosen et al. [1987].

Data and Analysis

Beginning in mid November 1985, forecasts of the zonal wind made by the MRF at 00 UTC each day for conditions at each successive 24-hour period out to 10 days have been accessed by us for the purpose of creating time series of M forecasts. We calculated M according to the relationship

$$M = \frac{2\pi a^3}{g} \int_{1000}^{100} \int_{\pi/2}^{-\pi/2} [u] \cos^2\phi \, d\phi \, dp, \quad (1)$$

where a is the mean radius of the earth, g acceleration due to gravity, ϕ latitude, p pressure, and [u] zonal-mean zonal wind which is available at every 2½° of latitude at the 10 standard pressure levels between 1000 and 100 mb.

To provide a measure of the skill of the MRF forecasts of M, we have computed values of a simple skill score, S, for each forecast lead time:

$$S = \frac{\sigma_p - \sigma_{MRF}}{\sigma_p} \times 100\%, \quad (2)$$

where σ_{MRF} is the root-mean-square error of the MRF forecasts over a given period of time, and σ_p is the root-mean-square error of persistence forecasts, i.e., forecasts in which forecasted $M(t_o + \Delta t)$ = observed $M(t_o)$ for each forecast lead time Δt. Hence, S measures the percentage improvement of the MRF over persistence forecasts so that $-\infty < S < 100\%$. Values of S have been computed for each full calendar month in our record, beginning therefore with December 1985, and for the entire period 1 December 1985 - 30 June 1987.

Skill of the MRF M Forecasts

Daily time series of the differences between values of M forecasted by the MRF or by persistence and the observed values of M are presented in Figure 1 for forecast lead times of 2, 5 and 10 days. As is evident from the figure, forecast errors, both for the MRF and persistence, increase as lead time increases, not a surprising result. Significantly, though, note that for a given lead time, the MRF forecast errors are typically smaller than those for persistence, a result evidenced in Figure 2 by the positive values of S for the MRF at all forecast lead times when averaged over the 18-month period from 1 December 1985 - 31 May 1987.

It is also worth noting that the MRF forecast errors contain a low-frequency component in addition to high-frequency noise for the longer lead

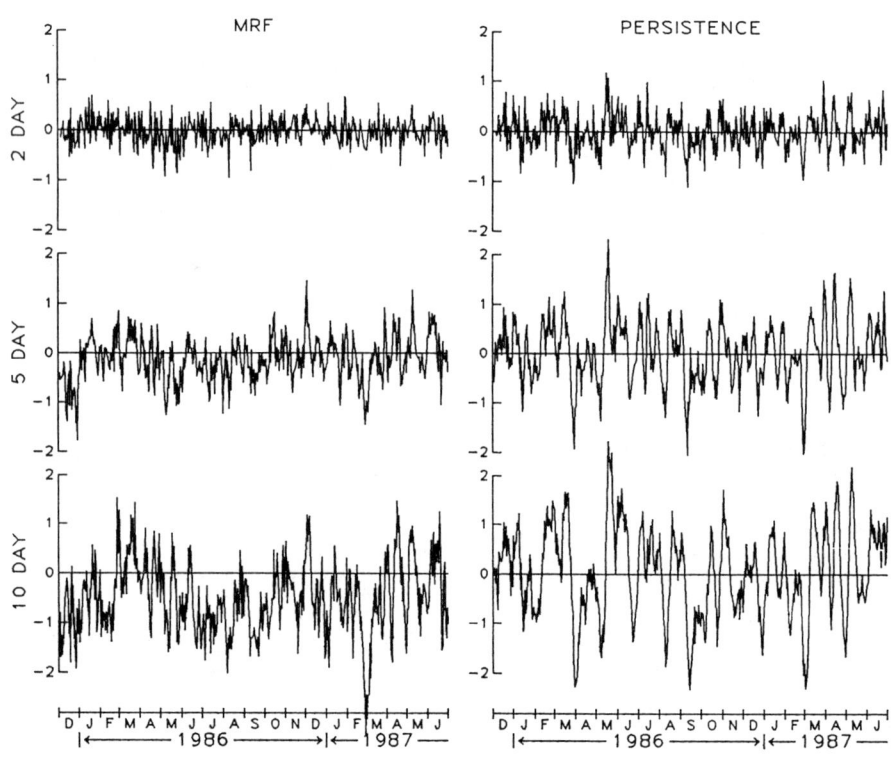

Fig. 1. Daily time series during 1 December 1985-30 June 1987 of the difference, MRF forecasted values minus observed values of M, for forecast lead times of 2, 5, and 10 days (left); and the difference, persistence-based M forecast values minus observed values of M, for the same forecast lead times (right). Units are 10^{25} kg m^{2}s^{-1}.

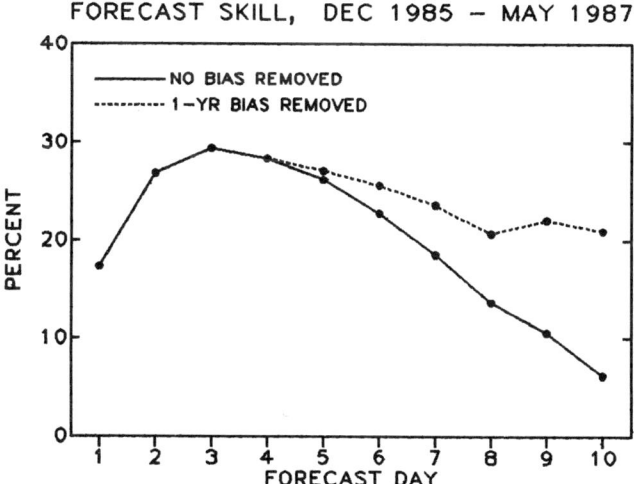

Fig. 2. Skill score, expressed as percentage improvement of the MRF over persistence, as a function of forecast lead time for the entire period 1 December 1985-30 June 1987 (solid line). The dashed line represents the skill for forecasts of 5 days and beyond obtained by removing the mean bias from forecasts made for the previous year.

time forecasts. Although this low-frequency component is not simply a uniform bias, we experimented with removing such a mean bias from the MRF forecasts for lead times of 5 to 10 days. The improvement in the MRF forecast skill proved to be fairly insensitive to the length of the period over which previous forecast errors were averaged, beyond a period of about 20 days; below this period the procedure was less beneficial. In Figure 2, we include the results of removing the average error from MRF forecasts that were made for as much of the previous 1 year as was available. An improvement in S for all the lead times between 5 and 10 days has occurred, with those for 9- and 10-day forecasts representing statistically significant reductions, at the 95 percent level of confidence, of up to 16 percent in σ_{MRF}.

Although the MRF is more skillful than persistence on average over our study period, inspection of Figure 1 does reveal periods when the MRF errors are larger than those for persistence. This result is evident in Figure 3, which depicts values of S for 2-, 5- and 10-day forecasts as a function of calendar month. A number of instances of negative skill are apparent, especially for the 10-day forecasts. Prior to February 1987 (and with the notable exception of December 1985), most of these instances involved marked decreases in the persistence error rather than significant changes in the MRF error, as shown by the plots of σ_{MRF} and σ_p in Figure 4. During February and March 1987, however, σ_{MRF} increased notably for the 10-day forecasts, for reasons not yet known to us. By April 1987, the MRF errors began to return to lower levels.

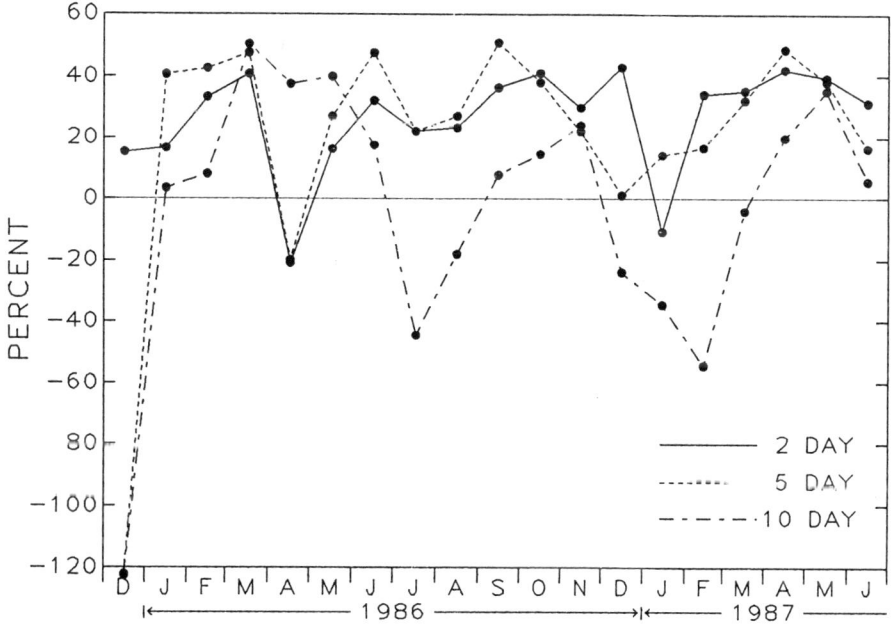

Fig. 3. Skill score, expressed as percentage improvement of the MRF over persistence, as a function of calendar month for forecast lead times of 2, 5, and 10 days.

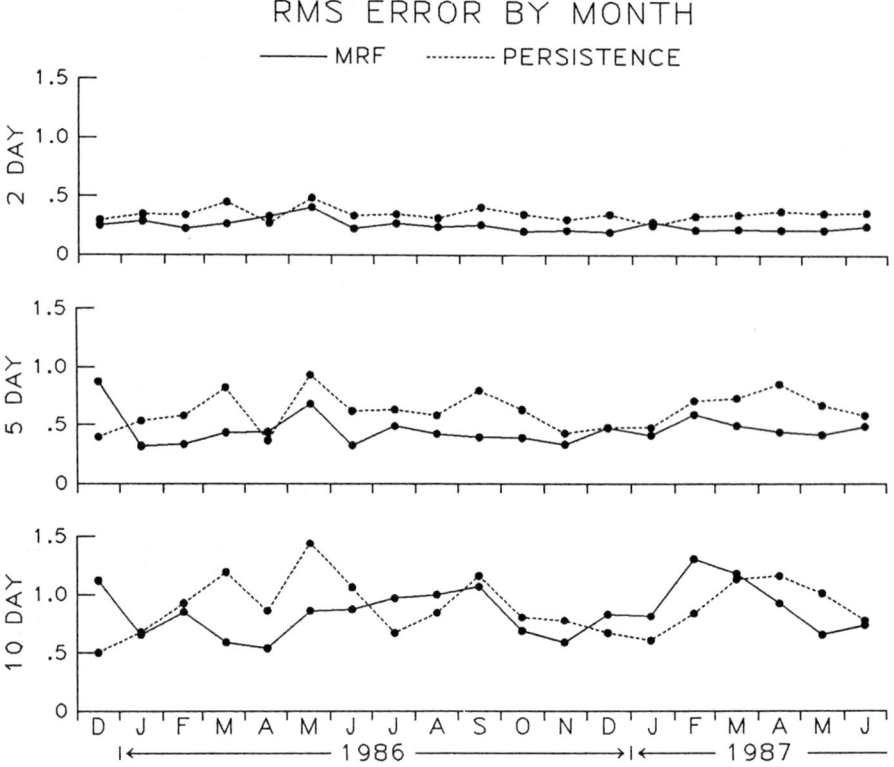

Fig. 4. The root-mean-square difference between observed values of M and those forecasted by the MRF, σ_{MRF}, (solid line) and between observed values of M and those forecasted by persistence, σ_p, (dashed line) for forecast lead times of 2, 5, and 10 days as a function of calendar month.

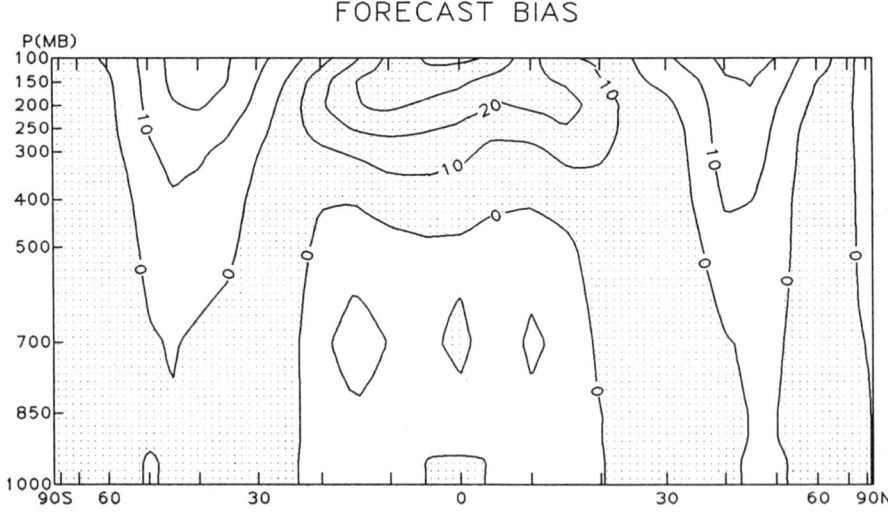

Fig. 5. The difference between 10-day forecasts of m and observed values of m, averaged over the period 1 July 1986-31 May 1987, as a function of pressure (marked along the ordinate in millibars) and latitude (marked along the abscissa in degrees). Units are 10^6 (kg m^{2}s^{-1})kg^{-1}; negative values are shaded.

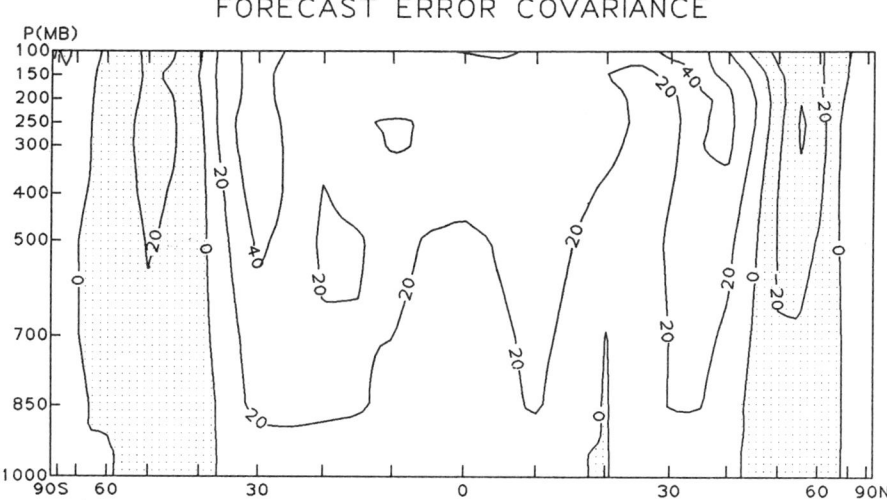

Fig. 6. The covariance between errors in 10-day forecasts of m and errors in 10-day forecasts of M, averaged over the period 1 July 1986-31 May 1987. Units are 10^{30} (kg m^{2}s^{-1})^{2}kg^{-1}; negative values are shaded.

Sources of MRF Forecast Errors

To identify the regions of the atmosphere that contribute most to the errors in the MRF forecasts of M, we have prepared latitude-height cross sections of the errors in the local values of the MRF angular momentum forecasts based on our archives of the [u] forecast fields at 10 pressure levels and every 5° of latitude. An example is shown in Figure 5 for the local mean biases during July 1986 - May 1987 in 10-day forecasts of m = [u] a cos ϕ, the angular momentum of the atmosphere per unit mass. Note that m has been defined so that the integral of the field displayed in Figure 5 over the mass of the atmosphere between 1000 and 100 mb yields the mean bias in the MRF 10-day forecasts of M for the July 1986 - May 1987 period.

Figure 5 indicates that the major source of the negative bias in the MRF forecasts of M evident in Figure 1 is located in the tropical upper troposphere, centered near the equator. This negative bias is mitigated to some extent by the tendency of the MRF to overforecast the strength of the westerlies somewhat poleward and above the location of the main zonal mean jet streams in each hemisphere.

With regard to the high-frequency noise in the MRF forecast errors, we present in Figure 6 a cross section of the covariance between m and M 10-day forecast errors averaged over the July 1986 - May 1987 period. The integral of this field over the mass of the atmosphere equals the variance in the M forecast errors about the mean bias error for this period. Figure 6 indicates that most of this high-frequency error component originates in the tropics and subtropics. It is worth noting, however, that this result largely reflects the strong weighting given to low latitudes in the formulation of m and M, and it need not imply that random errors in the forecasts of [u] are smaller at high latitudes than in the tropics.

Concluding Remarks

As noted by Rosen et al. [1987], we are pleased to discover that the NMC medium-range forecast model exhibits positive skill, on average, out to 10 days relative to persistence forecasts. Thus, dynamically based forecasts of Δl.o.d. appear to be a viable alternative to the empirical approaches currently in use. Work still needs to be done, however, in relating forecasts of M to Δl.o.d. on the very short time scales being considered here. Moreover, we need to recognize that the skill scores for the MRF fluctuate considerably from month to month, particularly for the longer lead times. Further analyzing the local sources of the MRF errors, along the lines depicted in Figures 5 and 6, could lead to improved skill scores, not only by suggesting better ways of removing bias errors than the simple approach used here, but also by indicating to meteorologists where some of the deficiencies in their forecast models lie.

Acknowledgments. The work at AER, Inc., reported here has been performed under Contract 957545 from the Jet Propulsion Laboratory, California Institute of Technology, sponsored by the National Aeronautics and Space Administration. The work at the Jet Propulsion Laboratory,

California Institute of Technology, represents the results of one phase of research carried out under contract with NASA.

References

Eubanks, T. M., J. A. Steppe, J. O. Dickey, and P. S. Callahan, A spectral analysis of the earth's angular momentum budget, J. Geophys. Res., 90, 5385-5404, 1985.

Hide, R., N. T. Birch, L. V. Morrison, D. J. Shea, and A. A. White, Atmospheric angular momentum fluctuations and changes in the length of the day, Nature, 286, 114-117, 1980.

Rosen, R. D., and D. A. Salstein, Variations in atmospheric angular momentum on global and regional scales and the length of day, J. Geophys. Res., 88, 5451-5470, 1983.

Rosen, R. D., and D. A. Salstein, Contribution of stratospheric winds to annual and semiannual fluctuations in atmospheric angular momentum and the length of day, J. Geophys. Res., 90, 8033-8041, 1985.

Rosen, R. D., D. A. Salstein, T. Nehrkorn, M. R. P. McCalla, A. J. Miller, J. O. Dickey, T. M. Eubanks, and J. A. Steppe, Medium-range numerical forecasts of atmospheric angular momentum, Mon. Wea. Rev., 115, 2170-2175, 1987.

FORECASTING SHORT-TERM CHANGES IN THE EARTH'S ROTATION
USING GLOBAL NUMERICAL WEATHER PREDICTION MODELS

Raymond Hide

Geophysical Fluid Dynamics Laboratory, Meteorological Office
(Met O 21), London Road, Bracknell, Berkshire, RG12 2SZ, U.K.

Introduction

The elucidation of tiny but detectable changes in the Earth's rotation is a key problem in geophysics and their prediction is of practical importance in geodesy and astronomy. Several years ago the theory of angular momentum exchange between the atmosphere and solid Earth was improved and then applied, using daily determinations of all three components of the total atmospheric angular momentum (AAM), to the dynamics of the "atmosphere-ocean-solid Earth" system. Short term changes in the Earth's rotation speed were thus shown to be largely of meteorological origin, so that Earth rotation measurements could be used as a proxy meteorological data set. By exploiting this finding it was established that the global atmospheric circulation fluctuates persistently if irregularly on time scales ranging from about 30 to 80 days. This intraseasonal fluctuation is clearly of potential practical significance in long-range weather forecasting, and it is now being studied intensively by research groups throughout the world seeking to understand underlying dynamical processes. Whether the phenomenon is basically a relaxation oscillation of the Hadley circulation in the Tropics, involving large-scale interactions with the oceans, or alternatively is produced by slow instabilities of mid-latitude flow, is not yet clear. The direct role of the stratosphere in the angular momentum balance and the indirect role of the oceans in modifying atmospheric excitation of polar motion are other relevant topics requiring further study.

Another response to the initial findings occurred in 1983 at the Hamburg IUGG General Assembly when, at the initiative of Professor H.Kautzleben, the International Association of Geodesy (IAG) organized a special working group (SSG 5-98) under the chairmanship of Dr. Jean O. Dickey charged with promoting further studies of all aspects of atmospheric excitation of changes in the Earth's rotation. The working group has urged meteorologists to forecast atmospheric angular momentum (AAM) changes on time scales of up to about 10 days from the output of reliable global numerical weather prediction (GNWP) models. One practical objective is to provide useful routine forecasts of changes in the length of day (and polar motion) for distribution through the newly formed International Earth Rotation Service (IERS) of the IUGG and the International Astronomical Union (IAU) to a variety of users, including geodesists concerned with spacecraft navigation and astronomers. AAM forecasts have now been started by three centres, the U.S. National Meteorological Center, the U.K. Meteorological Office and the European Centre for Medium Range Weather Forecasts. The initial results from these organizations are encouraging and it would be helpful if other meteorological centers would join in. Further advances in geodesy and geophysics will undoubtedly stem from this new venture, which should also deepen our understanding of the general circulation of the atmosphere and contribute to future improvements in the performance of global numerical weather prediction models.

Errors and Global Numerical Weather Prediction Models

Daily or twice-daily AAM determinations are now produced by several meteorological centers. The literature attests to the good use to which these data have been put by scientists in many countries, but errors are not always taken fully into account and a word of caution is in order here. In the calculation of AAM, wind and pressure data are needed over the whole globe and up to the highest possible levels in the atmosphere. Meteorologists made a special effort during two intervals of 3 months each in the year

Copyright 1990 by
International Union of Geodesy and Geophysics
and American Geophysical Union.

1979 to obtain good data coverage to high levels, well into the stratosphere, not only in extra-tropical regions in the northern hemisphere but also in the tropics and the extra-tropical southern hemisphere. But routine meteorological observations are still inadequate outside the troposphere in the extra-tropical regions of the northern hemisphere and there are concomitant systematic uncertainties in individual AAM determinations of up to about 10 per cent. Thus, although the AAM determinations are proving very valuable for certain purposes, including the refinement of "decade" length-of-day time-series, it is evident that some authors may not be fully aware of the limitations of the data sets currently in use. For example, errors in AAM determinations are much bigger than any direct contributions to length-of-day changes that the oceans might produce in a 10 year period.

The distribution of zonal winds with height and latitude is such that on average the atmosphere effectively super rotates relative to the solid Earth at about 10 ms^{-1}. The great mid-latitude jet streams make a dominant contribution to AAM. Short-term changes in the length of the day (l.o.d.) are associated with changes in the strength of the zonal winds and in the pattern of their distribution. One of the main tasks of global numerical weather prediction (GNWP) models is to integrate the governing mathematical equations with sufficient accuracy to produce useful "skill" in forecasting the evolution of the strength and distribution of winds.

In the mid-1970's and even earlier it was evident that progress in geodesy and meteorology could lead within a decade to close cooperation between practitioners of these two important areas of terrestrial physics. But it was also clear that GNWP models, then quite new, were not up to the task of providing useful AAM forecasts. The inadequate representation of dynamical effects due to surface orography was initially one of the principal sources of uncertainty in all GNWP models, in some of which the super rotation increased monotonically with time, in disagreement with observations. Fortunately, the improvement of the performance of GNWP models continues to be the object of a great deal of research and development work by meteorologists, and a few years ago, research in Canada and the United Kingdom on the representation of drag associated with internal gravity waves produced by the interaction of surface winds with orography led to new proposals for dealing with orography. Two meteorological centers have now implemented these improvements (the U.K. Meteorological Office (UKMO) nearly 3 years ago and the European Centre for Medium Range Weather Forecasts (ECMWF) more recently), and other leading centers in various countries, including the U.S. National Meteorological Center (USNMC), have either already followed suit or are expected to do so in the near future. There is evidence that time-dependent biases and other errors in AAM forecasts have been reduced to acceptable levels.

Concluding Remarks

In another paper in these proceedings, Dr. R. D. Rosen (1988) described his work with various colleagues on AAM forecasts using the USNMC model. I conclude my contribution to this symposium with an outline of comparable studies made at the UKMO and the ECMWF. It is impossible to name everyone involved indirectly in the operational AAM forecasts and analyses now being produced by these two centers, but direct contributions have been made by Dr. B. Barwell, Mr. M. Bell, Dr. R. Bromley, Mr. P. Trevelyan of the UKMO and Dr. A. Simmons and Dr. G. Sakellarides of the ECMWF. UKMO and ECMWF forecasts of up to 10 days of all three components of AAM are found to exhibit significant "skill", and it is intended in the very near future to supply the International Earth Rotation Service (IERS) with regular forecasts of the axial component of AAM, for use in making objective predictions of short-term changes in the length of the day. Full details of the methods and results of our attempts to forecast AAM changes using GNWP models will be reported elsewhere.

Reference

Rosen, R. D., D. A. Salstein, T. Nehrkorn, J. O. Dickey, T. M. Eubanks, J. A. Steppe, M. R. P. McCalla, and A. J. Miller, Forecasting atmospheric angular momentum and length-of-day using operational meteorological models, <u>Proceedings of the IUGG Interdisciplinary Symposium on Variations in the Earth's Rotation</u>, August 1987, Vancouver, B. C., Eds. Carter, W. E., D. McCarthy and P. Pâquet, AGU Publication Series, 1988.

GLOBAL WATER STORAGE AND POLAR MOTION

John W. Kuehne and Clark R. Wilson

The University of Texas at Austin, Dept. of Geological Sciences,
Center for Space Research, Institute for Geophysics.

Abstract. In order to determine the effect of water storage on Earth rotation variations, we have estimated the seasonal changes in water storage for 612 drainage basins covering all land except Antarctica. Estimates are derived from monthly mean meteorological data and the UNESCO Atlas of World Water Balance. Over all land, maximum water storage occurs in the late winter, and minimum storage in mid-summer. The use of drainage basins for the calculation permits the preparation of a monthly time series from 1900 to the present for comparison with the International Latitude Service polar motion series for the same period. Water storage time series for individual basins may be compared with river runoff and soil moisture observations available for some basins in order to verify the predictions derived from precipitation and temperature data.

Introduction

The rotation axis of the Earth moves with respect to the geographic coordinates. The motion has three main components: a nearly circular annual wobble, a 14-month wobble (Chandler Wobble), and a secular drift (polar wander). The annual wobble is surely forced by seasonal redistribution of air and water. The Chandler wobble is a resonant motion of the Earth excited by mass redistribution on and within the Earth. The cause of the secular drift is suspected to be post-glacial rebound.

Many investigations have confirmed the importance of air and water redistribution as the major causes of the annual wobble, but there have been far fewer studies of the Chandler wobble. Wilson and Haubrich [1976], and Wahr [1982] have shown that air mass redistribution makes an important contribution to Chandler wobble excitation, as indicated by evidence of coherence between atmospheric and polar motion time series near the Chandler frequency. However, the magnitude of air mass redistribution is inadequate, and additional sources of excitation must be sought. The work of Hinnov and Wilson [1987] which indicated correlation between globally integrated precipitation observations and ILS polar motion series motivated us to attempt a proper calculation of water storage effects by constructing a time series from 1900 to 1985 which accounts for the three components of water storage changes: precipitation, evapotranspiration, and runoff.

Using the newly computed water storage excitation time series, and a new air mass series, we have estimated the air mass and ground water contribution to polar motion excitation at both the annual and Chandler periods for comparison with polar motion excitation derived from the ILS data. We consider only loading effects, and not mass motion. For both air and water calculations, we use the inverted barometer hypothesis - that an increase in surface load over land results in a uniform decrease in load over the oceans.

Data

The availability of data controls the method of water storage computation. A complete water storage model might include precipitation, temperature, humidity, wind speed, light flux, soil moisture, river discharge, and snow depth, as a continuous function of space and time. Chandler wobble study requires at least 7 years of data (the 12 against 14 month beat period). However, these data are not available. Recent advances in remote sensing such as the direct detection of snow depth using passive microwave emmision, and perhaps future measurements of the divergence of moisture in the atmosphere, promise to ameliorate the problem. For now, we must use the available standard ground station parameters - monthly mean precipitation, temperature, and pressure - along with existing world water balance estimates and data.

All land (excluding Antarctica) was divided into 612 river drainage basins. Large river systems were subdivided into smaller basins which reflect the dendritic nature of the system. The average basin area is 200000 square kilometers. The basin map, as shown in figure 1, is digitized with a 1°x1° resolution. The basin is our fundamental unit of integration; we assume constant parameters over the entire basin.

For each basin we determine a time series from 1900 to 1985 of station pressure, temperature, and precipitation anomalies derived from all stations in that basin as recorded on the NCAR World Monthly Climatology tape [Spangler, 1986]. The anomaly is the simple average of all reported anomalies for a given month in that basin. Most of the 3500 stations on the tape began reporting after 1900, although only about 100 report for the entire 1900 to 1985 period. Some stations have reported less than 10 years. Most stations have some missing parameters, especially during wartime. Since the NCAR tape does not provide good station pressure coverage for China, we augmented it with about 75 years of monthly mean data for 23 Chinese stations that were on the NCAR tape but were missing large amounts of data. To our knowledge, these Chinese data have not been published.

As part of an international UNESCO effort to understand global hydrology, the Soviet Union has compiled maps of water balance elements for each continent [Korzoun, 1977]. Each of the 612 basins is assigned three values taken from contour maps at the basin centroid: annual evapotranspiration E^a; annual precipitation P^a; and a

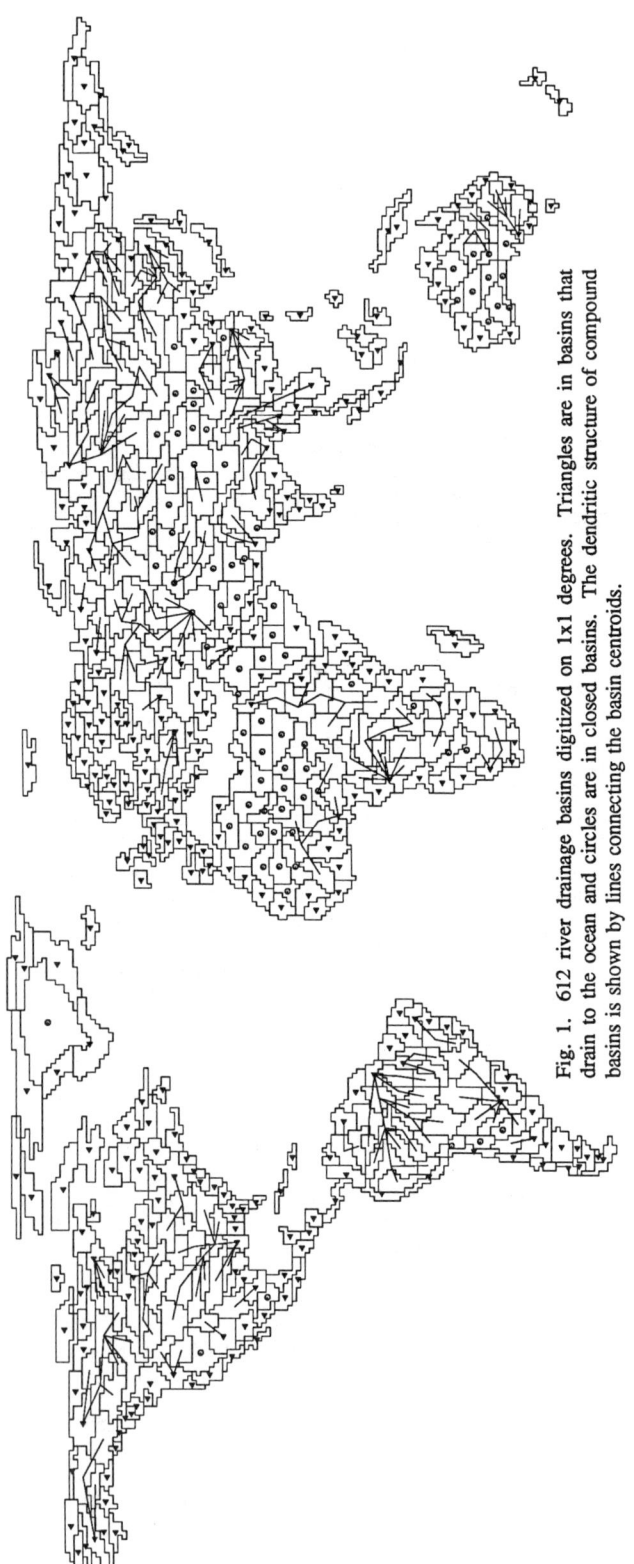

Fig. 1. 612 river drainage basins digitized on 1x1 degrees. Triangles are in basins that drain to the ocean and circles are in closed basins. The dendritic structure of compound basins is shown by lines connecting the basin centroids.

runoff coefficient ρ, which is the percentage of annual precipitation that went as runoff. Each basin is assigned: an annual distribution of runoff R_m, m∈{1,12}, which gives the mean monthly runoff as a percentage of annual runoff; the annual distribution of evapotranspiration E_m, which gives the monthly evapotranspiration as a percentage of annual evapotranspiration. There are 715 stations with annual distribution of runoff and 127 with annual distribution of evapotranspiration.

This data set contains monthly mean discharge values for 150 major rivers from about 1900 to 1964. Although not used in the work reported here, we intend to compare our predicted runoff with observed runoff for selected basins.

The National Meteorological Center has monthly mean soil moisture storage time series available for selected areas of North America. Although we have not yet made use of these, we intend to compare these with predictions derived from our meteorological data.

Basins with no station steal data from the nearest basin. Basins without an NCAR, R_m, or E_m station steal it from the nearest station in a neighboring basin. For pressure, temperature and precipitation, missing data are stolen from the nearest basin and are scaled by the ratio of the RMS departures from the mean for the two basins. Stolen monthly mean anomalies are flagged with the steal distance and basin number for later analysis. Basins steal anomalies across space, not time.

Although NCAR has discarded suspect values based on 4 or 5 standard deviations from the mean, we still found about 300 pressure values that exceeded 100 millibars from the station median pressure, and these were discarded.

Theory

Consider $S(\theta,\lambda,t)$ to be the surface time variable mass load per unit area on the Earth at colatitude θ, east longitude λ, and time t. We assume at any position on the Earth the average $S(\theta,\lambda,t)$ over the period 1900-85 is zero. Thus S is the departure from the mean. The Earth rotation excitation functions are given by the integrals

$$\Psi S(\theta,\lambda,t) = X_1 + iX_2 = \frac{a^4}{C-A}\int S(\theta,\lambda,t)\sin(\theta)\cos(\theta)e^{\lambda i}ds$$

$$X_3 = \frac{a^4}{C}\int S(\theta,\lambda,t)(\sin^2\theta - \frac{2}{3})ds$$

with ds=sinθdθdλ for a spherical Earth of radius a, C and A the principal moments of inertia of the Earth. The complex-valued excitation for polar motion refers to a coördinate system attached to the Earth with a complex plane defined by its origin at the geographic north pole, real axis along the Greenwich Meridian, and imaginary axis along 90° east. The units of X_1 and X_2 are radians. We will convert to milli-arc seconds by multiplying by 2.06×10^8, as is the convention in polar motion studies. X_3 refers to change in the rotation speed of the Earth induced by mass redistribution. Many studies have shown that motion (winds and currents) is probably most important in LOD changes, and we will not discuss X_3 further in this report.

The mass load of the atmosphere is observed directly on a barometer. In computing the excitation, we use the inverted barometer assumption, i.e., an increase in load over land results in a uniform decrease in load over the oceans. Hence if total atmospheric mass is constant the atmospheric load needs to be known only over land. Atmospheric mass is not conserved due to changes in water vapor content; rather the mass of all air and water is conserved. Thus by combining air and water excitations, each computed with the inverted barometer hypothesis, the mass due to water vapor is accounted for when the air and water excitations are added together.

Without a direct measure of the water load, we are forced to estimate the time rate of mass change in a basin as the difference between convergence and divergence, and then to integrate this with respect to time. We estimate the divergence as the runoff plus evapotranspiration. Runoff is taken to be general enough to account for any loss not due to evapotranspiration, including runoff through aquifers. Although the convergence must include runoff from another basin, we assume that integrating S over one month is sufficient time for all river runoff to reach the ocean (assuming of course that the basin is not closed), and that aquifers in a basin must discharge in the basin. Hence the convergence is due to precipitation only. We have not yet made special provision for closed basins like the Caspian Sea, whose storage must be controlled in part by upstream runoff.

Numerical Methods and Results

Consider precipitation or temperature in a basin. The average value for month $m \in \{1..12\}$ is computed as the arithmetic mean over all years and stations that report for that month. A particular value for a month (e.g. Jan 1900) is computed as the arithmetic mean over all stations reporting for that month. The monthly anomaly is the difference. For the pressure time series A_t, the mean and anomaly are computed as above for each station, and then the anomalies are averaged over all basins in the basin. After all basins have been processed, missing values are estimated and flagged as in 2.5.

For each basin we compute the average precipitation P_m. Using the notation of 2.3,

$$P^\alpha = \sum_{m=1}^{12} P_m \equiv \text{annual precipitation}$$

$$R^\alpha = \rho(P^\alpha) \equiv \text{annual runoff}$$

To conserve water, and because we felt that R^α was better determined than the Atlas value for E^α, we let

$$E^\alpha = P^\alpha - R^\alpha \equiv \text{annual evapotranspiration}$$

For basins with a stolen NCAR station, we use the Atlas value for P^α and scale P_m accordingly. Basins which do not drain to the ocean set R^α to zero, as all water must evaporate. In addition, we applied the Atlas precipitation correction to P^α. This corrects for any loss of precipitation in the gauge between time of accumulation and time of reading. Then

$$\Delta S_m = P_m - (R^\alpha R_m) - (E^\alpha E_m)$$

is the mean monthly change in water storage.

The runoff time series R_t is constructed from the precipitation time series P_t and the Atlas. The Atlas shows that for basins in the northern hemisphere higher than 45°, frozen precipitation is not released from the basin until spring, the effect being more pronounced with higher latitude. We assume that all frozen water melts over a year, and thus define 85 hydrologic years from July to June with yearly runoff R^y. R_t is $R^y \times \rho \times R_m$. For basins at latitudes below 45°N it is $P_t \times R_m \times \rho$.

The evaporation time series E_t is a perturbation

$$E^\alpha E_m + \frac{\partial E}{\partial T} \Delta T_t$$

where ΔT_t is the temperature anomaly for month t. We estimate $\partial E / \partial T$ as [Shaw, 1983; p.206]

$$\frac{\partial E}{\partial T} = \frac{\partial}{\partial T} \frac{22.5}{5} T = 5.08 mm/°C/month$$

The storage time series ΔS_t, is then $P_t - R_t - E_t$.

Our total global annual precipitation, runoff, and evapotranspiration are within 10 percent of estimates by Mather [1969] and Baumgartner and Reichel [1975] (table 1). Comparison by continent is more variable, but the ratios of runoff and evapotranspiration to precipitation are in general agreement.

TABLE 1. Global Annual Precipitation (P), Runoff (R), and Evapotranspiration (E) in thousands of cubic kilometers of water.

Author	P	E	R
Mather	106	37	69
Baumgartner,Reichel	111	40	71
Us	118	42	76

The Annual Excitation

Various estimates for the prograde and retrograde components of the annual excitation are shown in figure 2. The prograde component for the ILS series is better determined than the retrograde component because the signal to noise level is better at the prograde frequency. For the results derived from meteorological data, prograde and retrograde components are equally well determined. The noteworthy features of these diagrams are as follows.

A_t agrees reasonably well with the results of Wahr. Differences are likely due to: a different interpolation scheme, assuming constant pressure anomalies over the whole basin compared to the spline interpolation used by Wahr; different data sets, especially the additional Chinese data; and the added 13 years (1974-86) not used in Wahr's calculation.

S_t differs in phase by 60 degrees (about 2 months) from Van Hylckama's [1970] result. As much as 1 month of this may arise in the way we have assigned the time index to our results. We assigned the monthly mean to mid-month times, and performed integration of ΔS_t by a running sum. A better method would be to fit the annual sinusoid to the series of storage changes, and then integrate analytically.

We find that the retrograde annual component of ΨS_t is considerably larger than the prograde. If this is actually true, then water storage effects might not be very important in the annual wobble excitation, yet they might be important at other frequencies, relative to air mass redistribution.

The discrepancy in explaining the annual (prograde) part of the ILS series is large, at least as large as the estimate of the water contribution. Our water storage estimate STR does little to explain this discrepancy. However, the discrepancy in explaining the annual wobble remains as a measure of how well we can hope to fully account for air and water effects at the Chandler frequency.

The Chandler Frequency Excitation

The power and coherence spectra near the Chandler frequency f_c (.843 cpy) are used to determine whether an excitation series derived from meteorological data accounts for the excitation as determined from the ILS series. Only near f_c is the signal to noise level in the ILS data reasonably good. The coherence estimates are based on the discrete Fourier transform of the excitation data. To reduce spectral leakage, a cosine taper was applied to the first and last 10 percent of the excitation data. The annual sinusoid was removed, and each value in the frequency domain was averaged with the neighboring three values.

The coherence spectrum between our A_t series and the Wilson and Haubrich [1976] air mass series shows an average value near 0.8. This gives some measure of the error level in the series, and suggests that a degradation in coherence with the ILS data could be attributed partly to errors in the A_t series. There is little evidence of coherence between ΨP_t and ΨILS, or between $\Psi A_t + \Psi P_t$ and ΨILS.

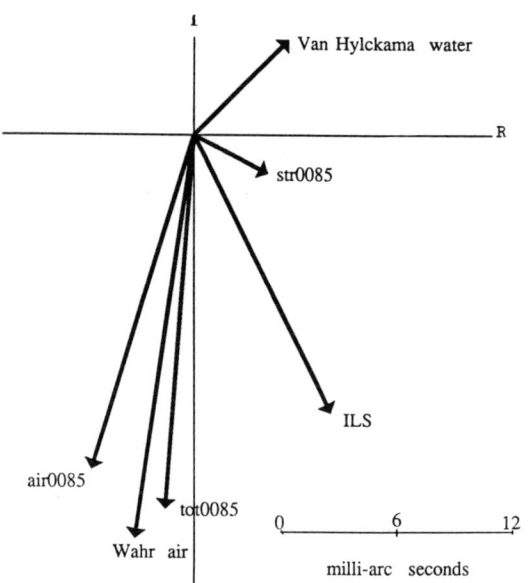 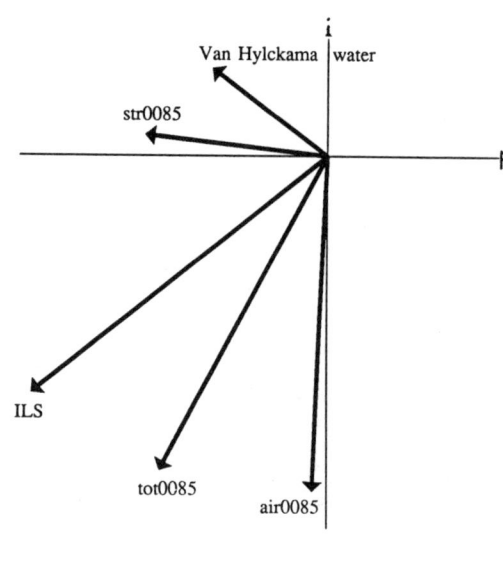

Fig. 2. Annual prograde excitation (left) and retrograde excitation (right). air0085 is the air-mass, str0085 is the water storage, tot0085 is water storage plus air-mass, Wahr and Van Hylckama are from other studies, and ILS is from 1900-85.

The power spectrum of ΨA_t is about 10 dB below the ILS spectrum near f_c, in agreement with earlier studies by Wilson and Haubrich and Wahr. The power spectrum of ΨP_t is within 3dB of the ILS spectrum, and follows it upward at low frequency, at a rate close to 12dB/octave. This is the spectral increase expected for integrated white noise. It suggests that errors in the series $\Psi \Delta S_t$, which would behave as white noise, could be dominant. Of course, part of this low frequency component could be real and contribute to low frequency polar motion, which is observed to have similar time scales (on the order of decades).

Discussion

There is probably a good chance that P_t is corrupted or dominated by noise. The low frequency behavior of the excitation has the spectral appearance of red noise. While the ILS spectrum behaves in this way also, the cause is thought to be due to viscous effects due to post-glacial rebound, and not to water redistribution in this century. The difficulty in estimating the water storage contribution is that convergence minus divergence involves subtracting numbers of about the same size, leaving the residual susceptible to errors. This is a fundamental problem with the method, and we can hope to improve the situation only by careful attention to the details of the calculation.

The time series and results given here are preliminary in that they are based on the simplest calculations and assumptions. There are many more variations possible, and more data to use, including the actual runoff time series and soil moisture measurements.

References

Baumgartner, A. and Reichel, E., *The World Water Balance*, Elseveir Scientific Publishing Company 1975.

Korzoun, V. I., *Atlas of World Water Balance*, The UNESCO Press, Paris 1977.

Mather, J. R., *The Average Annual Water Balance of the World - Symposium Banff*, AWRA Proceedings of Meeting 1969.

Shaw, E., *Hydrology in Practice*, Van Nostrand Reindhold 1983.

Spangler, W. and Jenne, R., *World Monthly Surface Station Climatology* (tape dataset), NCAR, Boulder, Colorado 1986.

Van Hylckama, T. E. A., Water Balance and Earth Unbalance, *International Association of Scientific Hydrology, Proc. Reading Symp. World Water Balance, Publ. 92*, AIHS-UNESCO 1970.

Wahr, J. M., The Effects of the Atmosphere and Oceans on the Earth's Wobble, I, *Geophys. J.R. astr. Soc.*, **70**, 349-372 1982.

Wahr, J. M. The Effects of the Atmosphere and Oceans on the Earth's Wobble and on the Seasonal Variations in the Length of Day, II, *Geophys. J.R. astr. Soc.*, **74**, 451-487 1982.

Wilson, C. R. and Haubrich, R. A. Meteorological Excitation of the Earth's Wobble, *Geophys. J.R. astr. Soc.*, **46**, 707-743 1976.

Wilson, C. R. and Haubrich, R. A. Atmospheric Contributions to the Excitation of the Earth's Wobble. 1901-1970, *Geophys. J.R. astr. Soc.*, **46**, 745-760 1976.

MAXIMUM LIKELIHOOD ESTIMATES OF POLAR MOTION PARAMETERS

Clark R. Wilson

Center for Space Research, Department of Geological Sciences
Institute for Geophysics, The University of Texas at Austin

R. O. Vicente

Faculty of Sciences, Department of Applied Mathematics, Lisbon, Portugal

Introduction

The frequency Fc and quality factor Qc of the Chandler wobble, the Earth's free nutation with a nearly 14 month period, are of geophysical interest because they provide information about the Earth's elastic and anelastic properties at a frequency well below the seismic band. Smith and Dahlen [1981] provide a thorough review of the relationships between Fc, Qc and the Earth's physical properties.

Estimates of Fc and Qc are made from observations of polar motion, the movement of the rotation axis with respect to geographical coordinates. In this study we use the monthly polar motion series of the International Latitude Service (ILS) [Yumi and Yokoyama, 1980], for the period January 1900 through December 1978, with supplementary data for the period 1979-85. The ILS data form the longest available series that has been reduced in a homogeneous way, and should provide the most reliable estimates. For the period 1979-85, both the Satellite Laser Ranging (SLR) data [BIH Annual Reports] and optical astrometry data [International Latitude Observatory of Mizusawa (ILOM)] are available. We smoothed and then interpolated both SLR and ILOM series with a cubic spline to obtain pole positions at the same intervals as the ILS series. The results from the combined ILS/SLR and ILS/ILOM series were essentially identical, and the remainder of this paper will refer to the results obtained with the ILS/ILOM series.

We use estimators that were developed by Jeffreys in two papers appearing in 1940 and 1968. The symbols Fc and Qc denote the true values of the polar motion parameters, while F and Q indicate estimates. We refer to the estimator from the 1940 paper by the Roman numeral I, and to estimates derived from it as F(I), Q(I). Similarly, F(II) and Q(II) refer to the estimators in Jeffreys' 1968 paper. Both I and II were developed from maximum likelihood arguments, assuming that a Gaussian random process is the cause of polar motion near the Chandler frequency. While this may not be correct, Monte Carlo experiments demonstrate that it is probably not a critical assumption, particularly for estimator II. The Monte Carlo experiments also permit an evaluation of estimator bias and variance in the presence of noise.

Copyright 1990 by
International Union of Geodesy and Geophysics
and American Geophysical Union.

Autoregressive Description of Polar Motion

The discrete polar motion equation [Wilson, 1985] is the foundation for understanding Jeffreys' maximum likelihood estimators. This equation relates time samples of pole position, M, to time samples of the excitation axis position, X. Using the complex pole coordinate description in which the real part is associated with motion along the Greenwich meridian, and the imaginary part with motion along 90° east longitude, the equation is

$$M_t = R \, X_{t-1/2} + S \, M_{t-1} \qquad (1)$$

where

$$R = \frac{\Omega T}{i \exp(-i\pi F_c T)}$$

$$S = \exp(i\Omega T)$$

$$\Omega = 2\pi F_c (1 + i/2Q_c)$$

Fc is the Chandler frequency expressed in cycles per year (cpy), and Qc is the dimensionless quality factor, proportional to the exponential decay time $Q_c/\pi F_c$. T is the time interval between observations, measured in years, and t is the time index which takes on integer values. An equation similar to (1) appeared in Jeffreys' 1940 paper, differing only in the value for the constant R, and the time index of X, the excitation series. The choice of R in (1) produces X in the same units and coordinate system as M. Equation (1) is a discrete version of the governing differential equation based on Euler's rigid body equations, and provides a very good approximation for the case of the ILS data where T is 1/12 year.

Estimating Fc and Qc requires that we find the central frequency and width of a spectral peak from a finite length time series, a problem that has received a great deal of attention in geophysical, time series, and electrical engineering literature. Although there exist many methods to solve this type of problem, some are clearly inappropriate. For example, if X is a broad band process, then polar motion has a continuous rather than discrete spectrum near the Chandler frequency, and will not be strictly sinusoidal in time. Thus, fitting sinusoids (harmonic analysis), which is a suitable method for line spectra, as found in tidal studies, for example, is inappropriate for the Chandler wobble.

Autoregressive spectral analysis, often called Maximum Entropy Spectral Analysis (MESA) [Ulrych and Bishop, 1975] seems particularly appropriate because equation (1) shows that the polar motion time series is an autoregressive process of order 1 (AR-1) if X is a Gaussian random process and the data are free of noise. Unfortunately, a noisy polar motion time series is no longer a simple AR-1 process [Box and Jenkins, 1970], and observed polar motion contains a drift and annual component which do not conform to the AR-1 model. Following Jeffreys we retain the AR-1 model by removing the drift and the annual term, and applying a correction for noise. This preserves the simplicity which permits analytical expressions for both the estimates and their uncertainty. Alternatively, one could allow the order of the AR process to increase to accommodate the additional variance introduced by noise and drift, which is equivalent to lengthening the Prediction Error Filter in the MESA [Vicente and Currie, 1976]; or one might introduce a more complicated Auto Regressive-Moving Average (ARMA) model to include the effects of noise and drift [Ooe, 1978 and Wilson, 1979].

Assuming the AR-1 model to hold, any linear combination of M is also an AR-1 process. The simplest linear combination is an average of N adjacent samples

$$\overline{M_t} = (N^{-1/2}) \sum_{k=t}^{t+N-1} M_k \qquad (2a)$$

The square root (N) normalization keeps the standard deviation of the noise in the average equal to that in the original data. Time samples of non-overlapping averages will obey equation (2b), which is similar in form to (1):

$$\overline{M_t} = R<X> + S^N \overline{M_{t-N}} \qquad (2b)$$

where $<X>$ denotes an appropriate average of the excitation over the interval. Equation (2) is the basis for estimator I.

Another linear combination of M is the Fourier coefficient at the frequency of +6/7 cpy (=0.8571 cpy) as determined from a 14-month segment of the monthly time series. The motivation for using this type of average is that the Fourier series coefficient is a narrow-band filtered version of the data, and thus will capture the signal of the Chandler wobble, while rejecting most of the broad band noise. For the Lth segment this coefficient is

$$A_L = (N^{-1/2}) \sum_{t=14L}^{14L+13} M_t \exp(-2\pi it/14) \qquad (3a)$$

where the square root normalization again provides the same standard error in A_L as in the original data. The sequence of these Fourier coefficients determined from adjacent non-overlapping 14-month segments obeys an equation similar to (2), and adjacent coefficients in segments, L, L-1 are related by

$$A_L = R <X> + S_L A_{L-1} \qquad (3b)$$

which is of the same form as (2) or (1), except that now $<X>$ is an appropriate average of the 6/7 cpy Fourier series coefficients of the excitation function. Equation (3) forms the basis for estimator II.

Maximum Likelihood Estimates

The annual motion and the slow drift must be removed because they do not conform to the AR-1 model. We use Jeffreys' [1968] procedure to remove the drift: the data are divided into subsets containing 7 years of data, which is exactly 7 annual periods and close to 6 Chandler periods; the mean value over each 7 year period is computed; a cubic spline interpolater is applied to each component of the 7 year means to obtain monthly values of the drift; finally, the interpolated drift series is subtracted from the original data. Figure 1 shows the drift computed in this way. After subtracting the drift, the annual component is determined by finding the best least squares fit sinusoid at a frequency of 1 cpy. In units of milli arc seconds (mas) the annual term is

$$(-45.4 \cos(+) - 84.2 \sin(+)) + i(72.5 \cos(+) - 31.7 \sin(+))$$

or in terms of prograde and retrograde components

$$(-38.6 + i78.3)\exp(i+) + (-6.8 - i5.9)\exp(i-)$$

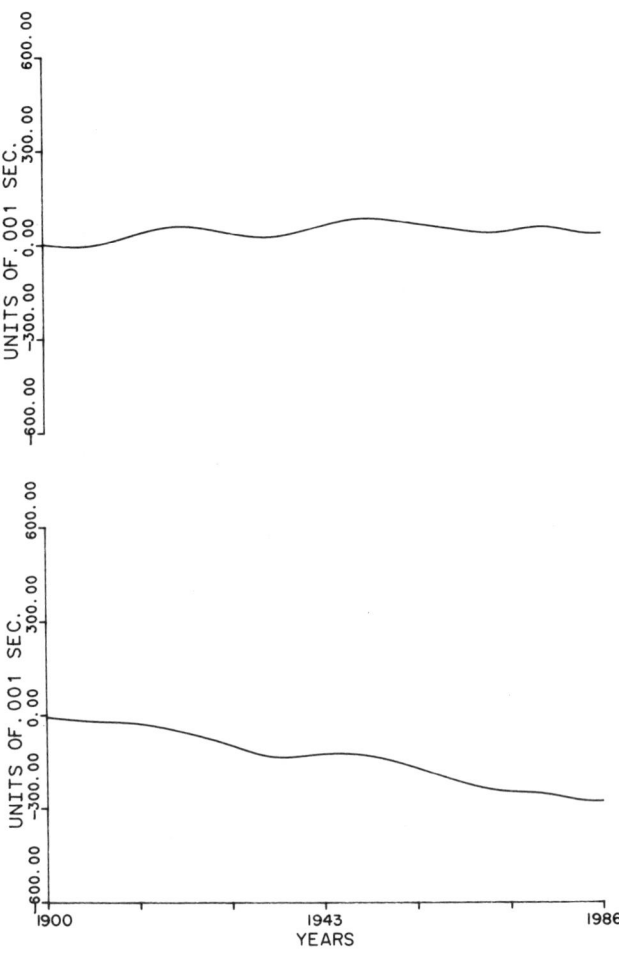

Fig. 1. The drift removed from the ILS/ILOM series as determined from means over 7 year intervals and cubic spline interpolated to monthly values. Upper curve is the component along 0° longitude; lower curve is the component along 90° East longitude.

The arguments of the trigonometric functions are radians after January 1, with the symbols (+) or (-) indicating the sign of the argument. After subtracting the drift and annual terms, the remaining series, shown in Figure 2, is assumed to be the Chandler wobble arising from a Gaussian random excitation, with added noise which is independent from month to month.

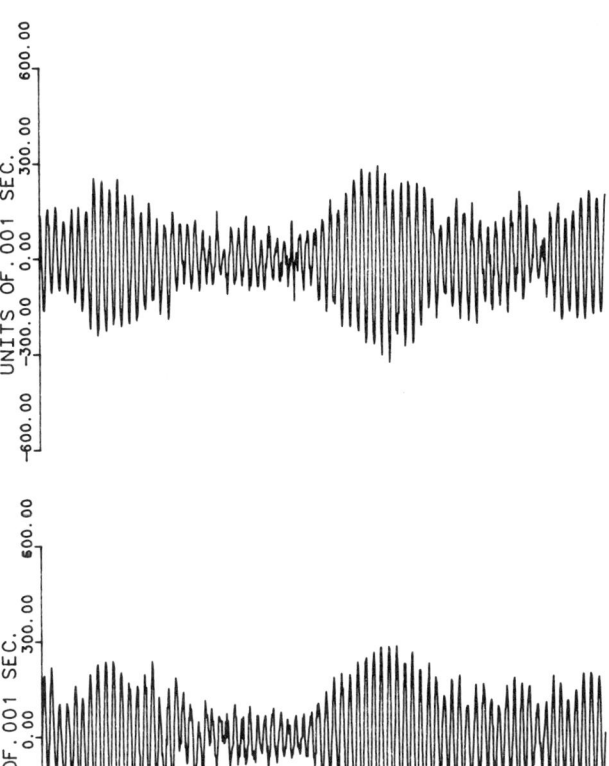

Fig. 2. The ILS/ILOM series after removal of drift and annual components. Upper curve is the component along 0° longitude; lower curve is the component along 90° East longitude.

In the maximum likelihood method (MLM), it is assumed that <X> forms a sequence of zero-mean Gaussian random numbers with independent real and imaginary parts. This is reasonable because even if individual values of X are not Gaussian, the averages, <X>, will tend to be by the Central Limit Theorem. Furthermore, independence among the sequence of non-overlapping averages should improve with increasing N, even if adjacent values of X are not independent. The MLM estimates correspond to those values of F and Q which minimize the variance of the series <X> determined from the data using equations (2) or (3), because least squares and maximum likelihood are equivalent for Gaussian random variables. Since the structure of equations (2) and (3) is identical, we present the explicit expressions using M to refer either to the simple average in (2) or to the Fourier coefficients in (3).

The estimates depend on the data through the variance U, and the covariances V and W.

$$U = \sum_{t=1}^{p} |M_t|^2 - 2ps^2 \quad (4)$$

$$V + iW = \sum_{t=2}^{p} M_t M^*_{t-1} \quad (5)$$

The number of independent values of M is denoted by p, and U has been corrected by subtracting the contribution of the noise, which has standard deviation s in each component of the complex datum. No noise correction is applied to V or W, because they incorporate products from non-overlapping intervals with presumably uncorrelated errors. Following Jeffreys, we define

$$a = V/U \qquad b = W/U$$

and the MLM estimates are

$$F = [1/(2\pi TN)] \tan^{-1}(b/a) \qquad 1/Q = \ln(a^2 + b^2)/(-2\pi FTN). \quad (6)$$

For estimator II, (equation 3), F is the correction to the trial frequency of 6/7 cpy. Standard errors for a and b are given by

$$s_a^2 = s_b^2 = (1-a^2-b^2)/2p \quad (7)$$

and may be used to calculate the corresponding errors of the estimates. Because 1/Q tends to be normally distributed, standard errors are first calculated in terms of 1/Q, and then the corresponding limits for Qc are determined from their reciprocals.

Table 1 shows estimates and standard errors obtained with estimator II, and estimator I with N=1,2,3. F(I), F(II), and associated standard errors are practically independent of the value assumed for s, but Q(I) and Q(II) depend strongly on s. The best modern optical determinations have standard error s of about 10 mas. For the ILS data, Jeffreys [1968] estimated s to be near 30 mas while Wilson [1979] found s to be about 28 mas. Estimates of Qc are shown for s = 10 and 30 mas. Table 1 shows that estimates obtained by methods I and II are generally inconsistent in that their confidence intervals do not overlap in most cases. Thus, one or both of the estimators must be at fault, and we turn to Monte Carlo experiments in the next section to discover the source of the inconsistency.

TABLE 1. Polar Motion Parameter Estimates
From the ILS/ILOM Series 1900-85

Method and N	F (cpy)	Q(s=10mas)	Q(s=30 mas)
I-1	0.8178 +/- .0116	10 (8,13)	57 (37,114)
I-2	0.8263 +/- .0067	15 (12,21)	35 (25,57)
I-3	0.8322 +/- .0106	18 (14,25)	26 (25,57)
II-14	0.8436 +/- .0022	123(70,471)	134(75,592)

Monte Carlo Evaluation of Estimators

Estimates obtained by methods I and II are complicated functions of the data, and involve only approximate corrections for the effects of noise. Thus it is not easy to analytically predict their performance, as measured by bias and variance, especially in the presence of noise. For this task, we turn to Monte Carlo studies using simulated polar motion data created with a random number generator. With simulated data, the true polar motion parameters are known exactly, and we may compare the estimates from I and II with the known values. The Monte Carlo experiments can also test the performance of the estimators when the data deviate from the AR-1 model.

The simulated polar motion data were generated from Gaussian, zero mean random numbers at 1 month intervals. The standard deviation of the excitation was chosen to be 15 mas in each component of X, based upon estimates of the excitation power spectrum near the Chandler frequency by Wilson and Haubrich [1976]. The monthly values of X were used to generate the simulated polar motion series using equation (1), with an initial pole position amplitude that was a uniformly distributed random number between 50 and 150 mas. Fc in equation (1) was fixed at 0.843 cpy, and Qc was set at 50, 100, or 200. The Gaussian noise added to the simulated polar motion series was given standard deviations varying between 0 and 40 mas in each coordinate.

The Monte Carlo experiments were performed on ensembles of 50 independent series of 1032 points, corresponding to 86 years of monthly data. The 50 estimates of Qc and Fc were used to determine estimator bias and standard deviation. The average predicted standard deviations from equation (7) were also computed. The most significant finding from these experiments was that Method II was far superior to I regardless of the noise level and the value of Qc. For example, standard deviations of F(II) were an order of magnitude or more smaller than those for F(I). For nonzero noise levels, Q(I) was very biased, yielding values that were an order of magnitude too small. This is consistent with the very low values for Q(I) shown in Table 1. Q(II) was also biased by noise, as described below.

Additional Monte Carlo experiments were conducted in which X consisted of a white noise series which had been integrated once or twice in time. This represents a departure from the AR-1 model, in that adjacent values of X are no longer independent. The performance of estimator II was found to be unchanged, and thus we conclude that the assumption that monthly values of X are Gaussian and independent is not critical.

While F(II) is unbiased by noise, Q(II) becomes biased as the noise level increases, as shown in Figure 3. The bias is approximately independent of the value of Qc. On the basis of this figure, we propose a correction for the estimate in Table 1 by the factor 1.33, and a corresponding increase in the upper limit on the confidence interval for Qc. To be conservative, we retain the original lower confidence limit on Qc. The Monte Carlo experiments also show that when s is 30 mas, equation (7) predicts standard errors for F(II) and Q(II) which agree reasonably well with the observed standard deviations in the Monte Carlo experiments.

Discussion and Conclusions

Table 2 summarizes the estimates obtained in this study by method II, including the adjustment for the bias in Q as discussed above. Table 2 also summarizes a number of other published estimates. Estimates obtained by method I are quite similar to those reported by Jeffreys [1940], and in both cases are inconsistent with method II results. This inconsistency is attributable to the poor performance of method I, demonstrated by the Monte Carlo experiments. The superior performance of method II is due to the fact that the Fourier coefficients derived from 14 month segments of the data at 6/7 cpy have vastly

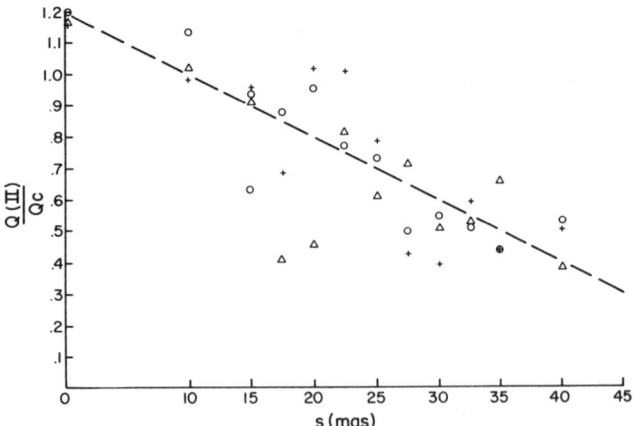

Fig. 3. The ratio of estimated Q by method II to the true value as a function of noise level. Each symbol represents the result of an experiment in which an ensemble of 50 simulated polar motion series were used to obtain average values of 1/Q(II), as described in the text.

TABLE 2. Polar Motion Parameter Estimates

Source	Method	F	Q
Jeffreys [1940]	I	0.8177 +/-.0127	46(37,60)
Jeffreys [1968]	II	0.8432 +/-.0043	61(37,193)
Ooe [1978]	ARMA	0.8400 +/-.0039	96(50,300)
Wilson & Vicente [1980]	ARMA	0.8430 +/-.0070	175(48,1000)
This Study (bias corrected)	II	0.8435 +/-.0022	179(74,789)

better signal to noise levels than an individual datum. The Fourier coefficient is effectively a narrow-band filtered version of the data which preserves that portion of the data with the best signal to noise level. Method II results obtained in this study are consistent with Jeffreys [1968] results and with other recent estimates.

Method II is probably suitable for more general spectral analysis problems in which the frequency and width of an isolated peak are to be estimated in the presence of noise. Method II constitutes a compromise between traditional Fourier analysis and autoregressive spectral methods, using the Fourier series as a narrow band filter to first reject the noise and then autoregressive analysis as a sensitive high resolution estimator of the central frequency and spectral line width.

Acknowledgement. This research was supported in part by the National Aeronautics and Space Administration Crustal Dynamics Project under grant NAG5-756.

References

Bureau Internationale de L'Heure, Annual Reports, Paris.
Box, G. and Jenkins, G., Time Series Analysis, Forecasting and Control, Holden Day, San Francisco, 1970.
International Latitude Observatory of Misuzawa, reports of daily polar motion using the observatories of the ILS.

Jeffreys, H., The Variation of latitude, Monthly Notices Royal Astronomical Society, 100, 139, 1940.

Jeffreys, H., The Variation of latitude, Monthly Notices Royal Astronomical Society, 141, 255-268, 1968.

Ooe, M., An optimal complex, AR.MA model of the Chandler wobble, Geophysical Journal Royal Astronomical Society, 53, 445-457, 1978.

Smith, M. and F. Dahlen, The period and Q of the Chandle Wobble, Geophysical Journal Royal Astronomical Society, 64, 223-281, 1981.

Ulrych, T. and T. Bishop, Maximum entropy spectral analysis and autoregressive decomposition, Reviews of Geophysics and Space Physics, 13, 183-200, 1975.

Vicente, R. and R. Currie, Maximum entropy spectrum of long period polar motion, Geophysical Journal Royal Astronomical Society, 46, 67,73, 1976.

Wilson, C. and R. Haubrich, Meteorological excitation of the Earth's wobble, Geophysical Journal Royal Astronomical Society, 46, 707-743, 1976.

Wilson, C., Estimation of the parameters of the earth's polar motion, in Time and the Earth's Rotation, ed. McCarthy, D. and J. D. H. Pilkington, D. Reidel, Dordrecht, 1979.

Wilson, C. and R. Vicente, An analysis of the homogeneous ILS polar motion series, Geophysical Journal Royal Astronomical Society, 62, 605-616, 1980.

Wilson, C., Discrete polar motion equations, Geophysical Journal Royal Astronomical Society, 80, 551-554, 1985.

Yumi, W. and K. Yokoyama, Results of the International Latitude Service in a homogeneous system, Central Bureau of the IPMS, Mizusawa, Japan, 1980.

INTERANNUAL AND DECADE FLUCTUATIONS IN THE EARTH'S ROTATION

Jean O. Dickey and T. Marshall Eubanks,
Jet Propulsion Laboratory, California Institute of Technology,
4800 Oak Grove Drive, Pasadena, California 91109-8099, U.S.A.

Raymond Hide
Geophysical Fluid Dynamics Laboratory, Meteorological Office (Met.0.21)
Bracknell, Berkshire RG12 2SZ, England, U.K.

The rotation of the solid Earth exhibits minute but complicated changes of up to several parts in 10^8 in speed [corresponding to a variation of several milliseconds in the length of the day (LOD)] and even larger variations in the direction of the rotation axis (polar motion). These changes occur over a broad spectrum of timescales, ranging from days to centuries and longer, reflecting the fact that they are produced by a wide variety of geophysical and astronomical phenomena. The most rapid "non-tidal" variations in the LOD, on timescales ranging from days up to a few years (up to about 1 millisecond in amplitude) have been shown to be largely of atmospheric origin. Routine determinations of daily fluctuations in the angular momentum of the Earth's atmosphere (AAM) are now available and they provide the basis for the elimination of meteorological contributions to the Earth rotation time series. The less rapid irregular "decade" variations of up to several milliseconds in the LOD are accepted by geophysicists to be manifestations of angular momentum transfer between the Earth's solid mantle and fluid core. Even longer period "secular" changes, due to tidal dissipation torques, produce a steady increase at a rate estimated from ancient eclipse records to lie between 1 and 2 milliseconds per century. Secular changes are also produced by internal sources, such as changes in the moment of inertia of the solid Earth resulting from the melting of ice after the ice ages. This summary outlines the main findings of a study of Earth rotation changes on the longer interannual and decade range of timescales, where the Earth's core takes over from the atmosphere in producing the most significant effects. Details of this study will be reported elsewhere (Dickey, Eubanks and Hide, to be published).

Torques at the core-mantle interface are due to time varying fluid motions in the core; several mechanisms have been forwarded as sources for the implied stress at the core-mantle interface [see for example, Lambeck (1980) and Hide (1986)]. They include: (1) tangential viscous drag in the Ekman-Hartmann boundary layer at the top of the core; (2) the action of normal pressure forces on topographic undulations on the core-mantle boundary ; and (3) electromagnetic coupling due to Lorentz forces between the conducting fluid core and partially conducting lower mantle. Clearly, the determination of the relative effectiveness of these mechanisms involves a variety of geophysical disciplines. The Earth rotation data presented in Figures 1-4 provide important constraints on models of the internal structure and dynamics of the core and lower mantle.

Viscous coupling is likely to be comparatively very weak and would only be significant under extreme assumptions about the viscosity of the core. Hence, most modern research is concerned with topographic and electromagnetic coupling. Difficulties, both of a qualitative and quantitative nature, are encountered when the electromagnetic coupling mechanism (first proposed by Bullard) is considered (Paulus and Stix, 1986). Matters of concern include assumptions made concerning the strength of the toroidal part of the geomagnetic field in the outer reaches of the core and the distribution of electrical conductivity in the lower mantle. According to the concept of topographic coupling [introduced by Hide (1969)] the magnitude of the stresses implied by the amplitude and timescale of the decade variations in the LOD might be accounted for if there were undulations on the core-mantle boundary with a height of less than one kilometer. These bumps could be supported by the viscous stresses associated with deep convention in the mantle. A method for obtaining estimates of the topographic core-mantle coupling has recently been proposed (Hide, 1986), which combines core motion models and core mantle boundary maps to produce estimates of these torques. Topographic maps of the core-mantle interface are based on the data of seismic tomography and the low degree harmonics of the geo-gravitational field together with various assumptions concerning the rheology of the upper and lower mantle. Also required in the topographic coupling calculation are pressure gradient fields at the core-mantle interface. Here, geomagnetic secular variation data are used to calculate the horizontal motion just below the core-mantle interface through the use of the frozen flux theorem of Alfven and additional assumptions used to close the system of

---
Extended summary of a paper delivered at Symposium U4 on "Variations in the Earth's Rotation", August 18-19, 1987, at the 19th General Assembly of the IUGG, Vancouver, B.C., Canada.

Copyright 1990 by
International Union of Geodesy and Geophysics
and American Geophysical Union.

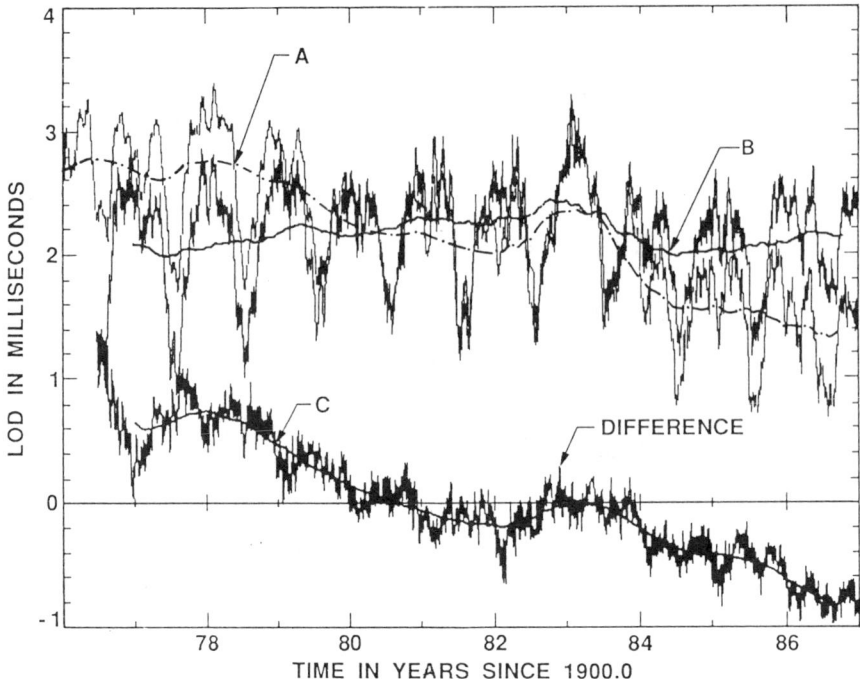

Figure 1(a). Determinations of residual length of day variations from 1976 to 1986 when periodic variations due to tidal changes in the moment of inertia of the solid Earth have been removed (defined to be LOD*). Curve (A) gives the time series of a "Kalman smoothing" of data obtained by means of the techniques of optical astrometry and space geodesy (Very Long Baseline Interferometry and Lunar Laser Ranging). Curve (B) is the LOD* as inferred from the atmospheric angular momentum as calculated by the National Meteorological Center. Curve(C) is the difference between these two series. Associated with each series are the results of eliminating the shortest period contributions and the seasonal effects by taking a "rectangular" 365-day running mean, leaving interannual and decade contributions. Note the enhancements in curve (C) during the El Nino periods (1977-8 and 1982-3).

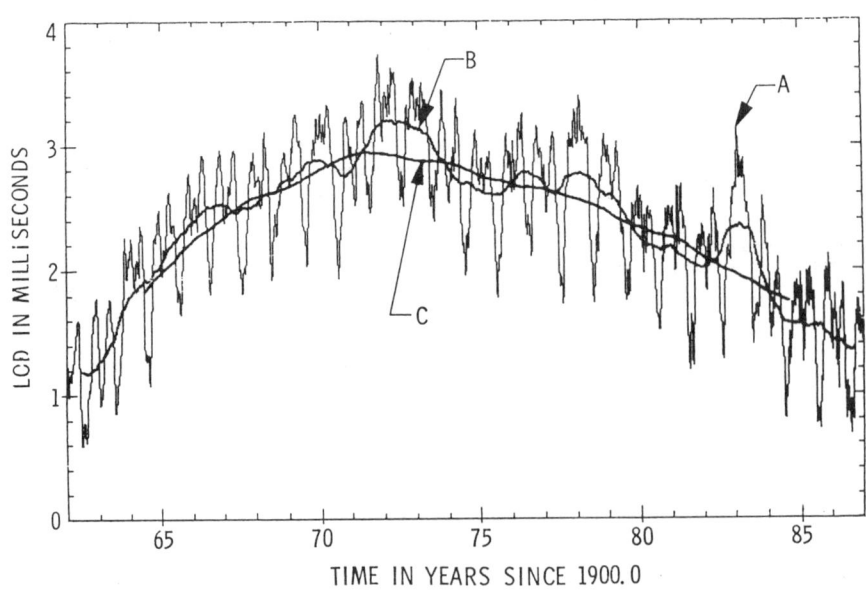

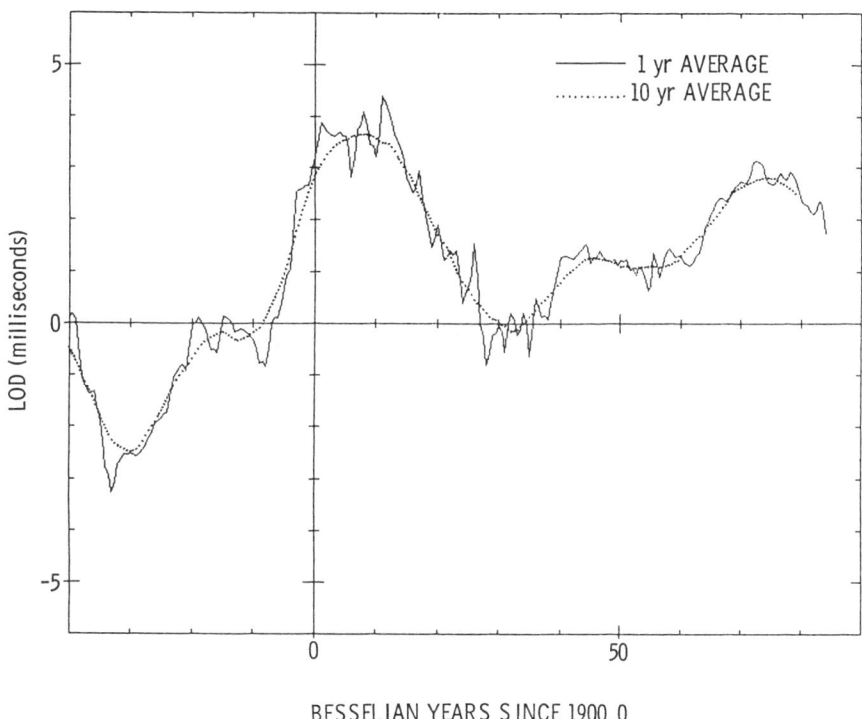

Figure 2. One year and ten year averages of length of day estimates; LOD* estimates obtained from McCarthy and Babcock (1986).

equations. The horizontal pressure gradients are computed from these velocity fields assuming geostrophic balance exists between the Coriolis acceleration and the horizontal pressure gradient in the outer case. The application of this technique is now being pursued in a related study by Clayton, Hager, Hide, Spieth and Voorhies.

The preliminary findings of the work are encouraging as a consistent picture of the core-mantle boundary is emerging based on the measurements of Earth rotation, and research into seismic tomography and geomagnetism. Earth rotation results (Fig. 3) indicate a torque of ~ 0.2 Hadley units in the axial component during the 1980 period. When torque calculations are produced with the combined results from seismic tomography and geomagnetism for the 1970-80 decade, axial torque estimates are an order of magnitude too large; this model has "bumps" on the core-mantle interface being at the 3 km level. When a D" layer is included in the analysis (a boundary layer in the bottom 100-300 km of the mantle, which is thought to have low viscosity), the undulations of the core-mantle boundary are ~ 500 m (Hager, Private Communication) and the magnitude of the torque becomes consistent those of Fig. 3. The size of these bumps is in accord with recent calculations based on corrections to the Standard IAU 1980 Nutation Model using Very Long Baseline Interferometry data (Eubanks et al, 1985; and Herring et al., 1986). The corrections are dominated by a large retrograde annual term of ~ 2 mas. These observations can be interpreted as implying a core-mantle boundary that is more elliptical than for an Earth in static rotational equilibrium; the total required deviation from the figure of equilibrium is 500 m (Gwinn et al., 1986). Collectively, these results indicate undulations at the core-mantle boundary of about 500 m in size and the existence of a D" layer in the mantle. The equatorial torque calculated from polar motion series contains a long term secular drift and as well as the "decade" variations. The longer term secular drift is presumably caused by the long term changes in the moment of inertia of the Earth, such as deglaciation effects, and has been removed from the polar motion series. These decade fluctuations in the equatorial torque (Fig. 4) are larger than the axial torque as is expected because of the effect of the equatorial budge.

Figure 1(b). Determinations of residual length of day variations from 1962 to 1986 when periodic variations due to tidal changes in the moment of inertia of the solid Earth have been removed (defined to be LOD*). Curve (A) gives the time series of a "Kalman smoothing" of data obtained by means of the techniques of optical astrometry and space geodesy (Very Long Baseline Interferometry and Lunar Laser Ranging). They indicate variations on time scales upwards of several days. Curve (B) is the results of eliminating the shortest period contributions to curve (A) by taking a "rectangular" 365-day running mean, leaving interannual and decade contributions. Curve (C) is the results of eliminating interannual variations from curve (B) by taking a 5-year "rectangular" running mean.

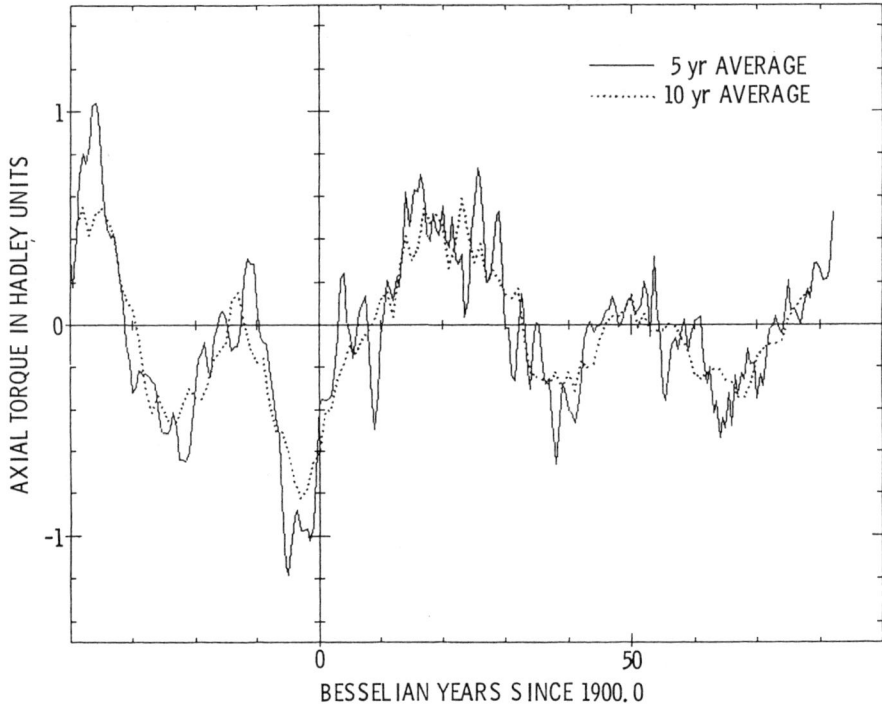

Figure 3. Five year and ten year average axial torque (1 Hadley = 10^{18} Nm) as inferred from length of data (LOD*) estimates by taking the derivative with respect to time (see Figure 2).

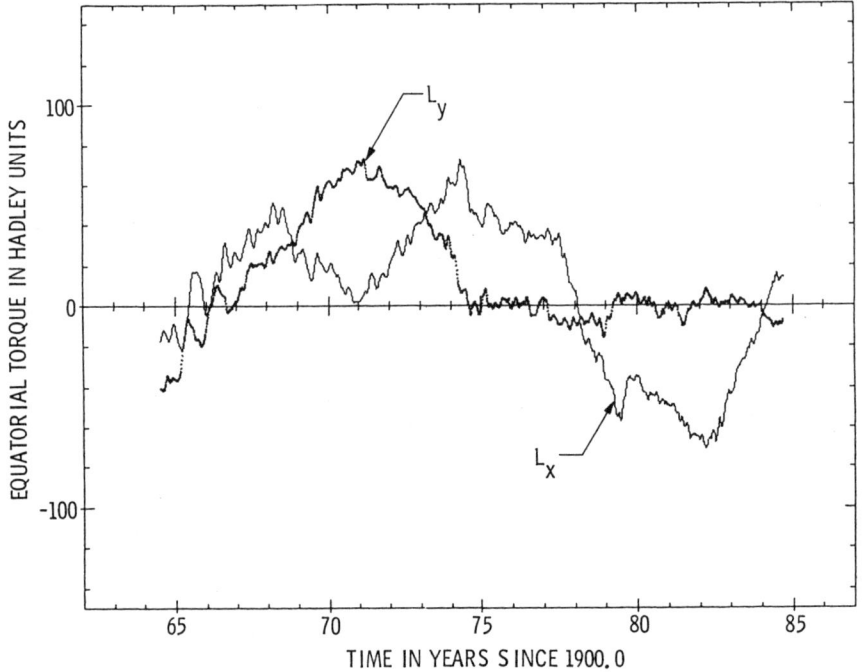

Figure 4. Equatorial components of the torque acting upon the solid Earth based on observations of polar motion with optical techniques from 1962 to 1982, supplemented by data from space geodetic techniques from 1970 to 1986. A linear trend (possibly due to changes in moment of inertia associated with the melting of ice after the last ice-age) has been removed from these data.

Acknowledgements. We acknowledged interesting discussions with R. Clayton, B. Hager, M. Spieth, and C. Voorhies and thank A. K. Babcock for providing us with length of day series beginning in 1656. This paper presents the results of one phase of research carried out at the Jet Propulsion Laboratory, California Institute of Technology, sponsored by the National Aeronautics and Space Administration.

References

Eubanks, T. M., Steppe, J. A., and Sovers, O. J., An Analysis and Intercomparison of VLBI Nutation Estimates, Proc. Inter. Conf. Earth Rotation and Terrestrial Reference Frame, Ohio State University, Columbus, Vol. 1, 326-340, 1985.

Gwinn, C. R., Herring, T. A., and Shapiro, I. I., Geodesy by Radio Interferometry: Studies of the Forced Nutations of the Earth 2. Interpretation, J. Geophys. Res., 91, 4755-4765, 1986.

Herring, T. A., Gwinn, C. R., and Shapiro, I. I., Geodesy by Radio Interferometry: Studies of the Forced Nutations of the Earth 1. Data Analysis, J. Geophys. Res., 91, 4745-4754, 1986.

Hide, R., Interaction between the Earth's liquid core and solid mantle, Nature, 222, 1055-1056, 1969

Hide, R., Presidential Address: The Earth's differential rotation, Quart. J. Roy. Astron. Soc., 27, 3-20, 1986.

McCarthy, D. D. and Babcock, A. K., The Length of Day Since 1656, Physics of the Earth and Planetary Interiors, 44, 281-292, 1986

Lambeck, K., The Earth's variable rotation, Cambridge University Press, 1980

Paulus, J. and Stix, M., Electromagnetic core-mantle coupling, Earth Rotation: Solved and Unsolved Problems, A. Cazenave, ed., D. Reidel Publishing Co., Dordrecht, 259-267, 1986.

SHORT PERIOD UT1 VARIATIONS FROM IRIS DAILY VLBI OBSERVATIONS

D. S. Robertson, W. E. Carter, and F. W. Fallon

National Geodetic Survey, Charting and Geodetic Services
National Ocean Service, NOAA, Rockville, Maryland 20852

Abstract. The superiority of very-long-baseline interferometry (VLBI) measurements over other modern techniques for determining the orientation of the Earth has been amply demonstrated [Robertson and Carter, 1985]. Nowhere is this superiority more evident than in the IRIS (International Radio Interferometric Surveying) determinations of UT1. One-hour observing sessions with the Westford-Wettzell interferometer routinely determine UT1 with an accuracy of 0.1 milliseconds of time, or about 4 cm displacement for a point on the Earth's equator [Robertson et al., 1985]. These observing sessions have been conducted on a daily basis since April, 1985. Detailed analysis of the resulting daily UT1 values indicate that the data contain more than enough accuracy to bound the anelastic effects of the Earth's mantle. Yet analysis of the mantle effects is hampered by process noise at the tidal frequencies. We believe that much of this noise originates in tidal motions of the water mass in the Earth's oceans [Robertson et al., 1987]. A preliminary analysis of the effects of a realistic tide model on the Earth's rotation rate is underway. The analysis is extremely computer-intensive; several months of CPU time on an HP-A900 minicomputer will be required to calculate the changing moment of inertia of the oceans, neglecting tidal currents (i.e. changing angular momentum of the oceans).

Introduction

Determinations of UT1 have long been hampered by the fact that UT1 has more high-frequency time variations than the other components of Earth orientation. To try to understand these time variations the IAU/IAG project MERIT [Wilkins, 1980] scheduled intensive observing campaigns in 1984 and 1985 with a goal of obtaining daily determinations of UT1. Although there was no technical reason why the IRIS project could not have conducted its 24-hour multi-station observing sessions every day instead of every 5 days, the costs of such an increase in duty cycle were prohibitive. However for monitoring only UT1 a much less costly observing program was found to be sufficient. In principle UT1 could be estimated from a single delay measurement of an equatorial source using one equatorially oriented baseline if the baseline coordinates, radio source coordinates, pole position, nutation, and clock parameters were well enough known. In practice, all of these parameters except the clock error parameters are sufficiently well known from the regular IRIS observations. Therefore all that is needed to monitor UT1 is a brief observing session designed to separate UT1 from clock errors parameters. This can be accomplished by observing a set of sources that cover a wide range in declination. In practice, the observing sessions employed for daily UT1 determinations use the Westford-Wettzell baseline and make eight observations on four sources that span about 70 degrees in declination. These observing sessions take about 45 minutes each day. Intercomparisons of the daily UT1 determinations with those from the 5-day, multi-station IRIS observing sessions consistently show agreement at the level of about 0.1 millisecond of time [Robertson et al., 1985]. This represents a bound on the combined error in both series, plus any real UT1 variations with periods less than about 10 days.

The important geophysical parameters that can be estimated from these high precision UT1 determinations are the coefficients of the elastic and anelastic response of the Earth's mantle to the disturbing (tidal) potential of the Sun and Moon. The tidal deformation of the Earth changes its moment of inertia, and thus changes it rotation rate. Yoder et al. [1981] give a detailed theoretical model of the tidal variations in UT1 for an elastic Earth with liquid oceans and core, in which the variations are proportional to the elasticity parameter k/C. Wahr has discussed the possibility of using the tidal variations in UT1 to determine the anelastic behavior of the mantle [Wahr, 1987; Wahr and Bergen, 1986]. Qualitatively, the anelastic effects are not difficult to understand: an anelastic Earth is a bit "softer" than a perfectly elastic one, and its response to tidal perturbations is slightly

Copyright 1990 by
International Union of Geodesy and Geophysics
and American Geophysical Union.

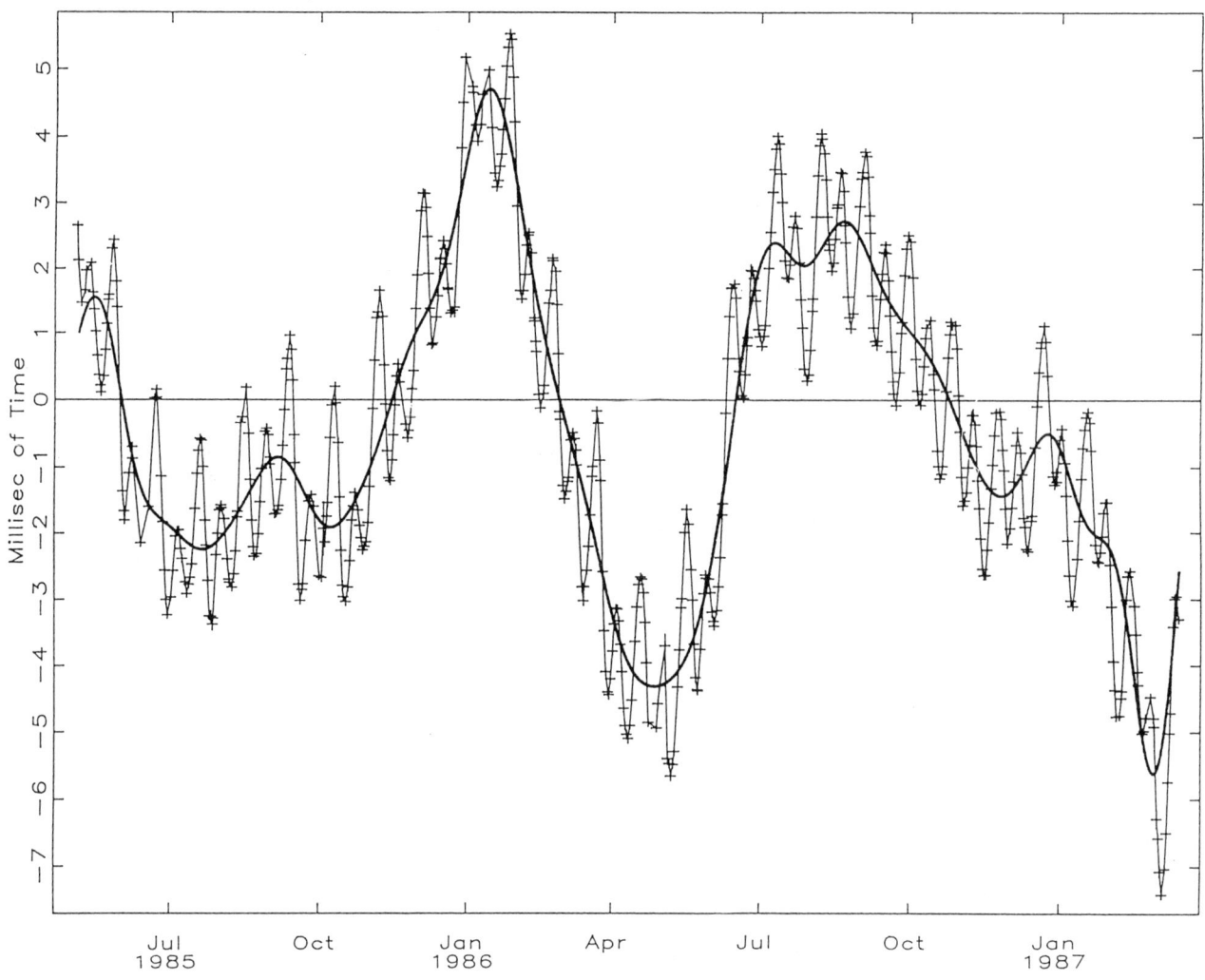

Fig. 1. Daily determinations of UT1, after removal of a linear trend and long-period sinusoids. The heavy line was generated by smoothing the observations with a low-pass boxcar filter with a cut-off at 10 cyc/year.

greater than the comparable effects for an elastic Earth. Quantitatively, the magnitude of the amplification is about 2% in k/C. Yoder's estimate for k/C is about 0.944, and Wahr's corrections for anelastic effects indicate that the value for fortnightly terms should be in the range of 0.97 to 1.06, and for monthly terms in the range of 0.96 to 1.00. This corresponds to an increase in the tidal UT1 variations by about 0.02 milliseconds of time. The need to measure such a small effect places a severe requirement on the measurement accuracies.

Data Analysis

Figure 1 shows the IRIS daily UT1 time series from May, 1985 through March, 1987, after removing a linear drift, three long-period sinusoids (6,12 and 24 month periods, respectively), and the effects of the tabulated variations in atmospheric angular momentum as compiled by the National Meteorological Center [Rosen and Salstein, 1983]. The high frequency tidal variations in UT1 are easily seen in these data. How accurately do the tidal variations seen in Figure 1 match the Yoder's theoretical model? Figure 2 shows periodograms of observed and theoretical time series at one-day intervals over the span of the VLBI observations. The tidal peaks at 27 and 14 days (13.5 and 26.1 cyc/year) are seen to match nicely, as do the lesser peaks at 9 and 31 days (11.8 and 40.6 cyc/year). The 9-day peak is outside the Nyquist limit for 5-day sampling, and can be observed only with the daily VLBI observations.

In order to use these data to evaluate the k/C tide coefficients with standard least-squares

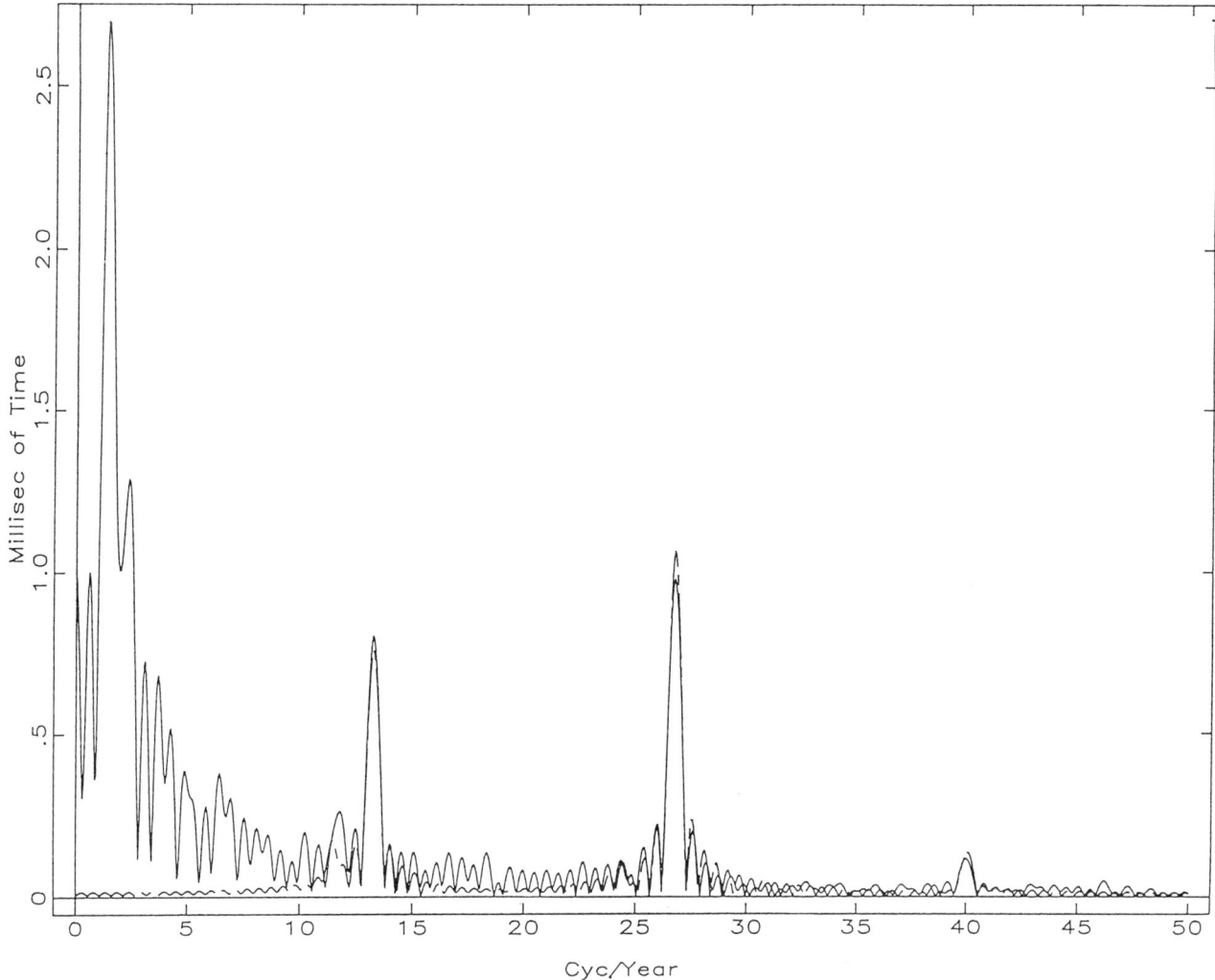

Fig. 2. The solid line shows a periodogram of the data shown in Fig. 3, and the dashed line is a similar periodogram of a daily tabulation of Yoder's tide model, evaluated over the same time interval as the VLBI data.

techniques it was necessary to remove the low-frequency variations. To do this the data were low-pass filtered using a filter with an absolute cut-off at 10 cycles/year (period of 36.524 days). The resulting function, shown in figure 1, was subtracted from the data. About 5 points at each end of the data set were not fit closely by this function, and were deleted in the ensuing analysis. A six-month sample of the differences, together with values calculated from the Yoder model, are shown in figure 3. To these data we fit three amplitude coefficients for the Yoder model, the first applied to the first 12 terms (with periods about 9 days), the second applied to the next 13 terms (with periods about 14 days), and the third applied to the next 16 terms (with periods about 27 days). The results of this fit are tabulated in table 1. The formal errors for the 14 and 27 day coefficients are seen to be substantially smaller than the precision required to determine the anelastic effects. Of course the formal errors are not the same as the total errors in the determinations. Rather, they represent the limiting accuracy which might be obtained in the absence of unmodeled systematic error. Their principal significance in this context is to demonstrate that the measurement errors are not a significant fraction of the true error budget.

To estimate the magnitude of the total error in the determinations of k/C we divided the data into subsets in two different ways and repeated the solutions, as shown in table 1. The resulting differences in the k/C determinations are generally much larger than the formal errors of the individual determinations. These differences provide much

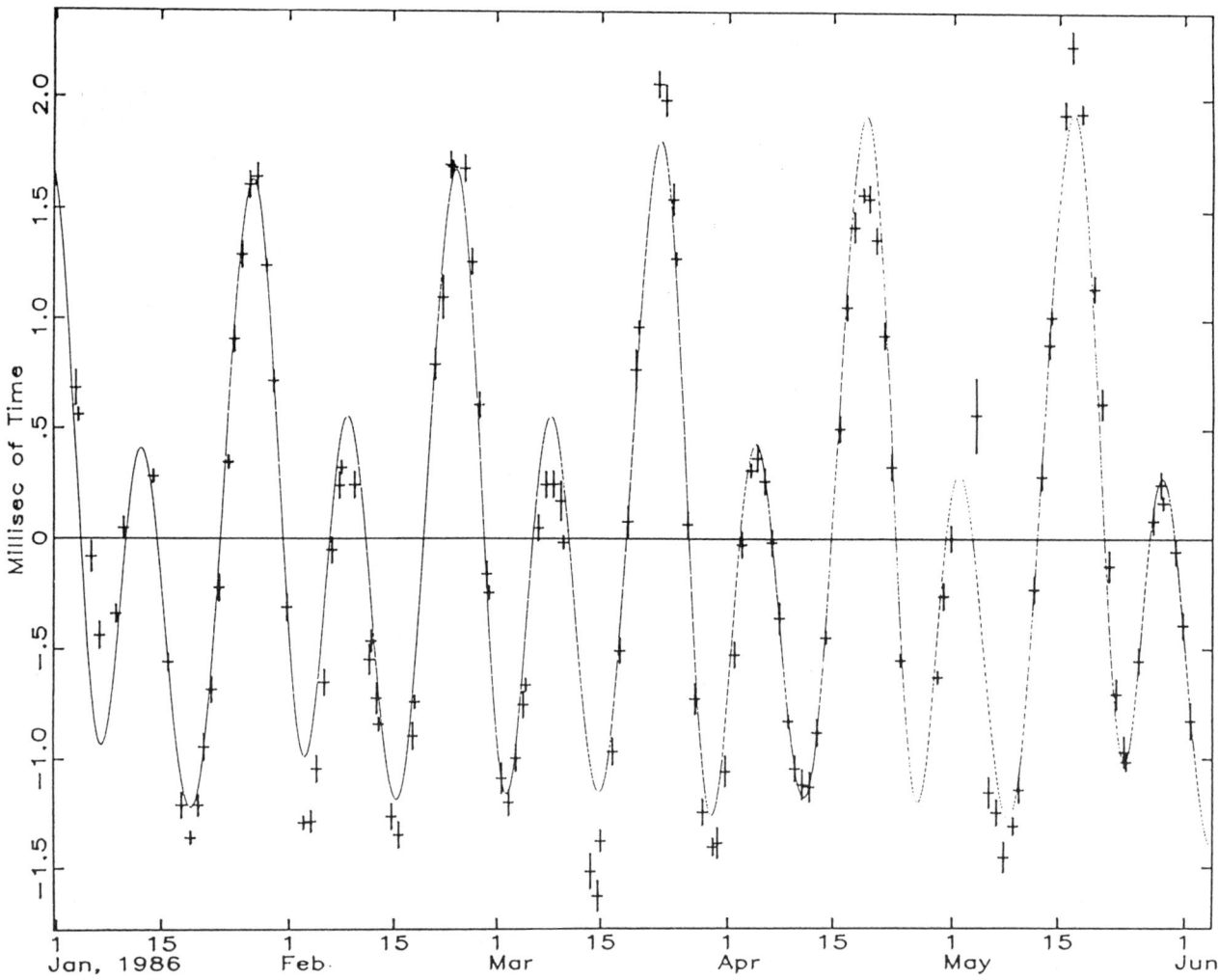

Fig. 3. Sample of the data shown in figure 1 after removal of the smooth curve from that figure. The solid line shows the Yoder tidal variation model.

TABLE 1. k/C determinations. The column labeled "sigma" tabulates the formal standard errors. Under "differences," the first column tabulates differences from interleaved subsets, and the second tabulates differences from the first and second halves of the data set.

Period	k/C	sigma	Differences 1	2
9 day	0.898	0.02	0.09	0.24
14 day	0.948	0.003	0.02	0.03
27 day	0.986	0.004	0.04	0.09

better estimates of the true uncertainties in the k/C determinations than do the formal errors. For the 27 day term and especially the 14-day term these uncertainty estimates approach the level needed for determining the mantle anelasticity. Clearly, however, there is some unmodeled source of systematic error or process noise interfering with the estimations of k/C. If it could be eliminated or substantially reduced then we might be very close to being able to observe the effects of the mantle anelasticity.

The set of possible sources of the observed process noise can be conveniently divided into 5 categories according to their distance from the center of the Earth: extra-terrestrial, atmospheric, oceanic, crust-mantle, and liquid-core effects. The two extreme categories can be largely dismissed out of hand: extra-terrestrial effects (luni-solar torques) are unlikely to hold surprises at any significant level, and the liquid core is unlikely to generate any response at periods shorter than 30 days. Of the other three effects, the ones that relate to the crust and mantle are the very ones we are trying to observe. That leaves the oceans and atmosphere as the principal problem areas.

The known effects of atmospheric winds and pressure variations have already been removed from the data, but the error levels in the wind values are poorly known. If there are large errors in the atmosphere data at fortnightly and monthly periods then they could prove to be a serious problem. Continuing studies should soon provide answers to this question.

The remaining likely source for errors in estimating k/C is oceanic effects. The oceans respond strongly at the tidal frequencies, and because of local resonances they are commonly significantly out-of-phase with the driving torques. The periodogram of the residuals to the UT1 data after fitting the k/C values, shown in figure 4, shows peaks at the tidal frequencies, exactly as might be expected from such out-of-phase oceanic effects. There are also peaks that are not at the tidal frequencies, which may or may not be related to oceanic effects. Yoder calculates that the ocean tides will cause UT1 variations at semi-diurnal and diurnal frequencies with amplitudes ranging from 0.02 to 0.07 milliseconds [Yoder et al., 1981]. Since this power is above the Nyquist frequency for the daily sampling employed here, it must alias into the frequency band that we are able to observe. There is no ready analytic cure for the process noise introduced by the oceans, but there may be ways to deal with it numerically. We are in the process of integrating the moment of inertia variations that are implied in the Schwiderski tide model [Schwiderski, 1980]. The integration is carried out over an assumed spherical Earth, as follows:

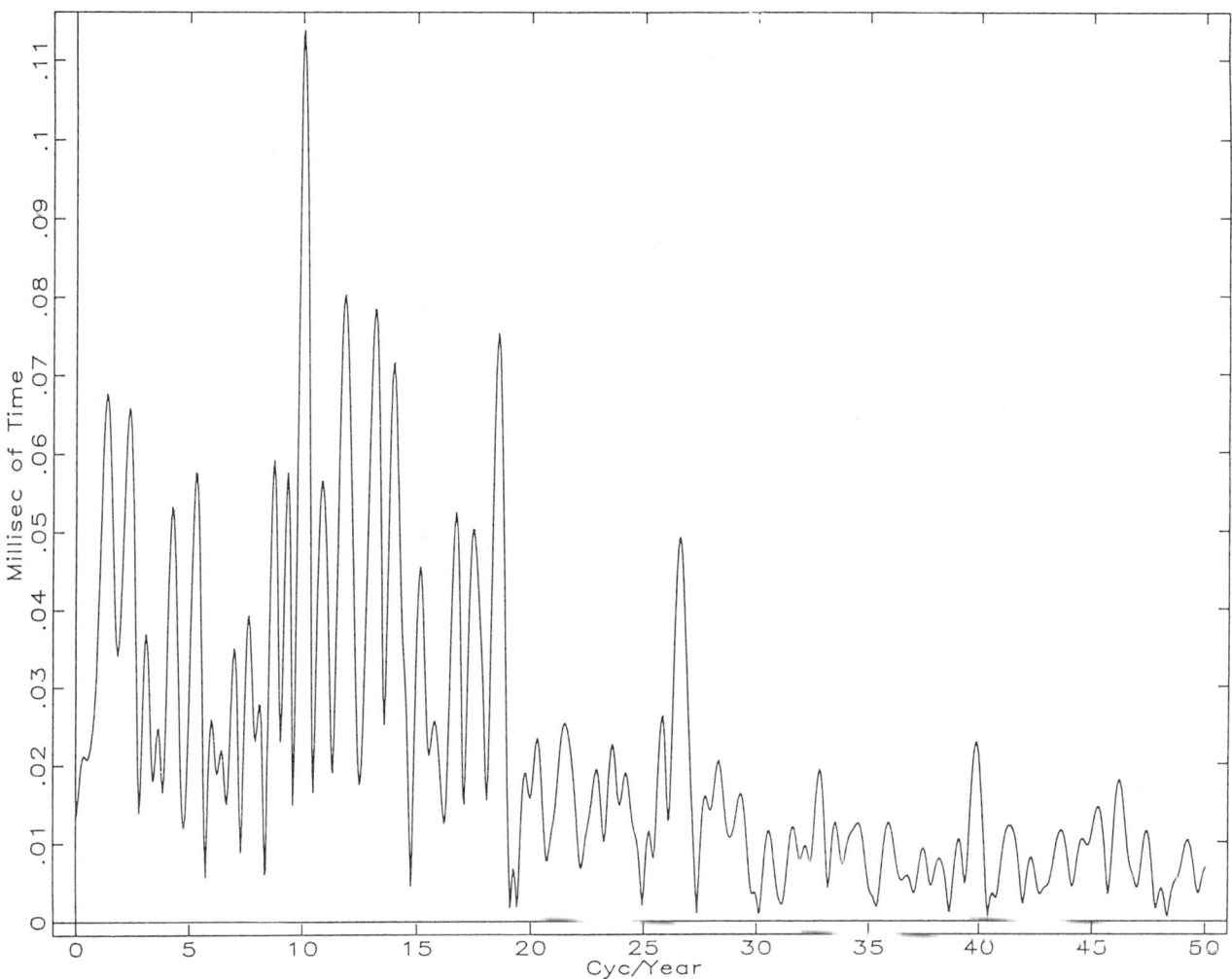

Fig. 4. Periodogram of the residuals of the UT1 determinations after fitting three k/C amplitudes as described in the text.

$$\delta LOD = K \int r^2 \, dm$$
$$= K \int \rho (R \cos \theta)^2 \, dV$$
$$= K \iint \rho (R \cos \theta)^2 \, \delta R \, (R \, d\theta) \, (R \cos \theta \, d\lambda)$$
$$= K \rho R^4 \iint (\cos \theta)^3 \, \delta R \, d\theta \, d\lambda$$
$$\cong K \rho R^4 \sum \sum (\cos \theta)^3 \, \delta R \, \delta \theta \, \delta \lambda$$

Where:

θ, λ = latitude, longitude
R = radius of the Earth
r = cylindric radius at latitude θ, = $R \cos \theta$
δR = tidal displacement at θ, λ
ρ = mean density of sea water, 1030 kg/m^3
K = conv. mom. of inert. in kg m^2 to δLOD in ms
   $\cong 8.64 \times 10^7 / C_{mantle}$, = 1.2138×10^{-30}

The numerical integration is being performed on a roughly 1° by 1° grid requiring more than 50,000 points between latitudes 89° N and 78° S, repeated at 1.5 hour intervals or 16 points per day. The corresponding variations in LOD for a two-month interval are shown in figure 5. The LOD variations are dominated by diurnal and semi-diurnal components with amplitudes of a few tenths of a millisecond, and there are hints of the fortnightly and monthly terms expected. Of course this calculation assumes that the oceans are coupled rigidly to the Earth, and it neglects changes in the angular momentum of the oceans resulting from tidal currents. Further study is needed to investigate the effects of these assumptions at the time scales of interest here.

Carrying out the necessary calculations to complete this numerical integration over the time period spanned by the VLBI UT1 determinations will take several more months. Whether these corrections will

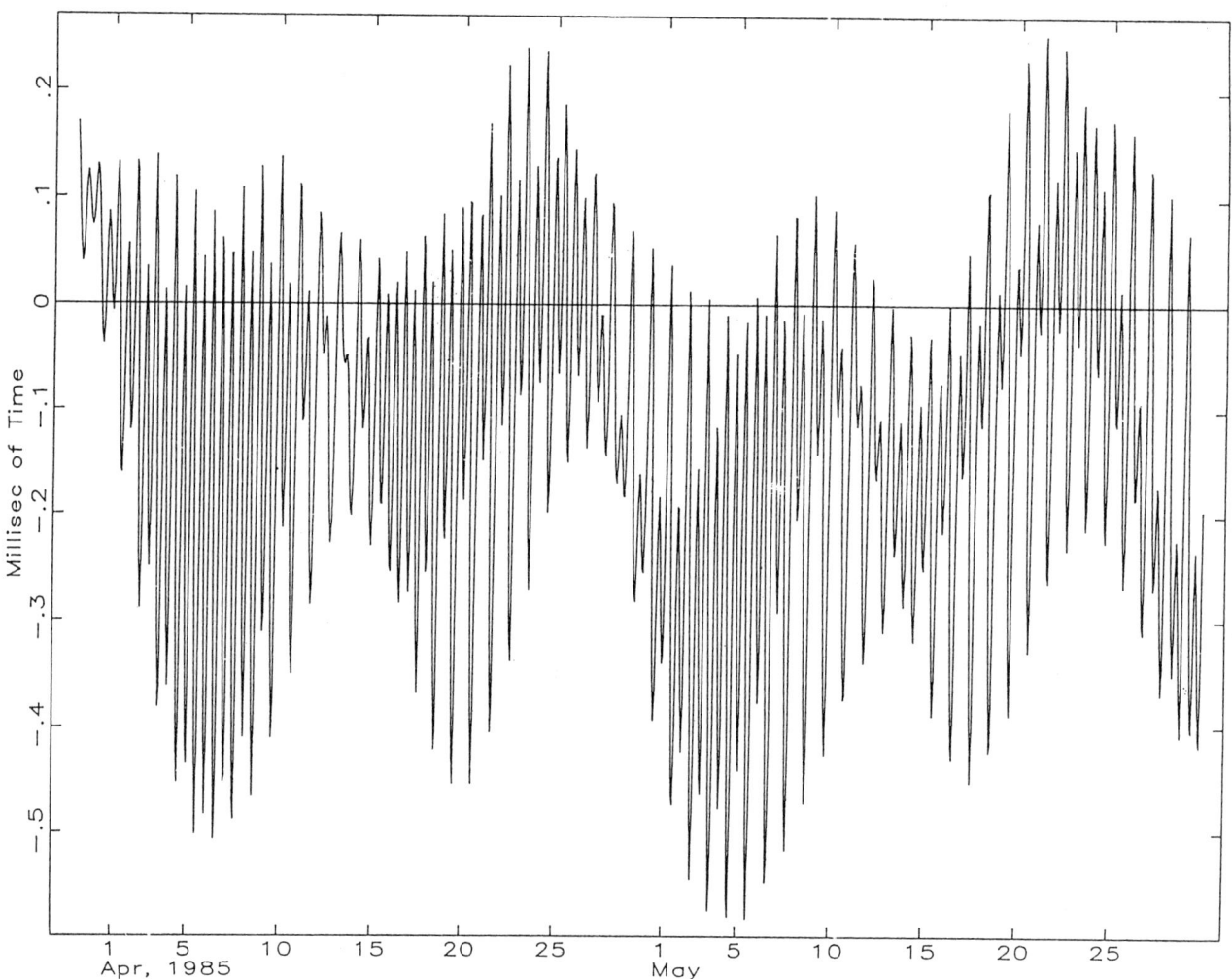

Fig. 5. Variations in LOD calculated by numerically integrating the variations in the moment of inertia of the oceans as implied by the Schwiderski tide model.

reduce the scatter in the estimates of k/C to the level that would make possible a useful determination of anelastic effects in the mantle remains to be seen. If the existing level of accuracy of ocean tide models is insufficient to reduce the scatter in the k/C estimates and the corresponding residuals in the UT1 observations, then these UT1 observations may have a role to play as a constraint on future ocean models.

If the semi-diurnal and diurnal tides cannot be modeled with sufficient accuracy to remove their effects from the UT1 values, it may be necessary to increase the sampling rate of the UT1 observations to at least the Nyquist frequency for the semi-diurnal tides, i.e., 4 observing sessions per day. VLBI is the only observing technique that can currently achieve such a sampling rate, and, indeed, sampling rates up to several times higher than this pose no particular technical problems.

Conclusions

The extraordinary power of VLBI for monitoring Earth rotation has been demonstrated by the quality of the IRIS results. The UT1 measurements have reached a level where the observational uncertainty is not a major part of the error budget, and the geophysical information that can be recovered is limited by our understanding of the complex interactions between the mantle, the oceans and the atmosphere. As these interactions are unraveled we can expect that the VLBI observations will play an important role in measuring anelastic effects in the mantle, and possibly in constraining models of ocean tide motions.

References

Robertson, D.S., W.E. Carter, J.A. Campbell, and H. Schuh, Daily UT1 Determinations from IRIS Very Long Baseline Interferometry, Nature, 316, 424-427, 1985b.

Robertson, D.S. and W.E. Carter, Earth Orientation Determinations from VLBI Observations, in Proceedings of the International Conference on Earth Rotation and the Terrestrial Reference Frame, Part II, Vol. 1, The Ohio State University, Columbus, Ohio, 296-306, 1985.

Robertson, D.S., W.E. Carter, and F.W. Fallon, Earth Rotation from the IRIS Project, in Proceedings of the IAU Symposium no. 129, J. Moran, ed., D. Reidel, Dordrecht, Holland, in press, 1987.

Rosen, R.D., and D.A. Salstein, Variations in Atmospheric Angular Momentum on Global and Regional Scales and the Length of Day, J. Geophys. Res., 88, 5451-5470, 1983.

Schwiderski, E.W., On Charting the Global Tides, Rev. Geophys. and Sp. Phys., 18, 243-268, 1980.

Wahr, J.M., Earth Orientation and Nutation, in Proceedings of the IAU Symposium no. 129, J. Moran, ed., D. Reidel, Dordrecht, Holland, in press, 1987.

Wahr, J.M., and Z. Bergen, The Effects of Mantle Anelasticity on Nutations, Earth Tides, and Tidal Variations in the Rotation Rate, Geophys. J. R. Astron. Soc., 87, 633-668, 1986.

Wilkins, G.A., (ed.), Project MERIT: A Review of the Techniques to be Used During Project MERIT to Monitor the Rotation of the Earth, Royal Greenwich Observatory, Herstmonceux, 77 pp., 1980.

Yoder, C.F., J.G. Williams, and M.E. Parke, Tidal Variations of Earth Rotation, J. Geophys. Res., 86, 881-891, 1981.

DAILY POLE POSITIONS MONITORED BY VERY LONG BASELINE INTERFEROMETRY

A. Nothnagel[1], G. D. Nicolson[1], H. Schuh[2], J. Campbell[2], and R. Kilger[3].

In January and February 1986 the Hartebeesthoek Radio Astronomy Observatory (HartRAO) and Wettzell Geodetic Fundamental Station performed a series of Very Long Baseline Interferometry (VLBI) experiments employing the Mark III system. Twenty-seven single baseline observing sessions of 2-hour duration and 6 sessions observed within multistation experiments of 24-hour duration were scheduled. The analysis described here considers only single baseline data extracted from the multistation experiments as well as the short 2 hour sessions. The measurements yielded accuracies of about ±2 mas for the x pole component and ±1 mas for the y pole component. The intensive series of pole positions agrees very well with the 5 day IRIS pole positions and indicates that periodic pole path fluctuations may exist.

Introduction

In January and February 1986 a Mark III Data Acquisition Terminal was installed temporarily at the Hartebeesthoek Radio Astronomy Observatory (HartRAO) on loan from the U S National Geodetic Survey (NGS), National Oceanographic and Atmospheric Administration of the U S Department of Commerce. Four experiments of 24-hour duration spread over 34 days were carried out including Westford Observatory in Massachusetts (USA), Richmond Station in Florida (USA), Wettzell Geodetic Fundamental Station in Bavaria (FRG) and HartRAO. In two additional experiments Wettzell was replaced by Onsala Space Observatory. A second group of experiments consisted of 27 sessions of 2-hours duration between the multistation experiments employing only Wettzell and HartRAO. Scheduling considerations and observations of these short sessions are described in Nothnagel et al. [1986].

The main purpose of these short daily sessions was to demonstrate the potential of the VLBI technique to monitor polar motion by relatively short and inexpensive experiments on a north-south baseline. Furthermore, this first daily pole position monitoring project of more than 1-month duration should allow for investigating short period fluctuations of the pole which were recently considered by several authors [Eubanks et al., 1986; Kolaczek, B. and Kosek, W., 1985].

Data Analysis

The complete data bases of the six multistation experiments in which HartRAO participated are included in the IRIS-A project solutions in order to provide, among other things, better positions for radio sources between 0 and -29° declination [Carter et al, 1987]. The NGS VLBI group, responsible for analyzing and disseminating the IRIS data, regularly runs global solutions in which all parameters are estimated quasi-simultaneously, resulting in a homogeneous set of radio source coordinates, station coordinates, Earth orientation parameters and corrections for the nutation angles Psi and Epsilon [Robertson and Carter, 1986]. Since the data collected in the 1986 HartRAO VLBI campaign are alone not sufficient for a comprehensive estimate of all parameters, we have to rely on some of the results of the above mentioned NGS global solutions for our analysis. As a suitable reference frame we selected the results from a global solution which included all IRIS experiments until October 1986 (Series 86 Oct 15). This has the advantage that the six long experiments including HartRAO are well imbedded between numerous other experiments and that the results, especially the nutation corrections, have stabilized for our observing period. From this global solution we adopted the station coordinates of the observing stations, the positions of the radio sources used, as well as smoothed nutation corrections for the observing period, and kept

---

[1]HartRAO, P O Box 443, Krugersdorp 1740, South Africa
[2]Geodetic Institute, Nussallee 17, D-5300 Bonn 1 Federal Republic of Germany
[3]Fundamentalstation Wettzell, D-8493 Kötzting, Federal Republic of Germany

Copyright 1990 by
International Union of Geodesy and Geophysics and American Geophysical Union.

these fixed in the least square fits of all data sets.

Then we analyzed the 27 short sessions between Wettzell and HartRAO and again the 6 experiments of 24-hour duration, but here only those observations on the long north-south baselines between Europe and Southern Africa, i.e., Wettzell – HartRAO and Onsala – HartRAO were used. This separation was done to complete the daily observations with data of the same single baseline origin and to better evaluate the results of the short experiments.

Standard least square fits were performed solving only for clock offset, clock rate, the two pole components and in the case of the 24-hour experiments one zenith atmospheric path delay at each station. The analysis of these sessions was done with the Bonn VLBI Software System (BVSS), which is very flexible for single baseline VLBI analysis. We fixed the third Earth orientation parameter, DUT1, to the daily values determined in the IRIS intensive series. Unfortunately, the series of daily DUT1 values has some gaps during our observing period, which may cause errors in the pole positions. However, a comparison between solutions using only the interpolated 5-day IRIS DUT1 values and solutions also incorporating the daily DUT1 values showed only a maximum change of 0.8 mas in the x component and 0.5 mas in the y component.

The initial analysis of the data was done with the nutation series adopted by the IAU in 1980. The post-fit residuals of the 24-hour experiments contain significant systematics (Figure 1) which originate from the inaccurate nutation series. Since we observed only radio sources between -30° and +40° declination a nutation offset in obliquity changes the apparent declination of a radio source depending on its right ascension and a nutation offset in longitude changes the apparent right ascension almost linearly. The respective interferometric delay corrections vary quasi-sinusoidal with increasing right ascension. The correction for any given source, however, remains nearly constant for the short observation period. If the sources are observed more or less in order of increasing right ascension, as it is the case on a long north-south baseline, a sinusoidal dependence is still discernible in a plot of delay corrections versus UT (Figure 2).

After applying nutation corrections taken from IRIS Bulletin A the post-fit residuals are much better distributed and the RMS of the delay residuals was reduced by 30 to 48 percent, down to a level of between 140 and 280 picoseconds (Figure 3). However, there is still room for improvement since the standard error quoted in IRIS Bulletin A is 1.3 mas for the nutation in longitude (Psi).

The daily short experiments can be divided into two groups. One group has been observed in the morning (start at 15:13 GST or 06:40 UT on Jan 28th, 17 successful sessions with 11 scans each), the second group in the evening (start at 04:02 GST or 19:20 UT on Jan 30th, 6 successful sessions with 10 scans each).

As can be seen from Figure 2 the effect of the

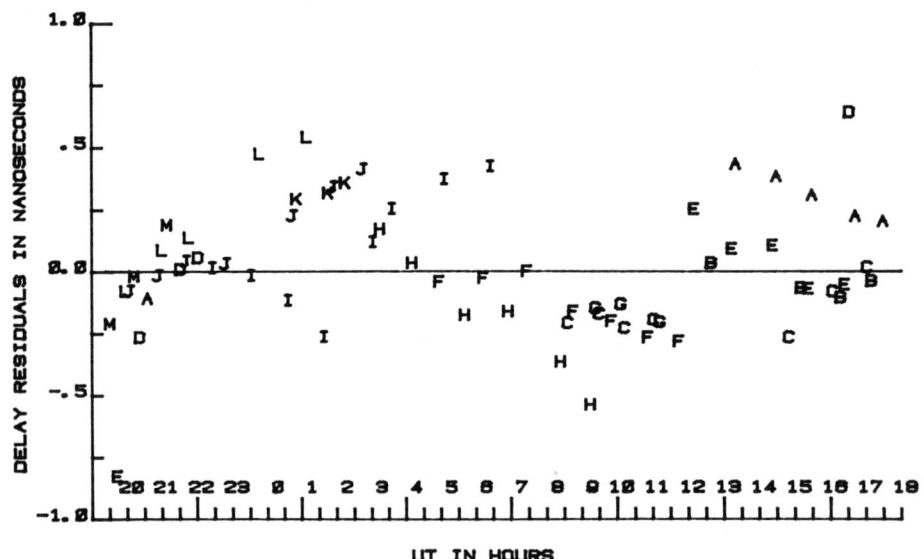

Fig. 1. Delay residuals of experiment 029/86 using IAU nutation for baseline HartRAO – Wettzell.
A = 0229+131   B = 2345-167   C = 3C454.3   D = 0552+398   E = 0106+013   F = NRAO530
G = 1921-293   H = 3C345   I = 3C273B   J = OJ287   K = 1034-293   L = 0727-115
M = 0454-234

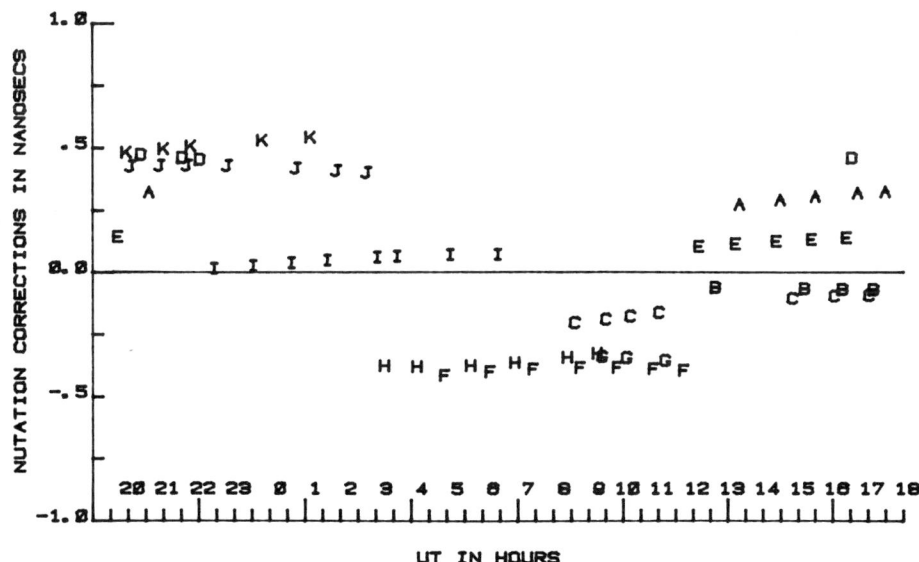

Fig. 2. Delay corrections for above observations, Nutation offsets dPsi = -1.3 mas and dEpsilon = +4.0 mas.

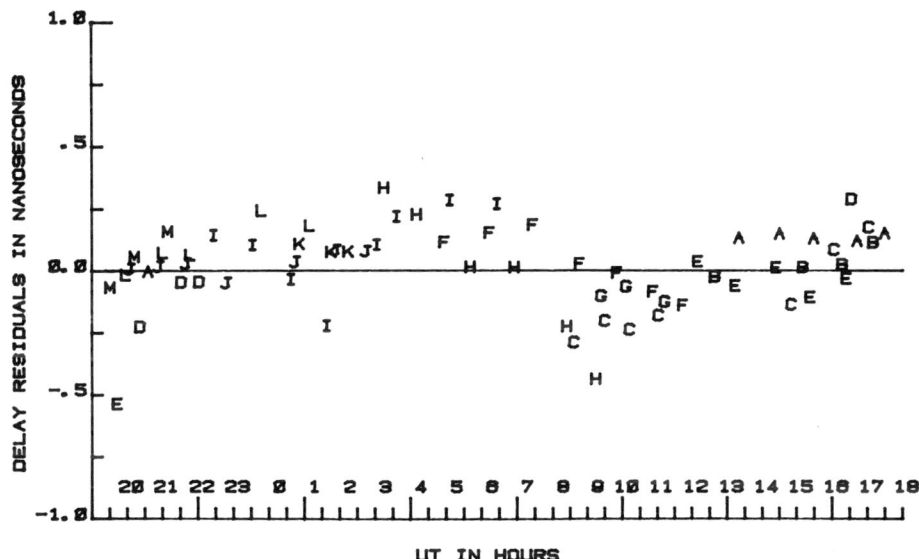

Fig. 3. Delay residuals of experiment 029/86 with nutation corrections applied.

incorrect nutation contaminates both groups in a completely opposite way. Positive corrections have to be applied to the observations in the morning experiments while the corrections to the observations performed in the evening are all negative. Resulting from these corrections the pole positions measured in the 2-hour sessions change by up to 2.5 mas. Accurate nutation corrections are therefore essential to be able to combine all subgroups consistently.

Results

Figure 4 shows the x and y pole components separately versus time with one pole position for each experiment, long and short, together with the IRIS pole positions. The trace of the y component does not show any inhomogeneities between the three different groups of morning, evening, and 24-hour experiments. Furthermore, the IRIS pole positions seem to be well integrated in the series of y pole

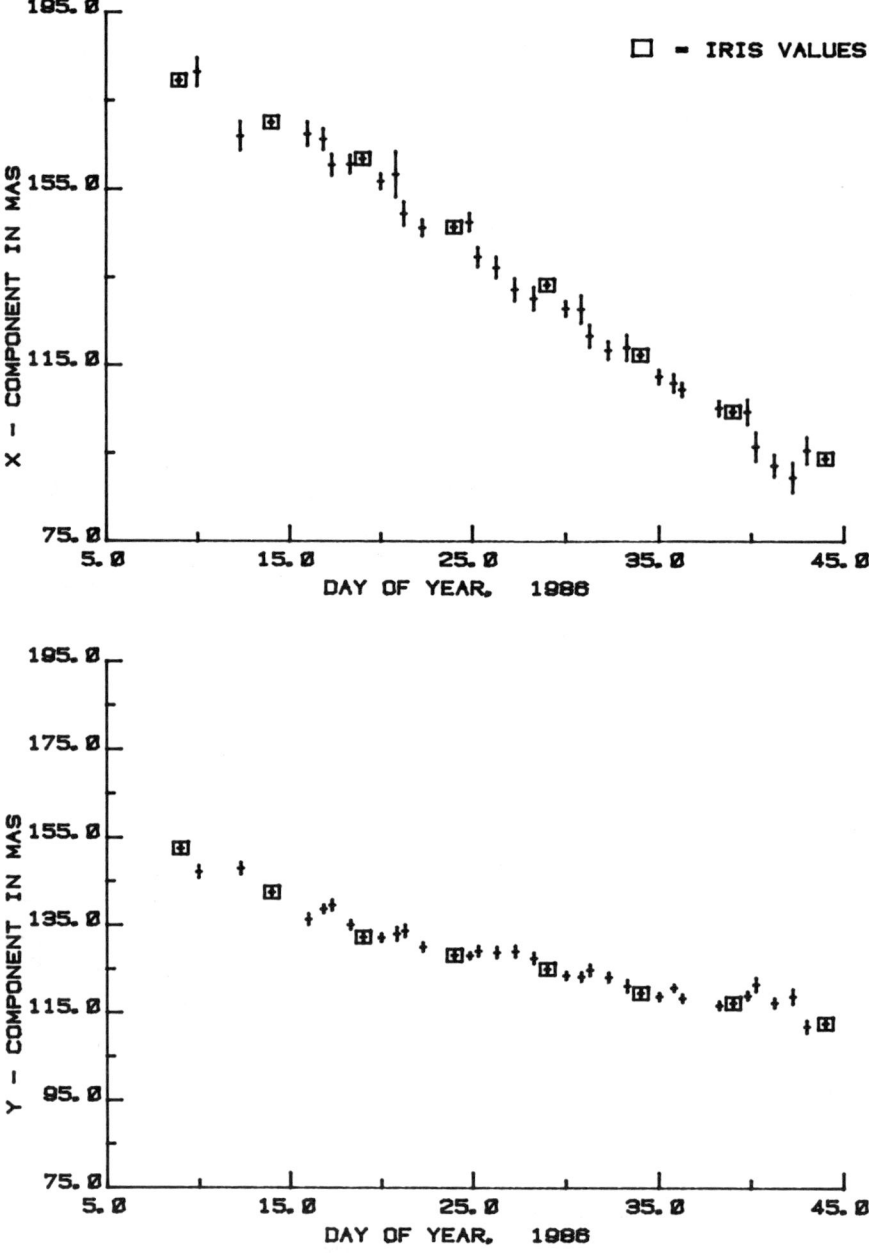

Fig. 4. Pole positions versus time

components and appear to be just the missing link in a continuous polar motion path. Further analysis of this y pole component series will show whether the apparent periodic variations are analytically detectable or not.

The series of the x components does not agree so well with the IRIS values. However, the bias of about −2.5 mas corresponds to only 8 cm which may easily be explained by inaccurate station coordinates of the HartRAO telescope. More important is the fact that the scatter of the x pole component series is less than ±2.5 mas.

In the 24-hour experiments the formal errors of the pole coordinates were based on observation weights adjusted so that the ratio of the variances a posteriori and the variances a priori was close to one (F-Test). The average adjustment values of the 6 experiments of 24-hour duration were taken as a measure for the 2-hour sessions because eleven observations do not permit a direct evaluation of the observation weights within a single observing session.

The most uncertain part in the analysis is the atmosphere. We used the CfA model [Davis et al., 1985] based on surface meteorological data and only in the six long experiments did we estimate atmospheric zenith path delays. The largest values within this group of experiments correspond to -11 cm excess zenith path at Wettzell and -3 cm at HartRAO. The average corrections were -7 cm and -2 cm, respectively.

However, these relatively large corrections are assumed to be due to the observing schedule used in the long experiments. A more realistic assumption is a general uncertainty of 3 cm in atmospheric path delay at both stations [Robertson et al., 1985]. We tested the influence of these errors on the solutions of the short experiments. The average change is 1.2 mas in the x pole component and 0.7 mas in the y component when modified atmospheric zenith path delays are used.

The combined effect of all these uncertainties discussed earlier can be estimated by forming the root sum squared (RSS) of the worst possible errors (Table 1).

The x component is in the worst case uncertain by 1.9 mas and the y component by 1.6 mas. This agrees with the formal errors of the least square fits.

TABLE 1. Root sum squared of errors.

	x_p	y_p	
Nutation	1.3	1.3	mas
DUT1	0.8	0.5	mas
Atmosphere	1.2	0.7	mas
RSS	1.9	1.6	mas

Conclusion

The series of daily pole position measurements on a single baseline shows remarkable accuracies. From the geometry of the baseline it was to be expected that the determination of the y pole component is better conditioned than that of the x pole component. It can therefore be assumed that short term periodicities are more easily detectable in the y pole component. In fact, the y pole position plot seems to show periodic variations.

More conclusive results are expected from a repetition series which was observed in January and February 1987 with an improved observing schedule and more redundancy. First results indicate that the formal errors of the least square fits are much lower and that the geometrical configuration is more stable.

References

Carter, W. E., D. S. Robertson, A. Nothnagel, G. D. Nicolson, H. Schuh, and J. Campbell, IRIS-S: Extending geodetic VLBI observations to the Southern Hemisphere, (submitted to JGR), 1988.

Davis, J. L., T. A. Herring, I. I. Shapiro, A. E. E. Rogers, and G. K. Elgered, Geodesy by radio interferometry: Effects of atmospheric modeling errors on estimates of baseline lengths, Radio Sci., 20, 1593-1607, 1985.

Eubanks, T. M., J. A. Steppe, and J. O. Dickey; Atmospheric Excitation of Rapid Polar Motions, Proceedings of IAU Symposium No.128, Earth Rotation and Reference Frames, D. Reidel, 1986.

IRIS Bulletin A, National Geodetic Survey, NOAA, N/CG 114, Rockville, Maryland, 1986.

Nothnagel, A., G. D. Nicolson, H. Schuh, J. Campbell, H. Cloppenburg, and R. Kilger, Radiointerferometric Polar Motion Determination Using a Very Long North-South Baseline, Proceedings of IAU Symposium No.128, Earth Rotation and Reference Frames, D. Reidel, 1986.

Robertson, D. S., and W. E. Carter, VLBI Determinations of Irregularities in the Earth's Nutation, Proceedings of IAU Symposium No.128, Earth Rotation and Reference Frames, D. Reidel, 1986.

Robertson, D. S., W. E. Carter, J. Campbell, and H. Schuh, Daily Earth Rotation Determinations from IRIS VLBI, Nature, 316, 424, 1985.

Kolaczek, B. and W. Kosek, On Short Periodic Oscillations of Pole Coordinates Determined by Different Techniques in the Merit Campaign, IAU General Assembly, New Delhi, India, 1985.

ERROR ANALYSIS FOR EARTH ORIENTATION RECOVERY FROM GPS DATA

N. Zelensky[1], J. Ray[2], and P. Liebrecht[3]

Abstract. Very precise satellite orbit determination, or terrestrial positioning using satellites, requires high-quality information on the Earth's orientation parameters: polar position (X,Y), and rotation rate (A1-UT1). For real time satellite tracking, errors in extrapolated values for these quantities can be a major source of uncertainty. The largest error comes from variations in the Earth's rotation rate which cannot be accurately predicted. This report considers the use of the GPS constellation of navigation satellites to determine the Earth orientation parameters in near real time.

The ORAN covariance analysis program has been used to simulate the full Block II constellation of 18 GPS satellites. The hypothetical GPS tracking network is coincident with the existing POLARIS VLBI network, operated expressly for the determination of earth orientation. Alternative network geometries have also been considered. As the reliability of any covariance analysis depends on the quality of the values assumed for the various tracking errors, considerable effort has gone into defining a reasonable error budget for these simulations. The simulations considered simultaneous solutions of Earth orientation together with the satellite orbits, ground clocks, station positions, and tropospheric scaling at each station. The principal contributors to Earth orientation uncertainty are found to be media effects and measurement noise.

Simulation results show that a POLARIS-like network should be able to determine the Earth orientation parameters about as well as the current VLBI program, using single-difference carrier phase observations of the GPS satellites. GPS offers the potential of lower costs and faster data turn-around, making it an attractive alternative to VLBI. GPS determinations could be nearly 30 times better for UT1 and perhaps 10 times better for pole position, compared with 10-day-old USNO predictions. Improvements of this magnitude are crucial for real time satellite tracking at the few meter level.

Introduction

Over the past few years, confidence in the value of the GPS system for high precision geodesy has grown, encouraged by improvements in the quality of results obtained using the available satellites. Baseline accuracy has improved by nearly two orders of magnitude in the last 3 years - from several parts per million to better than one part in 10 million. However the GPS capability for accurate determination of the Earth's orientation has only recently attracted attention [Abbot et al., 1987]. This paper evaluates the potential for recovering geodetic parameters, with emphasis given to Earth orientation, using the planned GPS Block II constellation of 18 navigation satellites.

This study is part of a larger NASA effort to evaluate the GPS capability to calibrate the required parameters for accurate near real time interferometric tracking of the geosynchronous Tracking and Data Relay Satellites (TDRS) (A. Au et al., Specifications for a TDRS-tracking system using interferometry, Interferometrics Inc., 1987).

Modeling Considerations

Covariance Analysis Program

The ORAN Program (ORbit ANalysis) (T. V. Martin and J. J. McCarthy, ORAN descriptive summary, EG&G/WASC, 1978) was used to evaluate the accuracy with which the Earth Orientation Parameters (EOP) can be determined from GPS tracking data in a simultaneous solution for the 18 satellite state vectors and other parameters. ORAN is a multiarc, multisatellite covariance analysis computer program which simulates a Bayesian least squares data reduction for orbital and geodetic parameters. It does not process actual data. Through the generation of normal equations, it has

---

[1] EG&G/WASC, 5000 Philadelphia Way, Lanham, MD 20706
[2] Interferometrics Inc., Code 621.9, NASA/GSFC, Greenbelt, MD 20771.
[3] Code 532.1, NASA/GSFC, Greenbelt, MD 20771

Copyright 1990 by
International Union of Geodesy and Geophysics and American Geophysical Union.

the capability of propagating the effect of measurement noise and the assumed level of systematic errors in any of the unadjusted parameters to compute a total uncertainty for each adjusted parameter. The uncertainty obtained is considered the accuracy with which a parameter can be recovered. Obviously, the validity of the accuracy estimate depends critically on the quality of the input error budget.

GPS Block II Constellation and Tracking Networks

The GPS Block II constellation, planned for full operation by the early 1990's, will consist of 21 satellites, three of which will be spares. In the simulations, 18 satellites are divided into 6 orbit planes, 3 satellites per plane, each at an inclination of 55 degrees.

As shown in Figure 1, three continental U.S. tracking networks are considered: 1) a small (1000-km baseline) conceptual network with 4 sites in the southwestern United States; 2) the POLARIS network consisting of dedicated geodetic radio observatories in Massachusetts, Florida, and Texas; 3) the POLARIS network plus the Owens Valley Radio Observatory in eastern California.

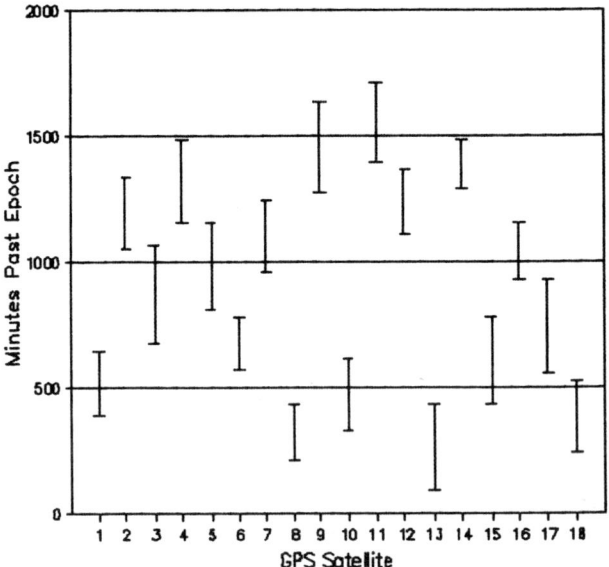

Fig. 2. Tracking schedule for the GPS Block II constellation using the POLARIS network.

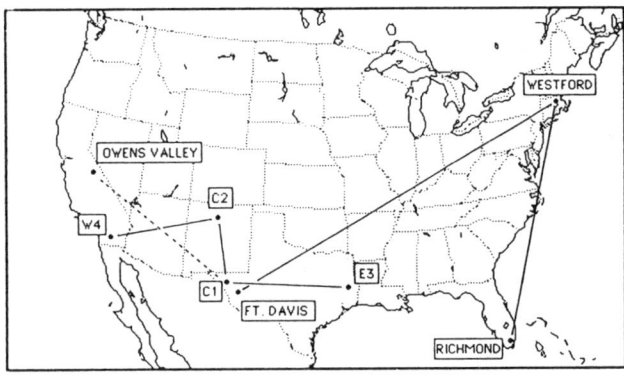

Fig. 1. GPS tracking network.

Figure 2 shows a typical 24 hour tracking schedule for the POLARIS network.

Measurement Modeling

The performance of three measurement types is evaluated: 1) P-code pseudo-range; 2) single differenced integrated Doppler (DID); 3) interferometric group delay.

The integrated Doppler (carrier phase) recorded simultaneously at two different sites is differenced for each satellite in common view to remove the effects of satellite clock variations, forming the DID observable. The DID and group delay data types are modeled similarly, except that the DID solution must adjust a constant bias parameter for each satellite/baseline/pass combination to account for the unknown number of integer cycle ambiguities inherent in the observables. Of the three data types, interferometric group delay has the advantage of being the least susceptible to dithering of the GPS emissions because no demodulation of the coded signal is involved. All receivers are assumed to track up to six satellites simultaneously. The nominal values assigned for the observation noise and sampling rate are based on the capabilities of existing receivers. The group delay measurement assumes a 2-meter steerable dish at each site. All measurements are assumed to be dual-frequency (L1 and L2) to permit direct calibration of ionospheric delays. Finally, it is assumed that the frequency standard at each site is a hydrogen maser clock. The recorded data can be transmitted over a dedicated telephone line and analyzed in near real time.

GPS Reference Frame and Error Budget

The GPS satellite orbits are used to define a dynamical reference system which will provide direct observations of the geodetic quantities of interest. For establishing such a reference frame, the simulations were considered to determine simultaneously the orbits, solar radiation pressure coefficients, site positions, EOP, and in addition, remove the troposphere and ground clock behavior effects from the data. The term "arc" refers to a specific data period over which a satellite's orbit is integrated; these data solely define the basis by which the satellite's position is adjusted.

The following terrestrial reference frame is

adopted: 1) the coordinates of one site are held fixed and 2) the EOP values assumed over the first day of data acquisition are held fixed. This set of initial terrestrial reference parameters have zero error assigned to them. An inconsistency between the assumed reference site position and initial axis orientation should not degrade the EOP solution accuracy using the high altitude GPS satellites.

The effect of independent ground clock behavior is modeled by holding one site clock fixed (with zero error) and adjusting the others. One clock offset spans the 2-day arc length, and 2 drift parameters account for each day of the arc. The troposphere delay is estimated using the Hopfield model, with one parameter per site per day. The combined effect of the measurement noise and the uncertainties in the unadjusted parameters are mapped into a 1-sigma total uncertainty for each adjusted parameter.

As in all covariance analysis, the results depend upon the assumptions made for the magnitude and character of the error sources. The assigned errors (Table 1) are believed to be conservative. However, some general assumptions are made concerning error sources for which no adequate models are available. These errors fall into three categories:

Multipath. The error due to multipath effects was not treated in this study. Although these effects can be significant for pseudo-range, it is anticipated that future improvements in antenna design and better positioning of the receivers should substantially reduce this error source.

Non-gravitational forces. The effect of solar radiation pressure acting on the complex satellite geometry coupled with the accelerations thought to be caused by unbalanced thermal radiation and absorption at the satellite ("Y-bias" effect) are important and difficult aspects to model. In a data reduction experiment to recover EOP using GPS [Abbot et al., 1987], the large errors at the end of a 6-day arc were attributed to mismodeling the "Y-bias" component. Indeed, simulations using a complex solar radiation pressure model show that even small errors in the assumed coefficients will cause a significant pertubation in the orbit after 2 days [Colombo, 1986]. Unfortunately, ORAN uses a simple "cannonball" model which cannot account for the character of the pertubation seen using a more complex model over longer arcs. It is not known to what extent this mismodels the true nature of the forces acting on the GPS satellites, but it can be assumed that limiting the duration of the arc span will minimize the effect of misrepresenting such force model errors. For this reason the arc length is limited to two days.

Media. The excellent coverage of the constellation which permits several lines-of-sight should allow a zenith offset parameter to accurately estimate the tropospheric delay. However, some atmosphere phenomena, such as horizontal gradients, may not be accounted for by a simple offset parameter. It is assumed that

TABLE 1. GPS Tracking Error Budget

Error Source	Uncertainty
Grav. const. (GM)	1 part per 10^8
Ionosphere (unmodeled)	0.5% per baseline
Gravity field error	45% (GEM10 - GEM7)
Solid Earth tides:	
dynamic (k_2)	1%
geometric (h_2, l_2)	10%
Ocean loading	100%
Polar motion:	
X	adjusted (fix 1st day)
Y	adjusted "
A1-UT1	adjusted "
Station locations	adjusted (fix 1 site)
Station clocks:	
offsets	adjusted (fix 1 clock)
drift rates	adjusted "
Troposphere	adjusted
Solar radiation pressure coeff.	adjusted
Total noise - 1 obs./min.: (corrected for ionosphere)	
Group delay	4.0 cm
Difference phase	1.4 cm
Pseudo-range	30. cm

systematic errors of this type will not exceed 0.5 cm at zenith.

The dual frequency GPS observations permit accurate removal of the dominant ionosphere effect. The ionosphere error which remains from neglecting higher order terms [Herring, 1983] is estimated to be under 0.2 percent of the total effect (under 2.0 cm for the worst conditions).

The residual troposphere and ionosphere errors are combined and represented as a fraction of the total ionospheric effect, assuming a very conservative error of 0.5 percent. ORAN uses the Bent model for the ionosphere. This combined representation is referred to as the media error.

Covariance Analysis Results

Nominal Solution Error Profile

The group delay solution using the POLARIS network, 2 days of GPS data, and an elevation cutoff angle of 10 degrees was selected as a standard by which other solutions can be compared. It can be seen (Figure 3) that this standard GPS simulation yields EOP accuracies which lie between the upper limit for quasar VLBI accuracy, based upon comparison with satellite laser ranging results [Robertson et al., 1985a, b], and the formal error for a typical 1986 VLBI solution (J.

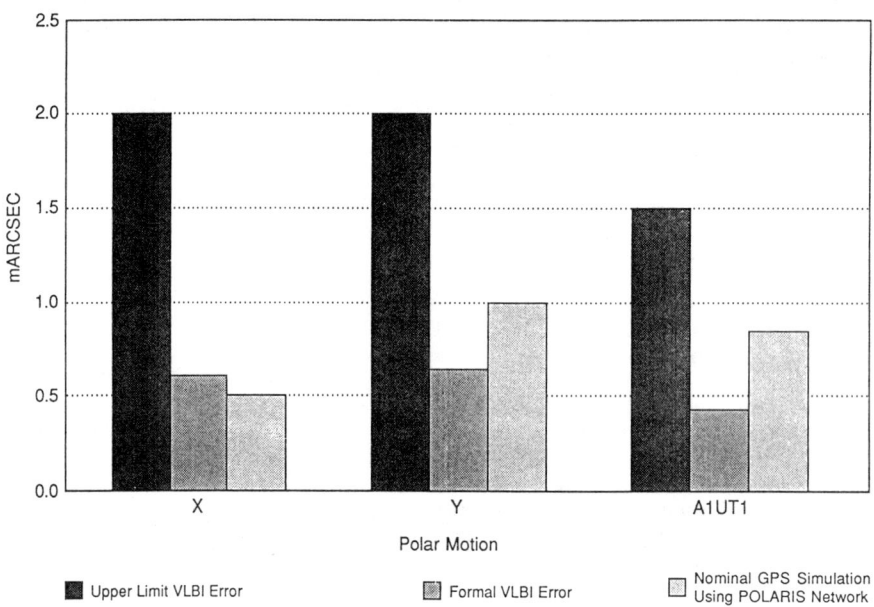

Fig. 3. Polar motion accuracy - VLBI versus GPS.

W. Ryan and C. Ma, <u>Crustal Dynamics Project Data Analysis - 1987</u>, NASA/GSFC, TM 100682, 1987). As shown in Figure 4, the unmodeled media effects constitute the largest GPS error source, followed by measurement noise.

Typical accuracies of the other parameters estimated in the GPS solution are given in Table 2. The estimated site position accuracy approaches that of quasar VLBI solutions, while all the other estimated parameter accuracies are

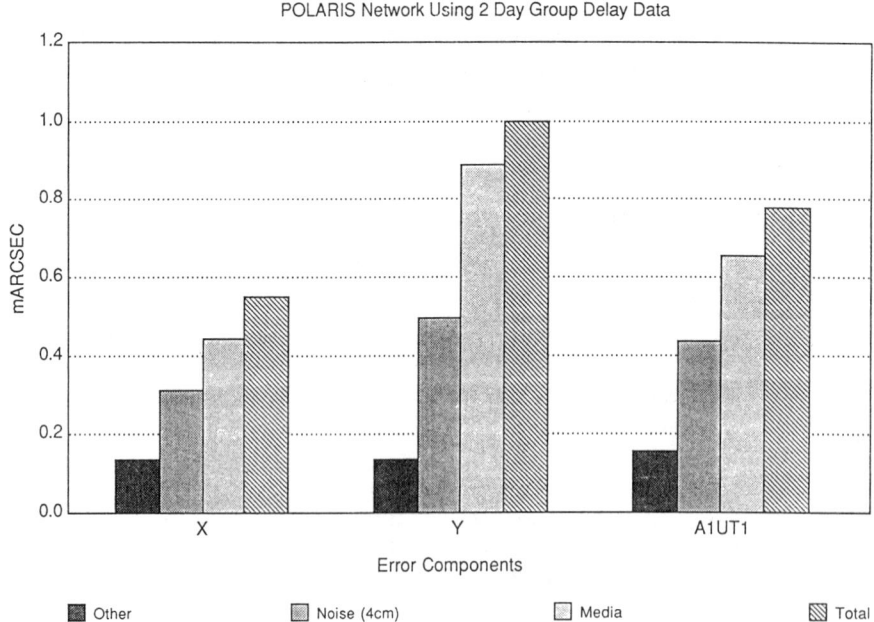

Fig. 4. Polar motion error profile for 2 days group delay data collected with POLARIS network.

TABLE 2. Geodetic Parameter Accuracies Using GPS
(For POLARIS Network with 2 Days of Data)

Measurement type	Calibrated parameters									
	Troposphere %	Clocks		Site position (cm)			Polar motion (marcsec)			GPS orbit error over 2 days (m)
		Offset (psec)	Drift (psec/day)	E	N	V	X	Y	A1-UT1	
Pseudo-range	0.7	300	170	3	4	9	2	1	1.5	1.0
Differenced phase	0.2	---	70	3	4	6	.5	1	.5	1.5
Group delay	0.2	70	70	2	3	5	.5	1	1.0	1.0

comparable to that of the VLBI solutions. The troposphere zenith estimate (0.2 percent or 0.4 cm at zenith) may seem optimistic, although that level of accuracy has been achieved in a set of special low-elevation angle VLBI experiments at GSFC. Systematic error which exceeds the error due to measurement noise for the adjusted parameters, is due, in most cases, to mismodeled media effects. Site position accuracy is degraded predominantly by the uncertainty in ocean loading on the Westford site. The low correlation coefficients (not shown) indicate all the parameters are highly separable in what appears to be a very robust solution.

EOP Solution Sensitivity

Solar radiation pressure (SOLRAD). The SOLRAD coefficients are typically estimated to an accuracy of 0.1 percent for the nominal 2-day solution. However, to evaluate EOP sensitivity to this effect, the SOLRAD parameter is not estimated, but rather an assumed systematic SOLRAD error is propagated through the nominal solution. Although ORAN employs a simple SOLRAD model, Figure 5 suggests that the EOP solution is not sensitive to this type of error over a 2-day arc.

Measurement type. Figure 6 compares the EOP accuracies found using the three measurement types: group delay, differenced phase, and pseudo-range. The POLARIS network is used with 2 days of data and an elevation cutoff of 10 degrees. The solution using pseudo-range also estimates CM (if CM is left unadjusted, it becomes a major error source). In the pseudo-range solution, GM is adjusted to an accuracy of two parts per billion. Although, the GPS EOP solution accuracy is comparable to quasar VLBI solutions

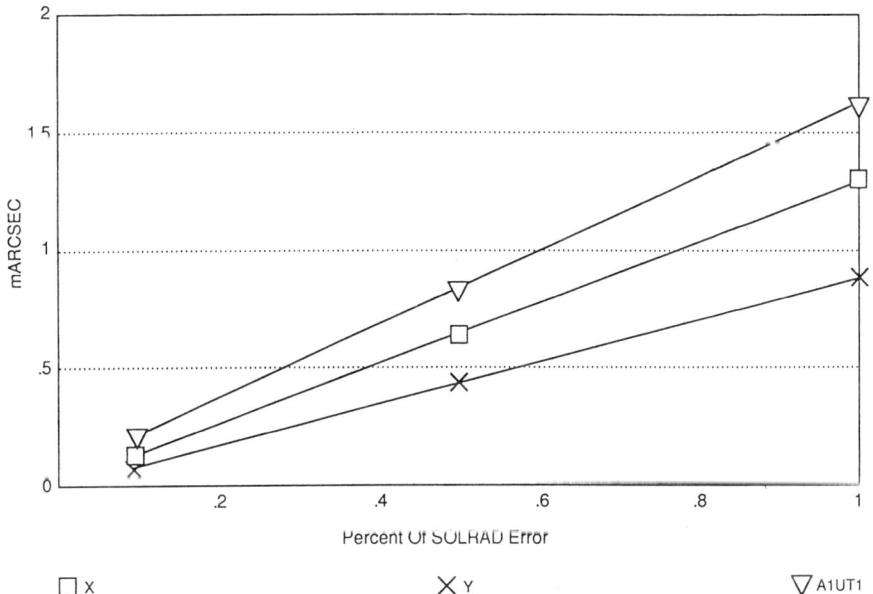

Fig. 5. Polar motion sensitivity - SOLRAD; POLARIS network with 2 days of group delay data.

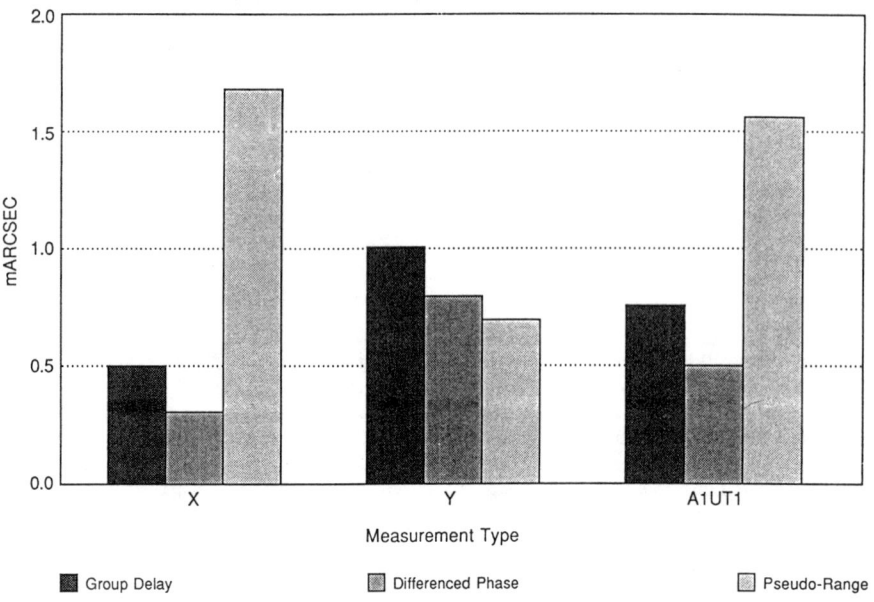

Fig. 6. Polar motion sensitivity - measurement type; POLARIS network with 2 days of data.

for all three GPS measurement types, this level of accuracy is questionable for the pseudo-range data type since multipath error is ignored here. In comparison with group delay, the slightly better accuracy seen for EOP recovery achieved using differenced phase data is due to the reduced sensitivity to media error. The accuracies for the other estimated parameters are very similar for both data types (Table 2).

Network geometry. As expected, the larger networks promise the better EOP solutions. Figure 7 shows the effect of varied network geometry (in all cases using group delay data, a 10 degree elevation cutoff, and a 2-day arc).

Elevation angle cutoff. The effect of the elevation cutoff angle was considered for tracking limited to 10, 20, and 30 degrees (using group delay data and a 2-day arc in all cases). In effect this reduces the duration of a typical pass from 6 to 4.5 to 3.3 hours, respectively. The strength of the orbit solution is thereby also reduced. Surprisingly, the EOP solution accuracies remain almost unaffected (Figure 8), and even improve slightly with the 20 degree elevation cutoff. The error due to propagation media is reduced in this solution, accounting for the improved total accuracy (Figure 9). ORAN models the effect of the ionosphere (media) by mapping the total electron content along the line of sight. Thus, the dependence of the ionosphere on the cosecant of the elevation angle contributes to a smaller net ionosphere error when using observations with larger elevation cutoff angles. As the solution weakens (for the shorter pass durations), contributions from other error sources begin to exceed the reduction in the ionospheric (media) effect. All the other adjusted parameter accuracies degrade significantly with the larger elevation cutoff angles. For example, the site position uncertainty increases from 7 cm for the 10 degree solution, to 42 cm for the 30 degree solution. The estimated troposphere zenith offset error increases from 0.2 to 0.5 to 2.0 percent, respectively. This test suggests that comparable EOP accuracies can be achieved using less stringent higher elevation tracking with the full Block II constellation.

Conclusions

This covariance analysis study evaluated the accuracy with which EOP and other parameters may be recovered from GPS data in near real time. The planned Block II constellation provided the satellite coverage. In the simulated solutions, the parameters of interest together with the 18 satellite state vectors were simultaneously adjusted. The accuracy of the recovered EOP was tested for sensitivity to network geometry, GPS measurement type, elevation cutoff angle, and solar radiation pressure error.

This study finds that all three components of Earth orientation can be recovered using 2 days of

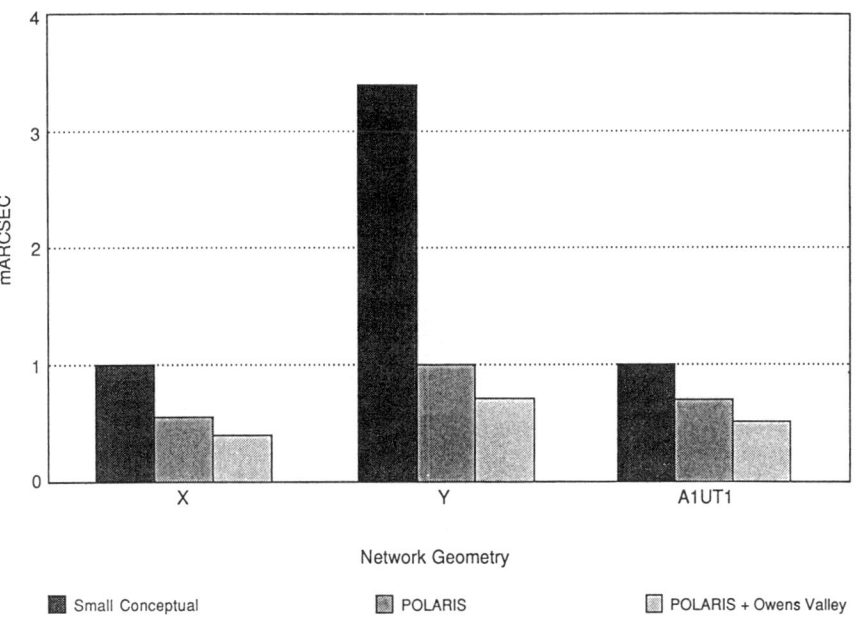

Fig. 7. Polar motion sensitivity - network; 2 days of group delay data.

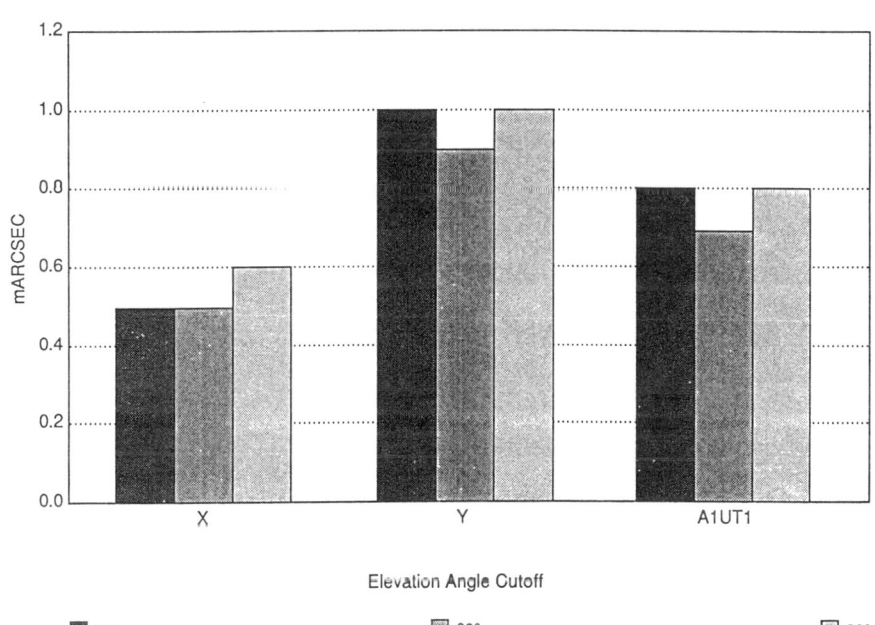

Fig. 8. Polar motion sensitivity - elevation angle; POLARIS network with 2 days of group delay data.

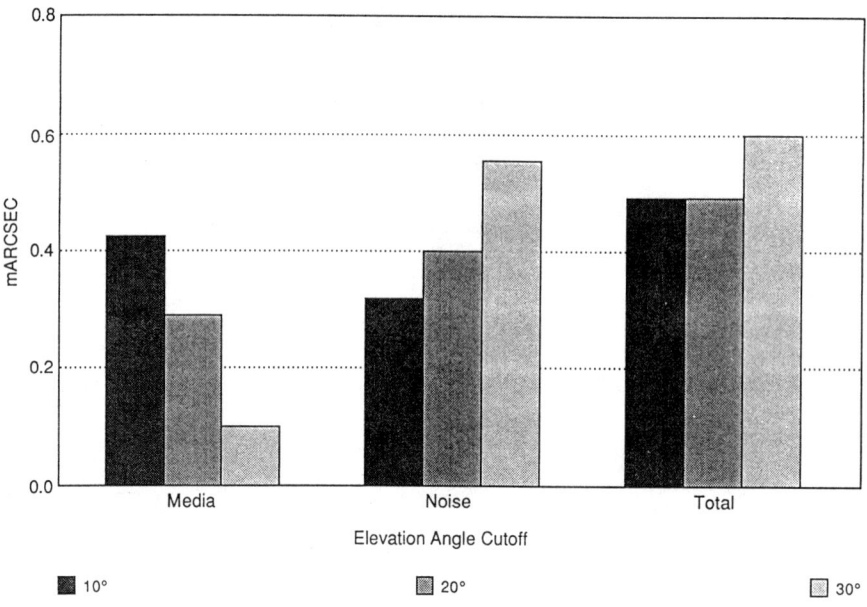

Fig. 9. Polar motion X-component error profile - elevation angle; POLARIS network with 2 days of group delay data.

GPS data with an accuracy comparable to quasar VLBI. The most accurate solutions were obtained using either differenced phase (differenced integrated Doppler) or group delay data, rather than pseudo-range, and the largest networks. The principal errors limiting polar motion recovery are unmodeled media effects and measurement noise. Polar motion insensitivity to elevation angle cutoff suggests that the more stringent requirements for low elevation tracking imposed for accurate satellite orbit or station position recovery can be relaxed for a network dedicated to Earth orientation monitoring. It is recommended that additional covariance studies be made using longer arcs, larger networks, and combinations of measurement types, to better evaluate the limits of the GPS capability for establishing a geodetic reference frame.

Acknowledgments. The authors wish to give their sincere thanks to Erricos Pavlis, Steve Klosko, Chopo Ma, Alan Schanzle, Mark Torrence, and Ron Williamson for the many helpful comments and useful critiques.

References

Abbot, R. I., R. W. King, Y. Bock, and C. C. Counselman III, Earth rotation from radio interferometric tracking of GPS satellites, in IAU Symposium 128, The Earth's Rotation and Reference Frames for Geodesy and Geodynamics, edited by A. K. Babcock and G. A. Wilkins, pp. 209-212, Kluwer, Dordrecht, 1987.

Colombo, O., Ephemeris errors of GPS satellites, Bull. Geod., 60, 64-84, 1986.

Herring, T. A., Precision and Accuracy of Intercontinental Distance Determinations Using Radio Interferometry, Ph.D. dissertation, Dept. of Earth and Planetary Science, Massachusetts Institute of Technology, Cambridge, Mass., 1983.

Robertson, D. S., W. E. Carter, B. D. Tapley, B. E. Schutz, and R. J. Eanes, Polar motion measurements: Subdecimeter accuracy verified by intercomparison, Science, 229, 1259-1261, 1985a.

Robertson, D. S., W. E. Carter, J. Campbell, and H. Schuh, Daily earth rotation determinations from IRIS very long baseline interferometry, Nature, 316, 424-427, 1985b.

SIMULATIONS TO RECOVER EARTH ROTATION PARAMETERS WITH GPS SYSTEM.

P. Pâquet and L. Louis

Royal Observatory of Belgium, 3 Av. Circulaire,
1180 Brussels, Belgium

Abstract. Results of simulations conducted to recover within 6 to 24 hours of observations the Earth Rotation Parameters (ERP) with the Global Positionning System (GPS) are presented. The whole GPS constellation supported by a maximum of 10 ground stations has been used as the basis system for data collection. The observed quantity is the integration of the Doppler frequency on dual frequency mode.

Perturbations are applied : on reference orbits for which the errors have a maximum variance of 1 meter both in along track and range components, on the coordinates of the tracking network whose the consistency is modified with a variance of 10 cm and on the tropospheric refraction on which an error of 2 percent is accepted.

Precisions of the subdecimeter level, both for Polar Motion and UT_1, are very promising and compete with those obtained by other technics like LASER ranging and Very Long Baseline Interferometry (VLBI). As a byproduct this simulation also shows that the interest of GPS for absolute positioning must be very limited and indicates that absolute positioning deduced from GPS data are not better than those obtained with the TRANSIT system.

Introduction

The Earth Rotation Parameters (ERP) are currently measured by space methods with a time resolution of 5 days. Presently, VLBI provides the best long term reference system and although it could be used for short term (≤ 1 day) measurements, such a network remains hard to manage in this particular window. In the present technical development of that type of observations, ideally VLBI could be used to provide, at regular time intervals, the best available absolute reference to which other more manageable techniques would calibrate their results obtained at higher frequency, from a few hours to 1 day for example.

Could GPS be a new tool to monitor the ERP within such short time intervals?

The GPS system is also extensively used and provides very good results for accurate differential positioning, but it is well known that it is not convenient for absolute positioning. The main reason is the poor geometry of the ground stations with respect to the GPS satellites constellation. Other applications or studies like the project POPSAT have demonstrated that for accurate absolute positioning the best satellite altitude is around 5000 to 6000 Km [Reigber et al., 1983].

With the support of 10 ground tracking stations, the capabilities of GPS were investigated both to monitor the ERP with a resolution of few hours and to determine absolute coordinates after 24 hours of observations. The error sources taken into account are related to the orbit, the refraction errors and the network consistency. The frequency error has not been introduced in the model with the consequence that the results are to be considered as optimistic. Nevertheless they are close to those obtained by Zelensky et al. in these proceedings [1988] and of the same order of precision as expected by Anderle [1982].

Data simulation

The observables

The simulated observed quantities are the integration over 60 seconds of the Doppler frequency as it will be observed on the broadcast frequencies generated by the 18 satellites of the full GPS constellation. The received

Copyright 1990 by
International Union of Geodesy and Geophysics
and American Geophysical Union.

signal is supposed to be free of ionospheric refraction, as indeed most of this effect can be removed in dual mode observation. On the basis of the ephemerides generated by numerical integration and of the coordinates of the conventional tracking network, the theoretical values of the observed quantities can be easily estimated.

To compute from the variance-covariance matrix the error of each unknown, the mean error m_0 of the observations must be estimated. It is deduced from simulated "observed quantities" computed by perturbing the reference orbit and / or the consistency of the tracking network and / or the tropospheric refraction correction.

Perturbations of the ephemerides

For the participating stations all different arcs observed are shifted, in along track and range by constant quantities extracted from two series of random numbers whose variance is given apriori. One of the series is associated with the along track component while the other series is associated with the range component. Most of the simulations were conducted with two apriori variances fixed to 0.5 and 1.0 meters.

The tracking network

The tracking network has been chosen to get a good world distribution in correspondance with the existing sites used for the tracking of TRANSIT satellites. Fig.1 shows the geographical distribution of the ground stations. Each station will observe all available passes for a maximum of 3 or 6 hours, symetrically distributed from rise time to set time. The minimun satellite elevation above the horizon is 10 degrees and the minimun elevation at closest approach is 15 degrees. To generate the simulated observed quantities, the network coordinates adopted to compute the theoretical values of the observed quantities are modified by adding to each component a number selected in a random series whose variance is given apriori. For most of the simulations, the apriori variance has been fixed to 10 cm which seems realistic according to the network errors estimated by Boucher [1987].

Perturbation of the tropsheric refraction

The meteorological parameters were kept the same for the whole network (Temp = 10^0, pressure = 1013 mb, humidity = 60 percent) and the adopted perturbation has

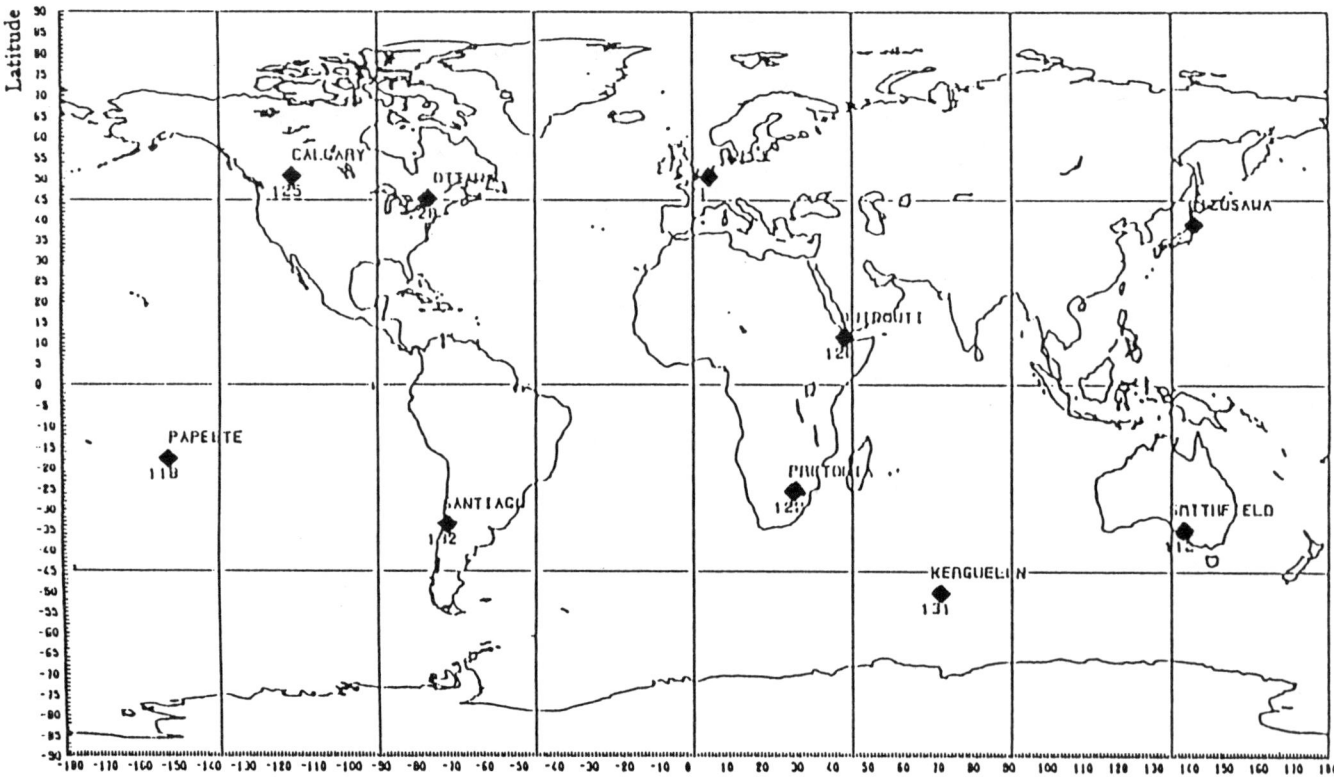

FIG. 1, STATION NETWORK FOR GPS SIMULATION
(10 STATIONS AND 18 SATELLITES)

been fixed in the range of 1 to 3 percent. At the zenith 2 percent corresponds to a range error of 3 cm while at 10^0 elevation it can reach 25 cm. The tropospheric refraction is estimated with the Hopfield's model [1971].

Precision of ERP and absolute positioning

Table 1 presents the sensitivity of ERP with respect to the various perturbations and the corresponding mean error m_0, while Table 2 gives the expected precision of ERP for different scenarios related to perturbations and length of the arc observed by each station. From Table 2 it can be seen that, after 24 hours of observations, the pole position could be given with a sub-decimeter precision while after 6 hrs it is limited to 10 to 15 cm. UT_1 would be determined with a precision of 0.15 ms to 0.30 ms.

Interesting results are given in fig.2 which represents the expected precision on each ERP versus the number of tracking stations. Starting from 2 to 10 stations, the site location has always been designed to keep homogeneous world distribution. It can be seen that the number of at least 10 stations is required to reach the precision of VLBI over time intervals of a few hours. As the results are based on simulations, generally considered as optimistic, for practical reasons and based on the experience of existing tracking network performances, we believe that a realistic tracking network would have to be extended to 12 to 14 stations.

Table 3 shows the sensitivity of absolute positioning for different variances of the perturbations applied to the orbit, network and tropospheric refraction; it also gives the global effect of all perturbations acting together. It confirms that under the defined constraints the precision of absolute positioning is of the order of 40 cm after 24 hrs.

TABLE 1

Sensitivity of ERP with respect to :
- orbit error
- network consistency
- refraction error

Observations : - 18 GPS satellites
- stations observe each arc maximun 6 hrs
- Doppler count of 60 seconds

Solution : - after 24 hrs of observations
- network of 10 stations
- number of equations 70589

Orbit error		Network error	Refraction error	ERP std. dev.			m_0
Al. Tr. (m)	Range (m)	(m)	(%)	X_p (m)	Y_p (m)	ΔUT_1 (ms)	(m)
0.30	0.30	-	-	0.034	0.031	0.10	0.038
0.50	0.50	-	-	0.044	0.040	0.13	0.050
1.00	1.00	-	-	0.062	0.057	0.18	0.070
-	-	0.10	-	0.002	0.002	0.01	0.002
-	-	0.30	-	0.004	0.003	0.02	0.004
-	-	-	1	0.005	0.005	0.02	0.006
-	-	-	3	0.016	0.014	0.05	0.018
0.50	0.50	0.10	2	0.050	0.045	0.15	0.051
1.00	1.00	0.10	2	0.070	0.063	0.20	0.071

TABLE 2

Sensitivity of the ERP with respect to :
- length of the arc observed by each station
- orbit errors
- refraction error 2 percent
- network of 10 stations, consistency 10 cm

Observations : - 18 GPS satellites
- Doppler count of 60 seconds

Solution : - after 6, 12, 24 hrs of observations

SOL hours	Al. Tr., Range → .50			Al. Tr., Range → 1.0			Nber Equ.
	ERP std. dev.			ERP std. dev.			
	X_p (m)	Y_p (m)	ΔUT_1 (ms)	X_p (m)	Y_p (m)	ΔUT_1 (ms)	

(a) Each arc observed a maximun of 3 hours :

06	0.141	0.122	0.42	0.195	0.170	0.59	10222
12	0.100	0.087	0.30	0.139	0.121	0.42	20444
24	0.071	0.061	0.21	0.098	0.085	0.30	40889

(a) Each arc observed a maximun of 6 hours :

06	0.100	0.090	0.20	0.140	0.125	0.40	16147
12	0.071	0.064	0.20	0.099	0.089	0.28	32294
24	0.050	0.045	0.15	0.070	0.063	0.20	64589

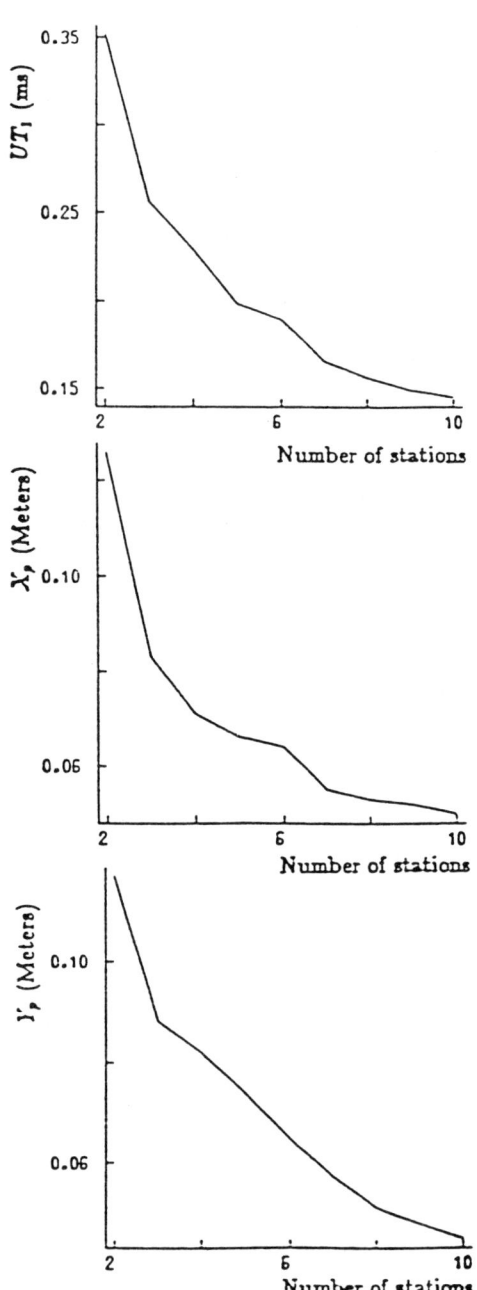

FIG. 2, PRECISION OF ERP ACCORDING TO THE NUMBER OF STATIONS INVOLVED IN THE GPS TRACKING NETWORK

TABLE 3

Sensitivity of Absolute Positioning with respect to :
- orbit error
- network consistency
- refraction error

Observations : - 18 GPS satellites
- stations observe each arc a maximun of 6 hrs
- Doppler count of 60 seconds

Solution : - after 24 hrs of observations
- network of 10 stations

Orbit error		Network error (m)	Refraction error (%)	Std. dev.		
Al. tr. (m)	Range (m)			X	Y (m)	Z
0.30	0.30	-	-	0.206	0.188	0.092
0.50	0.50	-	-	0.266	0.270	0.130
1.00	1.00	-	-	0.376	0.382	0.183
-	-	0.10	-	0.009	0.009	0.004
-	-	0.30	-	0.015	0.015	0.007
-	-	-	1	0.016	0.014	0.007
-	-	-	3	0.047	0.042	0.020
0.50	0.50	0.10	2	0.275	0.277	0.133
1.00	1.00	0.10	2	0.383	0.387	0.186

References

Anderle, R.J., L.K. Beuglass, J.T. Carr, Earth's Rotation and Polar Motion based on Global Positioning System Satellite data, Proc. High Precision Earth Rotation and Earth-Moon Dynamics, Ed. O. Calame, Reidel Publ., 1982.

Boucher, C., Definition and realization of Terrestrial Reference Systems for monitoring Earth Rotation, Proceedings of the IUGG Interdisciplinary Symposium on Variations in Earth Rotation, August 1987, Vancouver, B.C., Ed. Carter W.E., D. McCarthy and P. Pâquet, AGU Publ. Series, 1988.

Hopfield, H.S., Tropospheric effects on electromagnetically measured range: prediction from surface weather data, Radio Science, 6, (3), 1971.

Reigber, Ch., S. Hieber, E. Achterman, POPSAT - An active Solid Earth Monitoring System, Proceedings of the International Association of Geodesy Symposia, IUGG General Assembly, Hamburg, Ed. OSU Department of Geodetic Science, 1983.

Zelensky, N., J. Ray, P. Liebrecht, Error analysis for Earth Orientation recovery from GPS data. Proceedings of the IUGG Interdisciplinary Symposium on Variations in Earth Rotation, August 1987, Vancouver, B.C., Ed. Carter W.E., D. McCarthy and P. Pâquet, AGU Publ. Series, 1988.

STATION COORDINATES AND EARTH ROTATION PARAMETERS 1986

H. Hauck

Institut für Angewandte Geodäsie, Frankfurt am Main
Federal Republic of Germany

Abstract. Coordinates of the laser ranging stations and Earth rotation parameters from laser data of the year 1986 (Jan. - Oct.) are presented. The computations are based on normal points created from full-rate laser ranges to LAGEOS. Five sets of bimonthly solutions were transformed upon each other by a seven parameter transformation to get a mean solution. The bimonthly solutions are created in a terrestrial reference system close to the CIO system realized by fulfilling three condition equations. Consequently the terrestrial reference system where the mean annual solution is defined is close to the CIO system.

Earth rotation parameters are computed on the basis of the mean annual solution. Here the station coordinates are not completely fixed, but at the same time when earth rotation values are computed changes in the station coordinates with respect to the mean solution are also worked out.

Data Material and Evaluation Technique

The station coordinates were computed on the basis of compressed full rate laser ranging data (normal points) which were provided by the DGFI (Deutsches Geodätisches Forschungsinstitut) at München, FRG. They cover the time from January to October 1986. Only those normal points were used which had been created by more than nine full rate observations within a 2-minutes bin (Table 1).

The evaluation of the LAGEOS laser ranging data of 1986 by the analysis center at IfAG is based on the following procedure:

First a 60-day LAGEOS arc is computed holding a global set of very accurate a priori station coordinates fixed to get the state vector x_0 at the beginning of the arc and the best value for the empirical along track acceleration c_D. Hereby the satellite processor program UTO-

TABLE 1. Summary of normal points used to compute the station coordinates

Stat	Name	Location	Period	Npts
1181	GRDLAS	Potsdam	860108-861030	285
7086	MCLAS2	Ft. Davis	860104-861031	2632
7090	ML0501	Yarragadee	860101-861030	4731
7105	ML0702	Greenbelt	860102-861031	2231
7109	ML0802	Quincy	860103-861029	4801
7110	ML0402	Mt Laguna	860104-861031	4795
7121	HUAHIN	Huahine	860108-860412	404
7122	ML0601	Mazatlan	860102-861031	3089
7210	HOLLAS	Lure Obs.	860107-861031	2130
7510	MT1501	Askites	860528-860715	642
7512	MT1501	Kattavia	860919-861018	799
7515	MT1501	Dionysos	860730-860830	735
7517	MT1601	Roumeli	860530-860831	766
7520	MT1501	Karitsa	860329-860513	682
7525	MT1601	Xrisokellaria	860908-861018	277
7530	SAO3	Bar Giyyora	860904-861031	218
7540	MT1501	Matera	860114-860312	368
7541	MT1601	Matera	860110-860312	741
7550	MT1601	Basovizza	860331-860519	315
7810	ZIMMER	Zimmerwald	860305-861028	239
7834	WETLAS	Wettzell	860120-861030	1125
7835	GRASSE	Grasse	860829-861031	297
7838	SHOLAS	Simosato	860104-861031	2702
7839	AUSLAS	Graz	860112-861016	1087
7840	RGOLAS	Roy.Gr.Obs.	860103-861031	5083
7907	ARELAS	Arequipa	860108-861031	3251
7939	MATLAS	Matera	860102-861027	3401

PIA, which has been developed at the University of Texas at Austin, was applied [Schutz and Tapley 1982].

In a second step four successive 15-day arcs based on the state vector x_0 and the value c_D were predicted. During this prediction the range residuals O - C and the partial derivatives with respect to the parameters to be solved for were computed and stored. Then a program

Copyright 1990 by
International Union of Geodesy and Geophysics
and American Geophysical Union.

system developed at IfAG [Ehlert 1984] carried out an adjustment providing:

- a 60-day global solution
- four state vectors, each defining a 15-day arc
- 5-day mean polar motion values
- 5-day mean values for the change of the length of day (effects of zonal tides removed)
- 15-day values for the empirical along track acceleration and for the solar radiation

The adjustment procedure uses the 3-σ criterion for rejecting observations (normal points) and estimates station specific weighting factors attached to the observations.

Finally the independent bimonthly solutions were transformed upon each other by a seven parameter transformation recognizing the covariances of the estimated coordinates [Ehlert 1987].

Specifications

For the computation of the station coordinates in 1986 the parameters recommended in the MERIT standards (MERIT: Monitor Earth Rotation and Intercompare the Techniques of Observation and Analysis, MERIT standards, 1983) were used with the following exceptions:

1. 1950.0 instead of 2000.0 reference system
2. $GM = 3.98600440\ E+14\ m^3 s^{-2}$
3. ocean loading site displacements not applied
4. GEM-T1 gravitational field instead GEM-L2
5. no relativistic effects applied

The scale for the distances was based on the speed of light given value of 299 792 458.0 m/sec.

It was intended to define a terrestrial coordinate system which is very close to the CIO system. This was realized by computing 5-day polar motion values x_p and y_p using the a priori values $x_p(BIH)$ and $y_p(BIH)$ from the BIH table and fulfilling the two conditions

$$\sum x_p(BIH)_j - x_p(comp)_j = 0$$
$$\sum y_p(BIH)_j - y_p(comp)_j = 0$$

The parameter of rotation on the z-axis was defined by the condition equation

$$\sum \Delta \lambda_i = 0.$$

The subscript j refers to the 5-day time interval while i corresponds to the station number.

The Earth rotation parameters were computed on the basis of bimonthly solutions starting from the annual mean solution as a priori coordinates and $x_p = y_p = 0$ and DUT1R = const as a priori ERPs.

The coordinate system is chosen such that the following three condition equations are fulfilled [Bender and Goad 1979]:

$$\sum \Delta\lambda_i \cdot \sin\phi_i \cdot \cos\lambda_i - \Delta\phi_i \cdot \sin\lambda_i = 0$$
$$\sum \Delta\lambda_i \cdot \sin\phi_i \cdot \sin\lambda_i + \Delta\phi_i \cdot \cos\lambda_i = 0$$
$$\sum \Delta\lambda_i \cdot \cos\phi_i = 0$$

Here the subscript i refers to the station number. Only stations which have observed more than 25 percent of the average number of normal points per station are involved in the summation.

The effect of these condition equations is such that the computed bimonthly station solutions are given practically in that terrestrial reference system which is defined by the a priori station coordinates (mean annual solution).

Results

The five bimonthly solutions provided the following RMS values:

Jan-Feb86	±4.1 cm	Jul-Aug86	±5.0 cm
Mar-Apr86	±4.1 cm	Sep-Oct86	±4.7 cm
May-Jun86	±6.6 cm		

The mean station coordinates derived by the seven parameter transformation are shown in Tables 2 and 3.

TABLE 2. Solution 1986 in ellips. coordinates, $a = 6378137.0$ m, $1/f = 298.257$

Stat	height m	σ_h cm	longitude deg	σ_λ cm	latitude deg	σ_ϕ cm
1181	147.780	2.8	13.0652828	5.5	52.3802597	3.3
7086	1963.460	0.2	255.9841212	0.3	30.6769833	0.3
7090	244.505	0.3	115.3467416	0.3	-29.0465036	0.3
7105	22.334	0.2	283.1723074	0.3	39.0206047	0.4
7109	1109.477	0.2	239.0553025	0.2	39.9750021	0.2
7110	1842.204	0.2	243.5773330	0.2	32.8917369	0.2
7121	47.318	0.4	208.9588682	0.5	-16.7335196	0.5
7122	34.037	0.2	253.5409176	0.2	23.3428493	0.3
7210	3068.340	0.2	203.7440946	0.2	20.7072222	0.3
7510	183.934	0.8	25.5662186	0.9	40.9279972	0.7
7512	75.257	0.5	27.7808012	0.6	35.9514969	0.5
7515	511.903	0.6	23.9324626	0.7	38.0785711	0.6
7517	103.796	0.6	24.6941671	0.7	35.4042212	0.6
7520	599.906	0.5	20.6648325	0.6	39.7342290	0.5
7525	477.585	0.8	21.8776123	1.0	36.7914227	0.9
7530	776.438	0.8	35.0885494	0.8	31.7223942	1.0
7540	529.922	0.6	16.7044780	0.6	40.6485215	0.5
7541	529.676	0.5	16.7042825	0.6	40.6485043	0.4
7550	448.425	1.0	13.8755675	1.3	45.6429358	0.9
7810	951.065	1.0	7.4652158	1.1	46.8772269	1.0
7834	661.127	0.4	12.8780933	0.5	49.1449375	0.4
7835	1322.812	1.7	6.9211188	2.0	43.7546896	1.9
7838	101.587	0.3	135.9370392	0.3	33.5776945	0.4
7839	539.379	0.4	15.4933568	0.5	47.0671341	0.4
7840	75.385	0.3	0.3361207	0.4	50.8673792	0.3
7907	2492.268	0.3	288.5068772	0.4	-16.4657439	0.4
7939	535.789	0.3	16.7046827	0.4	40.6488309	0.3

TABLE 3. Solution 1986 in Cartesian coordinates

Stat	X m	σ_x cm	Y m	σ_y cm	Z m	σ_z cm
1181	3800621.46	3.8	882005.04	5.6	5028859.33	1.9
7086	-1330120.95	0.2	-5328532.25	0.3	3236146.73	0.2
7090	-2389007.59	0.3	5043331.77	0.2	-3078526.92	0.3
7105	1130720.30	0.3	-4831352.95	0.4	3994108.48	0.2
7109	-2517235.94	0.2	-4198558.28	0.2	4076571.70	0.2
7110	-2386279.12	0.2	-4802356.79	0.2	3444883.15	0.2
7121	-5345868.37	0.4	-2958248.34	0.5	-1824624.88	0.5
7122	-1660090.15	0.2	-5619103.17	0.2	2511639.16	0.2
7210	-5466006.93	0.2	-2404428.20	0.2	2242188.46	0.2
7510	4353447.34	0.7	2082667.37	0.9	4156505.62	0.7
7512	4573402.33	0.5	2409323.39	0.6	3723880.51	0.5
7515	4595218.86	0.6	2039436.31	0.7	3912628.35	0.5
7517	4728697.06	0.6	2174374.43	0.8	3674571.70	0.6
7520	4596045.02	0.6	1733477.64	0.6	4055719.83	0.4
7525	4745952.09	0.8	1905706.73	1.1	3799167.83	0.8
7530	4443968.47	0.9	3121946.72	0.8	3334695.51	1.0
7540	4641987.23	0.6	1393058.38	0.6	4133232.30	0.4
7541	4641992.99	0.5	1393042.84	0.6	4133230.69	0.4
7550	4336740.83	1.1	1071272.49	1.3	4537910.53	0.8
7810	4331283.69	1.1	567549.44	1.1	4633139.91	0.8
7834	4075530.10	0.4	931781.15	0.5	4801618.18	0.3
7835	4581091.80	1.0	556159.21	2.0	4389359.33	1.9
7838	-3822388.33	0.3	3699363.48	0.3	3507573.17	0.3
7839	4194426.72	0.5	1162693.77	0.5	4647246.51	0.3
7840	4033463.83	0.3	23662.24	0.4	4924305.04	0.2
7907	1942791.91	0.4	-5804077.68	0.3	-1796919.29	0.4
7939	4641965.07	0.3	1393069.81	0.4	4133262.20	0.2

The estimation of the formal errors given in Tables 2 and 3 are based on the standard deviations derived from the bimonthly solutions. They represent the internal accuracy (precision) of the laser ranges. A more realistic estimation of the external accuracy coming from the seven parameter transformation of the five sets of coordinates provides standard deviations which are 4 times bigger. Table 4 shows the computed earth rotation values.

References

Bender, P., and C. C. Goad, Probable LAGEOS contributions to a worldwide geodynamics control network, The Use of Artificial Satellites for Geodesy and Geodynamics, Vol. 2, (G. Veis and E. Livieratos, eds., National Technical University, Athens), 1979

Ehlert, D., EDV-Programme für die Ausgleichung vermittelnder Beobachtungen mit Bedingungsgleichungen zwischen den Unbekannten. Deutsche Geodätische Kommission, Reihe B, Heft Nr 268, Frankfurt am Main 1984

Ehlert, D, Die Analyse rezenter Erdkrustenbewegungen in statischen und kinematischen Modellen. Deutsche Geodätische Kommission, Reihe B, Heft Nr 279, Frankfurt am Main 1987

Schutz, B., and B. Tapley, UTOPIA, University of Texas Orbit Processor, Departement of Aerospace Engineering and Engineering Mechanics Austin, Texas, 1982

TABLE 4. Earth rotation values in 1986

MJD	X_p arcsec	Y_p arcsec	Lodr msec	MJD	X_p arcsec	Y_p arcsec	Lodr msec	MJD	X_p arcsec	Y_p arcsec	Lodr msec
46434	0.1803	0.1654	-1.2805	46534	-0.0630	0.1643	-1.4553	46634	-0.0348	0.3611	-0.4293
46439	0.1688	0.1562	-1.2041	46539	-0.0695	0.1736	-1.3798	46639	-0.0309	0.3672	-0.4581
46444	0.1587	0.1462	-1.2229	46544	-0.0779	0.1813	-1.4061	46644	-0.0251	0.3714	-0.3746
46449	0.1496	0.1385	-1.4093	46549	-0.0815	0.1948	-1.4210	46649	-0.0177	0.3745	-0.3967
46454	0.1346	0.1320	-1.4068	46554	-0.0857	0.2018	-1.5014	46654	-0.0083	0.3811	-0.5057
46459	0.1232	0.1271	-1.4281	46559	-0.0905	0.2126	-1.5446	46659	-0.0016	0.3839	-0.5430
46464	0.1033	0.1230	-1.6282	46564	-0.0928	0.2231	-1.4897	46664	0.0039	0.3872	-0.5413
46469	0.0940	0.1219	-1.5874	46569	-0.0968	0.2314	-0.9202	46669	0.0138	0.3929	-0.5648
46474	0.0828	0.1169	-1.5488	46574	-0.1002	0.2430	-0.9555	46674	0.0219	0.3960	-0.7136
46479	0.0682	0.1153	-1.3500	46579	-0.0982	0.2553	-0.9350	46679	0.0279	0.3987	-0.6728
46484	0.0565	0.1159	-1.2739	46584	-0.0990	0.2667	-0.5527	46684	0.0359	0.3970	-0.7835
46489	0.0408	0.1157	-1.2551	46589	-0.1003	0.2746	-0.5489	46689	0.0448	0.4001	-1.2078
46494	0.0261	0.1179	-1.2041	46594	-0.0965	0.2859	-0.4819	46694	0.0507	0.3991	-1.2245
46499	0.0130	0.1199	-1.1699	46599	-0.0906	0.2955	-0.5554	46699	0.0549	0.3964	-1.2584
46504	-0.0014	0.1209	-1.1643	46604	-0.0821	0.3088	-0.5869	46704	0.0623	0.3940	-1.4734
46509	-0.0136	0.1244	-0.9278	46609	-0.0733	0.3186	-0.5562	46709	0.0676	0.3949	-1.4848
46514	-0.0259	0.1318	-1.0385	46614	-0.0690	0.3201	-0.5380	46714	0.0721	0.3930	-1.4817
46519	-0.0360	0.1371	-1.1994	46619	-0.0608	0.3434	-0.5658	46719	0.0800	0.3911	-1.5348
46524	-0.0473	0.1424	-1.3993	46624	-0.0531	0.3525	-0.5302	46724	0.0903	0.3909	-1.4922
46529	-0.0583	0.1516	-1.4266	46629	-0.0485	0.3522	-0.4132	46729	0.0926	0.3904	-1.5152

REFERENCE FRAME OF LLR

Jin Wen-jing and Wang Qiang-guo

Shanghai Observatory, Shanghai, China

Abstract Since 1985, lunar laser ranging program has been operated at three stations of two nations on a regular basis. The situation of running only one lunar laser ranging station was over. Recently the precision of ranging distance has been improved from 30cm to 8cm. The results of comparing the station coordinates obtained from SLR with those from LLR are presented in this paper. Some parameters related to lunar motion, lunar physical condition, lunar tidal acceleration etc. are also shown and discussed.

Introduction

The station coordinates are necessary parameters for national defence and economic constructions. As it is well known that these coordinates of a station determined by new techniques such as VLBI, SLR, LLR are different. In order to compare the systematic errors of new techniques, collocation observations at Harvard Radio Station (USA), Wettzell (FRG), Orroral Valley (Australia) were suggested [Muller, 1983]. Unfortunately this plan is implemented slowly. Although only collocation observations of SLR and LLR were carried out at several lunar laser stations, comparison of terrestrial reference frame of SLR and LLR can still be made. Usually the dynamical reference frame can be obtained through the observations of bodies in the solar system, such as the sun, the moon, the minor planets etc., but the special research subjects related to the moon still need to observe the moon, for example to determine the coordinates of retro-reflectors, lunar harmonious constants of gravity etc. Of course such relative parameters can be also obtained from lunar laser ranging with high precision.

Data Used and Its Precision

The LLR data of shooting reflectors, Apollo 15, 14, 11, Lunakhod 2 obtained from McDonald 2.7m telescope, McDonald laser ranging system (MLRS), Haleakala (Hawaii), CERGA (Grasse, France) were used during 1972-1986. The physical model adopted was described in the paper of Dr. Jin et al. [1985].

From analysing the LLR data given by McDonald 2.7m telescope, the sidereal month term 27.32 day and annual term of 365.25 day are found during Julian Date 2441317.5-2444816.5. The experimental term is

$$25.7\sin((JD-2441317.9)/27.32)+ 18.9\sin((JD-2441261.7)/365.25) \text{ ns},$$

which was mentioned by Dr. Calame [Calame, 1978]. The cause of this may be induced by ephemeris.

The number of normal point N obtained from each station every year, the internal consistency, the precision of normal point and the external consistency i.e., the fluctuation of normal points after adjusting 60 parameters such as station and retro-reflector coordinates, solar and lunar orbit parameters etc. are listed in Table 1.

As shown in Table 1, in recent years, the precisions whether internal or external one are upgraded gradually, so the possibility exists to discuss the systematic errors of new techniques using the collocation observation of LLR and SLR.

Collocation Observation of LLR and SLR

There are three stations running both LLR and SLR programs at the same instrument or at the same yard of observatory such as McDonald, Lure, CERGA observatories. The station coordinates determined by LLR are listed in Table 2 in which the precision of coordinate is about 13cm. The comparisons of station coordinates between the determined values of SLR, processed by CSR (Austin, USA) with the data of MERIT campaign [Feissel, 1985] and of those of LLR are made here. At the same instrument of MLRS station the differences of coordinates x, y, z, are listed at the first three lines of Table 3.

If LLR and SLR program have been carried out at the separated instruemnts, the comparisons of the distances determined by SLR and LLR with those determined by geodetic survey [Altamini, 1986] are also shown in Table 3.

TABLE 1. Precision of LLR   unit: ns

	year	1972	1973	1974	1975	1976	1977	1978	
	N	359	435	327	361	315	242	222	
	E_{in}	0.9	0.9	0.8	1.0	1.0	0.7	1.0	
	E_{ex}	3.6	3.0	3.1	3.0	2.5	2.5	3.3	
McDonald 2.7m	year	1979	1980	1981	1982	1983	1984	1985	1986
	N	272	292	198	12	28	74	49	
	E_{in}	0.8	0.7	0.7	0.7	0.9	1.5	1.2	
	E_{ex}	2.5	2.3	6.6	11.4	3.5	3.6	3.0	
CERGA	N				62	232	356	715	120
	E_{in}				1.0	1.5	1.5	1.2	1.3
	E_{ex}				6.0	4.9	3.9	2.1	2.5
Haleakala	N						6	209	125
	E_{in}						0.7	0.4	0.2
	E_{ex}						1.6	1.2	0.9
MLRS	N					44	236	247	76
	E_{in}					1.2	0.8	0.4	0.1
	E_{ex}					2.4	2.3	1.0	3.8

TABLE 2. Earth-Fixed Rectangular Coordinates of Stations   unit: m

station	x	y	z
MCD	-1330781.180 ±0.221	-5328755.932 ±0.139	3235697.681 ±0.050
MLRS	-1330121.011 ±0.216	-5328532.574 ±0.136	3236146.256 ±0.073
CER	4581692.818 ±0.075	556195.672 ±0.219	4389355.007 ±0.103
HAL	-5466008.573 ±0.044	-2404427.185 ±0.269	2242190.341 ±0.065

TABLE 3. Difference of Coordinates Determined by SLR and LLR [1]   unit: m

		Results of LLR and SLR	Geodetic Survey	Difference
$MLRS_1 - MLRS_2$ [2]	x	-0.048	0	0.039
	y	-0.492	0	-0.005
	z	-0.548	0	-0.364
$MCD - MLRS_2$	x	-660.217	-660.329	0.199
	y	-223.850	-223.299	-0.064
	z	-449.123	-449.020	0.081
MG-SG [3]	x	+0.940	+0.577	-0.045
	y	+36.249	+36.474	-0.303
	z	-4.235	-4.438	0.283

1. At Haleakala station, the corresponding value of geodetic survey has not been found.
2. Here coordinates of $MLRS_1$ and $MLRS_2$ are determined by LLR and SLR, respectively.
3. MG, SG indicate the instrumental coordinates of LLR and SLR at CERGA respectively.

TABLE 4. Coordinates of Four Lunar Reflectors [1]   unit: m

Reflector	X	Y	Z	dX	dY	dZ [2]
0	1591973.198 ±25.448	69049.632 ±57.853	20990.959 ±4.300	-38.976	143.634	-15.351
2	1652724.397 ±19.038	-52094 5.998 ±60.114	-109741.930 ±4.547	62.160	149.649	-14.290
3	1554704.061 ±5.448	98144.575 ±56.487	764998.544 ±4.149	17.793	140.529	-11.538
4	1339372.135 ±29.894	801916.256 ±48.627	756352.355 ±3.608	-41.644	122.900	-9.252

1. The notations of 0, 2, 3, 4 indicate Apollo 11, 14, 15 and Lunakhod 2.
2. dX, dY, dZ indicate the difference of comparing with those published at MERIT standards [Melbourne, 1983].

From these values, the transform factors of two reference frames ΔZ, K, Q_z, are obtained and the values are 0.475 ± 0.149 m, $-0.899 \times 10^{-7} \pm 0.233 \times 10^{-7}$, $-0''.0013 \pm 0''.0048$, respectively. After reducing the station coordinates to the same reference frame with these factors, the differences are listed at the last column of Table 3 [Zhu Wen-yao, 1986].

The differences of station coordinates determined by different techniques are caused by the measuring error and different physical model of motion such as disturbing force considered, so the transform coefficients determined by the observing data of three stations are not sufficient to discuss the relation of two reference frames. The suggestion of resuming the LLR observations of Orroral Valley situated in the southern hemisphere and carrying out observations at Wettzell station soon are very important. It is also useful to determine three parameters x, y, UT1-UTC of ERP and improve the network of LLR.

Lunar Physical Parameters

As it is well known that dynamical reference frame is determined by observing natural bodies of the solar system such as the sun, the moon, the major planet and the asteroid. In addition to the above mentioned purpose, the data of observing the moon can be used to research the lunar gravity field, physical libration, tidal acceleration and so on.

Using the LLR data of fifteen years, the coordinates of retro-reflectors have been obtained by the authors and listed in Table 4. Because the variation of the distance from the station to the reflector is not sensitive to the variation of space position of the reflector, the measuring error of reflector coordinate is slight large. It is less useful for the cartography of lunar surface as the control points in a global selenodetic network.

The information on lunar moment of inertia parameters of β, γ and the harmonics of the lunar gravitational field of C_{30}, C_{31}, ... S_{32}, S_{33} with the error are obtained and listed in Table 5.

In this solution the correction of the precession constant and tidal secular acceleration of the lunar orbital longitude are also obtained, the values are $-0''.0018$/yr, $-24''.5$/century^2, respectively. They are of significance for the research of the lunar secular motion and dynamcial frame.

TABLE 5. Lunar Gravity Harmonic Coefficients

	Coefficient	Precision	Difference[1]
C30	-0.48367×10^{-5}	$\pm 0.1104 \times 10^{-5}$	0.37893×10^{-5}
C31	0.30580×10^{-4}	$\pm 0.2188 \times 10^{-5}$	-0.13047×10^{-6}
C32	0.48071×10^{-5}	$\pm 0.9506 \times 10^{-8}$	-0.26885×10^{-7}
C33	0.14589×10^{-5}	$\pm 0.4460 \times 10^{-7}$	0.22930×10^{-7}
S31	0.35815×10^{-5}	$\pm 0.1021 \times 10^{-5}$	-0.20292×10^{-5}
S32	0.17089×10^{-5}	$\pm 0.8907 \times 10^{-8}$	0.24934×10^{-7}
S33	-0.29782×10^{-6}	$\pm 0.1828 \times 10^{-7}$	0.45685×10^{-7}
β	0.63166×10^{-3}	$\pm 0.1406 \times 10^{-6}$	-0.16311×10^{-7}
γ	0.22800×10^{-3}	$\pm 0.1429 \times 10^{-7}$	-0.10319×10^{-9}

1. The harmonic coefficients obtained in this paper minus the values of MERIT standards.

Acknowledgements. Partial work of the paper was finished by Dr. Jin when she stayed at Astronomy Department, the University of Texas, Austin, in January 1987. The authors are indebted to Dr. Shelus and Mr. Ricklefs for the helpful discussion.

References

Muller, I.I., Zhu, S.Y., and Bock, I., Reference frame requirement and the MERIT campaign, Dept. of Geodetic Science and Survey No. 329, Ohio State University, Columbus, 1982.

Jin Wen-jing, Wang Qiang-guo, Determination of ERP with laser ranging and discussion of the influence of the adopted parameters, Proceedings of the International Conference on Earth Rotation and the Terrestrial Reference Frame, 1, 287-295, 1985.

Calame, O., and Guinot, B., Earth rotation by lunar distances (EROLD), BIH Annual Report for 1978, D-27, 1978.

Feissel, M., Earth rotation from laser ranging to Lageos, Observational Results on Earth Rotation and Reference Systems, B-67, 1986.

Altamimi, Z., Boucher, C., and Feissel, M., Directory of sites participating to the realisation of the BTS for 1986, BIH Notes Internes, 1986.

Melbourne, W., et al. Project MERIT standards, U.S. Naval Observatory Circular No. 167, 1983.

Zhu Wen-yao, The coordinates system of MEDOC and its influence on the measuring polar motion, Science Bulletin, No. 5, 360-364, 1986.

DEFINITION AND REALIZATION OF TERRESTRIAL REFERENCE SYSTEMS FOR MONITORING EARTH ROTATION

Claude Boucher

Institut Géographique National
Saint-Mandé, France

Abstract. A terrestrial reference system is one of the key elements in the modelling and monitoring of Earth rotation. Improvement in the concepts are necessary to derive proper models to match as accurately as possible the advanced measurement techniques, such as VLBI, satellite or lunar laser ranging, supraconducting gravimeters, laser gyro ... Possible definitions are reviewed in a relativistic and continuum mechanichal background. Realizations of such systems are also investigated and their current implementation as a standard of the coming International Earth Rotation Service is outlined.

The level of accuracy reached by a variety of techniques sensitive to the Earth's rotation has pushed the necessity to improve the concept and realization of a terrestrial system.

Many concerns have been denoted recently to these problems, mainly through the MERIT-COTES campaigns (G.A. Wilkins, I.I. Mueller, 1986).

An improved concept of terrestrial system is a key element of the proper modelling of data provided by various techniques, already developed or under development : optical astrometry, VLBI, satellite or lunar laser ranging, radio-electric tracking of satellites, supraconducting gravimeters, laser gyroscopes ...

For more about basic definitions and terminology, please refer to J. Kovalevsky and I.I. Mueller, 1981, for instance.

A terrestrial system is the actual implementation, in a specific data analysis, of an ideal terrestrial system, which is a quasi cartesian system with a given origin, scale and orientation. But in order to ensure a proper realization of such a reference system, one selects a reference frame, characterized by a finite set of points for which one provides coordinates as a function of time, $x(t)$.

Due to the very nature of terrestrial reference systems, there exists a large variety of them. Their number is regularly increasing through the multiplicity of measurement techniques and data analysis.

The Terrestrial Reference Frames

For terrestrial reference frames, several aspects must be specified :
- the type of point (either reference points located on the Earth's crust, such as geodetic control marks, tracking instrument reference points; or moving points such as an artificial satellite);
- the type of coordinate system associated to the quasi cartesian reference system, such as cartesian, spherical, cylindrical, geographical, astronomical, ...
- the quality of the estimation of x.

The current terrestrial frames, as used in geodetic or astronomical communities, are :
a) Sets of tracking stations used in space geodesy, together with their cartesian (X, Y, Z) or geographical (λ, ϕ, h) coordinates, for which an ellipsoid has to be specified. The secular time variations (as we shall discuss further below) have also to be given, either as time derivative at a reference epoch t_o (i.e. $\dot{X}$, $\dot{Y}$, $\dot{Z}$ or $\dot{\lambda}$, $\dot{\phi}$, $\dot{h}$), or as a sequence of estimations at various epochs t_k, usually regularly spaced - monthly, yearly values - (X_k, Y_k, Z_k or λ_k, ϕ_k, h_k).
b) Sets of optical instruments, with their astronomical coordinates (Λ, Φ) and their time variations.
c) Sets of geodetic marks with their geographical (λ, ϕ, h) or grid (E, N, h) coordinates. Here, h is obtained through geoid undulation N and orthometric height H :

$$h = N + H \quad (1)$$

d) Sets of cartesian coordinates of the centre of mass of an artificial satellite, at regular epochs (ephemerides)

The case (a) is the most widely used in space geodesy. (b) was used by the BIH up to 1983 for its system, and is also (implicitly) used by geodetic agencies for the orientation of terrestrial control networks using triangulation or traverses, through Laplace azimuths. The resulting frame is traditionally expressed as (c). Finally, (d) is also used, for instance, Transit Doppler point positioning with Broadcast or Precise Ephemerides.

Relative Origins, Scales and Orientations

For all systems, the physical model adopted, either newtonian or relativistic, enables us to define the concept of origin, scale and orientation. The general relationship between two quasi cartesian systems $(X_k^{(i)})_{k=1,2,3}$ and $i=1,2$ is therefore

$$X_k^{(2)} = X_k^{(1)} + T_k^{(1,2)} + Q_{k,\ell}^{(1,2)} X_\ell^{(1)} + \delta X_k^{(1,2)} \quad (2)$$

where

$T_k^{(1,2)}$ are the translation components

$$Q_{k,\ell}^{(1,2)} = \begin{bmatrix} D^{(1,2)} & -\varepsilon_3^{(1,2)} & \varepsilon_2^{(1,2)} \\ \varepsilon_3^{(1,2)} & D^{(1,2)} & -\varepsilon_1^{(1,2)} \\ -\varepsilon_2^{(1,2)} & \varepsilon_1^{(1,2)} & D^{(1,2)} \end{bmatrix} \quad (3)$$

$D^{(1,2)}$ is the scale factor and

$\varepsilon_k^{(1,2)}$ are the rotation angles around axis (k),

$\delta X_k^{(1,2)}$ is a correction term to take non-linearities and relativistic terms into account.

Such a general relationship is valid for the transformation between two terrestrial reference systems, providing they are nearly geocentric and vaguely aligned on the equatorial system with a common zero meridian (so called "Greenwich").

One has furthermore assumed a same direct orientation of the vector basis. Consequently, (2) is a first order expansion of a three-dimensional euclidian affinity, which is the general relationship between two orthogonal, isotropic and direct affine frames.

Depending on the way the various terrestrial systems and frames are determined, there are various discrepancies in either origin, scale or orientation, which can be predicted directly from the adopted definitions.

It is important to actually determine these discrepancies, in order to compare them with the values estimated from a direct comparison of the frames. This can show systematic errors in the estimated values, coming from bias in the coordinates of the frames and the usually non global coverage of the colocation stations used in the computation.

We can select some examples :
a) In VLBI, the origin of the frame is defined by fixing coordinates of a station. Consequently, the origin shift of this frame, with regard to any other system, will be given by colocating an instrument belonging to the other system at this VLBI reference station, and comparing related coordinates.

b) Techniques like VLBI or LLR derive currently information from a solar system barycentric frame, in a relativistic framework. The terrestrial geocentric frame is obtained usually by a Lorentzian boost and a spatial rotation. When comparing with a local geocentric Lorentz frame, as used by dynamical satellite techniques, such as SLR, there remains a scale discrepancy coming mainly from the solar gravitation at the Earth's level, or

$$\left\langle \frac{U_\odot}{c^2} \right\rangle = \frac{GM_\odot}{Ac^2} = 1.5 \times 10^{-8}$$

c) In most techniques, such as VLBI or SLR, the scale is defined through the adoption of the velocity of light, C. This value is fortunately fixed in the new definition of the meter to :

$$C = 299\ 792\ 458\ m \times s^{-1} \quad (CIPM\ 1983)$$

The previous recommended value (IUGG 1957) of $299\ 792\ 500\ m \times s^{-1}$ created a scale discrepancy of

$$\frac{\Delta C}{C} = 1.40 \times 10^{-7}$$

It must be noticed that this is only true for dynamical techniques when GM is adjusted. If this parameter is held fixed, it also influences strongly the scale of the terrestrial system.

Time Variations

Another important aspect has to be taken into account for a refined realization of a terrestrial system, namely the time dependent effects.

On a general basis, the cartesian coordinates of a point are functions of time X(t). For the corresponding frame, it is then certainly interesting to model all or a part of the time variations of the coordinates. This is particularly feasible for geodetic or astronomical points lying on the crust. The major phenomena which provide such time variations are
- solid earth tides (30cm),
- ocean loading (a few cm),
- tectonic plate motion (10cm/year),
- land uplift
- local deformations due to ground water changes
- deformations due to atmospheric pressure (a few cm)

- direct luni-solar effect on astronomical vertical

Some of these effects can be modelled with a good accuracy. A review of current models can be found in W. Melbourne et al., 1983 (MERIT Standards). Two models are of particular interest for terrestrial frames :
- solid earth tide correction for ground station positions, especially important for the vertical component

In particular

$$\Delta h = -0.121 \left(\frac{3}{2}\sin^2\phi - \frac{1}{2}\right) m \qquad (4)$$

(MERIT Standards p. A5-7)

is the permanent tidal deformation, where ϕ is the latitude of the station. It must be specified if this correction is applied or not.
- tectonic plate motion correction, acting on horizontal components. The usual ones, such as the series of Minster-Jordan models, are defined through a set of angular velocity vectors $\bar{\Omega}_p$, one for each plate, and expressed in the terrestrial system, so that the velocity of a point of coordinates $\bar{X}$ is :

$$\dot{\bar{X}} = \bar{\Omega}_p \wedge \bar{X} \qquad (5)$$

Two absolute motion models are usually adopted in data analysis :
AMO-2, derived from the RM-2 model by applying a no global rotation condition,
AM1-2, which minimizes the motion of a set of hot spots, also derived from RM-2 (J.B. Minster and T.H. Jordan, 1978).
AMO-2 depends only on the adopted contour of plate boundaries, whereas AM1-2 depends on the selection of the hot spots which are more subject to uncertainties. On the other hand, AMO-2 corresponds to the type of law of evolution one one wants to give to terrestrial frames (see below), and has been consequently adopted by MERIT Standards (Update 1, December 1985). Nevertheless, AM1-2 leads to a system linked to the mantle which is needed to express a geopotential model without secular variations due to a residual rotation of the system. It is therefore favoured by groups which perform dynamical analysis of satellite tracking data.

In a given system (i), a ground point has therefore the following expression :

$$\bar{X}^{(i)}(t) = \bar{X}_o^{(i)} + \dot{\bar{X}}_o^{(i)}(t - t_o) + \bar{L}^{(i)}(t) + \Delta\bar{X}_{tid}(t) \qquad (6)$$

where

$$\dot{X}_o^{(i)} = \Omega_p^{(i)} \wedge X_o^{(i)} + \dot{h}\frac{\bar{X}_o^{(i)}}{|\bar{X}_o^{(i)}|} \qquad (7)$$

This modelling, although arbitrary, seems a good compromise between the model and stochastic components. The modelled part takes into account
- tidal variation
- plate motion
- secular vertical motion.

As both points on the crust and reference systems are moving, it is important to understand the problems related to time evolution.
From (2), we see that if (6) is true in (i), we get in (j) :

$$\bar{X}_r^{(j)} = \bar{X}_{or}^{(i)} + \dot{\bar{X}}_{or}^{(i)}(t - t_o) + \bar{L}_r^{(i)}(t) + \Delta\bar{X}_{tid}(t) + T^{(i,j)}(t) + Q^{(i,j)}(t)\bar{X}_{or}^{(i)} + \delta X \qquad (8)$$

The assumption that (6) is also valid for (j) means that T and Q can be expanded to first order with a sufficient accuracy, and :

$$\begin{cases} \bar{X}_{or}^{(j)} = \bar{X}_{or}^{(i)} + T_o^{(i,j)} + Q_o^{(i,j)}\bar{X}_{or}^{(i)} \\ \dot{\bar{X}}_o^{(j)} = \dot{\bar{X}}_o^{(i)} + \dot{T}_o^{(i,j)} + \dot{Q}_o^{(i,j)}\bar{X}_{or}^{(i)} \\ \bar{L}_r^{(j)} = \bar{L}_r^{(i)} \end{cases} \qquad (9)$$

The question of the definition of the terrestrial system is then divided into two aspects :
- to define it at a reference epoch (t_o), through seven conditions for origin, scale and orientation;
- to assess an evolution law. For this, several options are conceivable.

For the origin, either a station gets assessed coordinates (and time variations), like in VLBI or terrestrial system, or the system is modelled as geocentric, like in LLR or dynamical satellite techniques. This holds also for time evolution.

For the scale, a choice of constants (c, GM) determines it and its time evolution (no evolution).

For the orientation, one usually fixes or constrains Earth rotation parameters to a specific a priori series, or fixed value at a reference epoch, and applies a no-net rotation condition (see below).

One could also define a terrestrial system and its evolution through unambiguous statements, such as principal axes of momenta for the origin (e.g. selected at geocenter) ...

But these definitions are hardly operational and sensitive to the redistribution of masses with the Earth.

A rather tempting approach is the concept of Tisserand axes. Such axes are defined by minimizing the kinetic energy. They are characterized by nul linear and angular momentums.

Let $\vec{X}$ and $\vec{V}$ be position and velocity related to an external frame, and $\vec{x}$, $\vec{v}$ for a terrestrial frame :

$$\begin{cases} \vec{X} = \vec{r}_o + \vec{x} \\ \vec{V} = \vec{V}_o + \vec{v} + \vec{\omega} \wedge \vec{x} \end{cases} \qquad (10)$$

The kinetic energy is :

$$T = \frac{1}{2} \int_c v^2 dm \qquad (11)$$

whereas the linear momentum is

$$\vec{p} = \int_c \vec{v} \, dm \qquad (12)$$

and the angular momentum :

$$\vec{h} = \int_c \vec{x} \wedge \vec{v} \, dm \qquad (13)$$

If we apply a variational expression to the terrestrial frame, we get :

$$\begin{cases} \vec{0} = \delta \vec{r}_o + \delta \vec{x} \\ \vec{0} = \delta \vec{V}_o + \delta \vec{v} + \delta \vec{\omega} \wedge \vec{x} + \vec{\omega} \wedge \delta \vec{x} \end{cases} \qquad (14)$$

Whence :

$$\delta T = \int_c \vec{v}.\delta \vec{v} \, dm = -\vec{p}.\delta \vec{V}_o - \vec{h}.\delta \vec{\omega} + \vec{p}.(\vec{\omega} \wedge \delta \vec{r}_o) \qquad (15)$$

so that

$$\delta T = 0 \iff \vec{p} = \vec{h} = \vec{0} \qquad (16)$$

One has some freedom to select the domain of integration (C). Usually, one chooses the crust (M.L. Smith, 1980).

A Tisserand terrestrial system is defined by its position at a reference epoch and the following laws, from (6) :

$$\int_c \left(\dot{\overline{X}}_o^{(i)} + \frac{d\overline{L}^{(i)}}{dt} \right) dm = \overline{0} \qquad (17)$$

$$\int_c \overline{X}_o^{(i)} \wedge \left(\dot{\overline{X}}_o^{(i)} + \frac{d\overline{L}^{(i)}}{dt} \right) dm = \overline{0} \qquad (18)$$

the tidal terms vanishing.
If we apply (7), we get :

$$\int_c \overline{\Omega}_p \wedge \overline{X}_o^{(i)} dm + \int_c \dot{h} \frac{\overline{X}_o^{(i)}}{|\overline{X}_o^{(i)}|} dm + \int_c \frac{d\overline{L}^{(i)}}{dt} dm = \overline{0} \qquad (19)$$

$$\int_c \overline{X}_o^{(i)} \wedge (\overline{\Omega}_p \wedge \overline{X}_o^{(i)}) \, dm + \int \overline{X}_o^{(i)} \wedge \frac{d\overline{L}^{(i)}}{dt} dm = \overline{0} \qquad (20)$$

Analysis centres which process various types of data (either one specific type, or a combination) not only have to select models and constants, but also the parameters which are solved for in the (usually least squares) estimator. A variety of choices has been done by these centres.
Among the numerous possibilities which exist regarding the adoption of such models for station positions, we shall only mention 4 possibilities which are actually used :
a) constant solved-for positions : $\overline{X}_o$
b) solved-for position and velocity at a reference epoch t_o :

$$\overline{X}_o + \dot{\overline{X}}_o (t - t_o)$$

c) solved-for position at t_o, plus correction models :

$$\overline{X}_o + \overline{\Omega}_p \wedge \overline{X}_o (t - t_o) + \overline{\Delta X}_{tid}(t)$$

d) values solved-for at regular epochs (dayly, monthly, yearly) as step functions :

$$\overline{X}_k = \overline{X}_o + \overline{L}_k \quad t \in [t_k, t_{k+1}[\quad t_{k+1} = t_k + \Delta t$$

In the cases (a) to (c), the system and its time evolution are fixed by constraints on origin, scale and orientation as seen before. In the case (d), it can be done in the same way or through other constraints, such as discretized Tisserand condition :

$$\begin{cases} \sum_r m_r (\overline{X}_{k+1,r} - \overline{X}_{k,r}) = 0 \\ \sum_r m_r \overline{X}_{k,r} \wedge (\overline{X}_{k+1,r} - \overline{X}_{k,r}) = 0 \end{cases} \qquad (21)$$

The main remaining problem is to estimate reliable and accurate transformation parameters between systems. Even if they can be totally or partially derived from a priori considerations, it will be desirable to compute through a (usually least squares) estimation of the numerical value of the (usually) seven transformation parameters occuring into (2), taken as model.
This will lead to a comparison of two systems or a combination of several ones, including the estimation of their relationships (see C. Boucher, M. Feissel, 1984) using available frames and colocation.

An example : BIH Terrestrial System (BTS)

Since 1984, the BIH Terrestrial System has been redefined using data provided by new geodetic space techniques. Its realization, based on a study by Boucher and Feissel (1984), consists of a combination of sets of station coordinates derived from VLBI, LLR, SLR and Doppler and, in addition, the corresponding series of Earth rotation parameters.
For this purpose, a global adjustment has been elaborated in a combined least square estimation requiring colocation sites in the different networks. The selected sites are those for which station coordinates are available in at least two networks. A reference point per site is chosen and station coordinates are brought to their reference point value by means of the known geodetic ties.
The origin of the BTS is taken from dynamical solutions such as SLR and LLR. The scale is the one of SLR and the orientation is obtained by

rotating the individual systems by the angles derived from the comparisons of the corresponding series of ERP with the BIH series.

As an output of the global adjustment, three types of results, referred to the BTS, are obtained :
- a set of cartesian coordinates for each site
- a set of transformation parameters between each individual system and the BTS
- a series of ERP at 0.05 year interval for the whole year of its realization

Three annual realizations of BTS have been achieved for 1984, 1985 and 1986, whose results are published in the corresponding BIH Annual Reports. For more details, see also Boucher and Altamimi, 1985 and 1986.

About the data used in the combination, we select here those of BTS 86 realization :
- VLBI : SSC(NGS) 87 R 01, SSC(GSFC) 87 R 01 and SSC(JPL) 83 R 05
- LLR : SSC(JPL) 87 M 01
- SLR : SSC(CSR) 86 L 01 and SSC(DGFI) 87 L 01
- Doppler : SSC(DMA) 77 D 01

In BTS86 realization, we have made use of AMO-2 plate tectonic absolute motion model derived from the global RM2 model (Minster and Jordan, 1978). The choice of AMO-2 is due to the fact that it is derived with the condition of no net rotation of the Earth's surface and then recommended by MERIT Standards (Melbourne et al. update 1, 1985).

BTS86 realization contains 51 colocated sites, distributed over six plates, whose coordinates are related to the epoch 1984.0, whenever the date of station coordinates in the individual networks is available. Complement of analysis and results of BTS86 realization is given by Boucher and Altamimi (1987) in BIH Annual Report for 1986, page D-63.

The Terrestrial System used by the International Earth Rotation Service

Improvements have to be undertaken in the future for new solutions, such as BTS87 realization, as well as in the framework of the International Earth Rotation Service. These improvements concern, in particular :
- increasing of colocation sites
- accuracy and reliability of local ties between stations which can be obtained by local surveys or GPS baselines

The IERS must also select a clear and consistent definition of the terrestrial system and derived frame.

A possible candidate is to adopt the BTS, with necessary refinements. This would also have the practical consequence to create no discontinuity between BIH and IERS Earth rotation parameters.

The important issue is also to clearly specify all these aspects into standards (update of MERIT Standards into IERS Standards), in order to fulfill both theoretical and practical applications of such a new conventional terrestrial system, satisfactory for present and forthcoming years.

References

Boucher, C., and Altamimi, Z., Towards an improved realization of the BIH terrestrial frame, 3rd MERIT Workshop, Columbus, July 1985

Boucher, C., and Altamimi, Z., Status of the realization of the BIH terrestrial system, IAU Symposium 128, Coolfront, U.S.A., 1986

Boucher, C., and Altamimi, Z., Complement of analysis of BIH terrestrial system for 1986, BIH Annual Report for 1986

Boucher, C., and Feissel, M., Realization of the BIH terrestrial system, International Symposium on Space Techniques for Geodynamics, Sopron, Hungary, 9-13 July 1984

Dickey, J.O. et al., Reference frames : determinations and connections, IUA, Symposium 128, Coolfront, U.S.A., 1986

Hothem, L.D., Vincenty, T., and Moose, R.E., Relationship between Doppler and other advanced geodetic system measurements based on global data, International Geodetic Symposium on Satellite Doppler Positioning, Las Cruces, USA, 1982

Hothem, L.D., Vincenty, T., and Hoyle, D.B., Analyses of Doppler, Satellite Laser, VLBI and terrestrial coordinate systems, IUGG / IAG XVIIIth General Assembly, Hamburg, FRG, August 1983

Kovalevsky, J., Mueller, I.I., Comments on conventional terrestrial and quasi inertial reference systems, Proc. Symp. "Reference coordinate systems for Earth dynamics", Warsaw pp. 375-384, 1981

Melbourne, W. et al., Project MERIT Standards, USNO, Circular n° 167, Washington, 1983

Minster, J.B., and Jordan, T.H., 1987, Present day plate motions, J. Geophys. Res. 83, 5331-5354 1987

Smith, M.L., The theoretical description of the nutation of the Earth, IAU Coll. 56, Varsaw, 1980

Wilkins, G.A., and Mueller, I.I., On the rotation of the Earth and the terrestrial reference system, Joint summary Report of the IAU/IUGG Working Groups MERIT and COTES, 1986, Bull. Géod. 60, pp. 85-100, 1986

A CORRELATION STUDY OF THE EARTH'S ROTATION WITH EL NIÑO/SOUTHERN OSCILLATION

B. Fong Chao

Geodynamics Branch
NASA Goddard Space Flight Center
Greenbelt, Maryland 20771

Abstract. The El Niño/Southern Oscillation (ENSO) is the single most prominent interannual fluctuation in the atmosphere and ocean. In order to conserve the total angular momentum, its associated motion and mass redistribution will cause variations in the (solid) Earth's rotation, in both length of day (LOD) and polar motion. This paper examines the importance of ENSO in the excitation of Earth rotation variations by means of cross correlation in the time domain and complex coherence in the frequency domain. A Monte Carlo simulation is conducted to obtain estimates for the coherence confidence threshold that are needed, particularly as short spectral averages and tapering window functions are employed.

ENSO's effect on LOD is represented by the sea-level (air) pressure difference between Tahiti and Darwin, Australia, similar to the Southern Oscillation Index. It is found that the cross correlation function between this ENSO representation and LOD change on the interannual time scale, roughly from 1 to 10 years, has a maximum value of 0.68 at the LOD phase lag of 2 months during 1972-86. This indicates the following: (i) Most of the interannual LOD variation is caused by ENSO. This extends the atmosphere-LOD relationship into the interannual time scale. (ii) The transfer of ENSO's axial angular momentum to the solid Earth lags behind the (Tahiti − Darwin) pressure variation by about 2 months. This provides gross constraints on the atmospheric/oceanic modeling of the evolution of ENSO. The corresponding coherence spectrum, in addition, shows minimum correlation around biennial periods, suggesting the influence in LOD of the stratospheric quasi-biennial oscillation.

ENSO's effect on the polar motion is less certain. In this paper, it is simply modeled as two varying pressure masses centered at Tahiti and Darwin. Its coherence spectrum with the observed Chandler wobble excitation for 1962-86 shows a near-zero phase and a squared coherence peak that falls between the 85 per cent and 90 per cent confidence thresholds. Considering the simple assumptions made and the limited spectral resolving power of the available data, this correlation suggests that an observable portion of the Chandler excitation energy was provided by ENSO during 1962-86.

A paper based on the present study, entitled "Correlation of interannual length-of-day variation with El Niño/Southern Oscillation, 1972-1986", is in press, J. Geophys. Res., 1988 (Paper #7B1047).

STATISTICAL INVESTIGATIONS ON ATMOSPHERIC ANGULAR MOMENTUM FUNCTIONS AND ON THEIR EFFECTS POLAR MOTIONS

Aleksander Brzezinski[1]

Institute of Theoretical Geodesy, Technical University, Graz, Austria

Abstract. The short term atmospheric excitation of polar motion was analyzed using the available three series of the atmospheric "effective angular momentum" (EAM) vector. It was shown that the pressure terms X^p_1 and X^p_2 at periods less than 100 days can be adequately modeled by a 3rd order autoregressive process, whose striking feature is pseudoharmonic oscillation with a period of about 9 to 13 days and a very short relaxation time of 2 to 4 days. Different estimates of the same components of the equatorial EAM functions are compared in order to compute their signal and noise statistics. The wind terms X^w_1 and X^w_2 are found to contain more observational noise at short periods than the pressure terms but the comparisons are limited by the fact that there are differences in the epochs to which the meteorological centers refer their data. Finally, the atmospheric effects on polar motion are discussed using various statistical models for the excitation process.

(This paper was published in manuscripta geodaetica, 12/3, 1987.)

---

[1] On leave from Space Research Centre, Polish Academy of Sciences, Warsaw, Poland.